IUTAM Symposium on Theoretical, Computational and Modelling Aspects of Inelastic Media

IUTAM BOOKSERIES
Volume 11

Series Editors

G.M.L. Gladwell, *University of Waterloo, Waterloo, Ontario, Canada*
R. Moreau, *INPG, Grenoble, France*

Editorial Board

J. Engelbrecht, *Institute of Cybernetics, Tallinn, Estonia*
L.B. Freund, *Brown University, Providence, USA*
A. Kluwick, *Technische Universität, Vienna, Austria*
H.K. Moffatt, *University of Cambridge, Cambridge, UK*
N. Olhoff *Aalborg University, Aalborg, Denmark*
K. Tsutomu, *IIDS, Tokyo, Japan*
D. van Campen, *Technical University Eindhoven, Eindhoven, The Netherlands*
Z. Zheng, *Chinese Academy of Sciences, Beijing, China*

Aims and Scope of the Series

The IUTAM Bookseries publishes the proceedings of IUTAM symposia under the auspices of the IUTAM Board.

For other titles published in this series, go to
www.springer.com/series/7695

B. Daya Reddy

Editor

IUTAM Symposium on Theoretical, Computational and Modelling Aspects of Inelastic Media

Proceedings of the IUTAM Symposium held at Cape Town, South Africa, January 14–18, 2008

B. Daya Reddy
Centre for Research in Computational and Applied Mechanics
University of Cape Town
7701 Rondebosch
South Africa

ISBN-13: 978-1-4020-9089-9   e-ISBN-13: 978-1-4020-9090-5

Library of Congress Control Number: 2008936007

© 2008 Springer Science+Business Media, B.V.
No part of this work may be reproduced, stored in a retrieval system, or transmitted
in any form or by any means, electronic, mechanical, photocopying, microfilming, recording
or otherwise, without written permission from the Publisher, with the exception
of any material supplied specifically for the purpose of being entered
and executed on a computer system, for exclusive use by the purchaser of the work.

Printed on acid-free paper

9 8 7 6 5 4 3 2 1

springer.com

Table of Contents

Preface ix

List of Committee Members and Sponsors xi

List of Participants xiii

Multiscale Modeling and Simulation of Microstructure

On Computational Procedures for Multi-Scale Finite Element Analysis of Inelastic Solids 3
D. Perić, D.D. Somer, E.A. de Souza Neto and W.G. Dettmer

Material Characterization Based on Microstructural Computations and Homogenization 15
P. Wriggers and M. Hain

Relaxed Potentials and Evolution Equations for Inelastic Microstructures 27
K. Hackl and D.M. Kochmann

Towards Effective Simulation of Effective Elastoplastic Evolution 41
C. Carstensen and R. Huth

Numerical Approximation Techniques for Rate-Independent Inelasticity 53
A. Mielke

Damage and Fracture

A Computational Methodology for Modeling Ductile Fracture 67
A.A. Benzerga

Multiscale Methods for Fracturing Solids 79
S. Loehnert and D.S. Mueller-Hoeppe

A Regularized Brittle Damage Model Solved by a Level Set Technique 89
N. Moës, N. Chevaugeon and F. Dufour

Gradient and Other Non-Local Theories

A Counterpoint to Cermelli and Gurtin's Criteria for Choosing the 'Correct' Geometric Dislocation Tensor in Finite Plasticity 99
A. Acharya

On Stability for Elastoplasticity of Integral-Type 107
F. Marotti de Sciarra

On the Mathematical Formulations of a Model of Strain Gradient Plasticity 117
F. Ebobisse, A.T. McBride and B.D. Reddy

Uniqueness of Strong Solutions in Infinitesimal Perfect Gradient-Plasticity with Plastic Spin 129
P. Neff

Algorithms and Computational Aspects

SQP Methods for Incremental Plasticity with Kinematic Hardening 143
C. Wieners

Simulation of Forming Processes Using Overlapping Domain Decomposition and Inexact Newton Methods 155
S. Brunssen, C. Hager, F. Schmid and B. Wohlmuth

Variational Formulation of the Cam-Clay Model 165
M. Hjiaj and G. de Saxcé

Anisotropic Modelling of Metals in Forming Processes 175
S. Reese and I.N. Vladimirov

Inelastic Media under Uncertainty: Stochastic Models and Computational Approaches 185
H.G. Matthies and B.V. Rosić

Automated Computational Modelling for Solid Mechanics 195
K.B. Ølgaard, G.N. Wells and A. Logg

Generalised Functions for Modelling Singularities: Direct and Inverse Problems 205
S. Caddemi and I. Caliò

Discontinuous Galerkin Methods

A Discontinuous Galerkin Method for an Incompatibility-Based Strain
Gradient Plasticity Theory   217
J. Ostien and K. Garikipati

Some Applications of Discontinuous Galerkin Methods in Solid Mechanics   227
A. Lew, A. Ten Eyck and R. Rangarajan

Some Aspects of a Discontinuous Galerkin Formulation for Gradient
Plasticity at Finite Strains   237
A. McBride and B.D. Reddy

Computational Dynamics

Energy-Momentum Algorithms for the Nonlinear Dynamics of
Elastoplastic Solids   251
F. Armero

Internal Variable Formulations of Problems in Elastoplastic Dynamics   263
M.A.E. Kaunda

Time-FE Methods for the Nonlinear Dynamics of Constrained Inelastic
Systems   275
R. Mohr, S. Uhlar, A. Menzel and P. Steinmann

The Potential for SPH Modelling of Solid Deformation and Fracture   287
P.W. Cleary and R. Das

Effect of Material Parameters in the Izod Test for Polymers   297
V. Tvergaard and A. Needleman

Experimental and Computational Aspects

The Response of "Large" Square Tubes (Width/Thickness Ratio > 45) to
Opposite Lateral Blast Loads Followed by Dynamic Axial Load   309
S. Chung Kim Yuen and G.N. Nurick

Modelling the Behaviour of Fibre-Metal Laminates Subjected to Localised
Blast Loading   319
D. Karagiozova, G.S. Langdon and G.N. Nurick

On the Measurement and Evaluation of the Width of Portevin–Le Chatelier
Deformation Bands with Application to AA5083-H116 Aluminium Alloy   329
A. Benallal, T. Berstad, T. Børvik, O. Hopperstad and R. Nogueira de Codes

Shakedown and Limit Analysis

Direct Evaluation of Limits in Plasticity and Creep Deformation 341
A.R.S. Ponter

On Recent Progress in Shakedown Analysis and Applications to Large-Scale Problems 349
D. Weichert, A. Hachemi, S. Mouhtamid and A.D. Nguyen

Viscoelasticity

Local and Global Regularity in Time Dependent Viscoplasticity 363
H.-D. Alber and S. Nesenenko

Hamiltonian Theory of Viscoelasticity 373
A. Hanyga and M. Seredyńska

Author Index 385

Subject Index 387

Preface

Inelastic media constitute a rich source of interesting and important problems in theoretical, experimental and computational mechanics.

Significant insights have been gained through studies of the mathematical characteristics of new models. New constitutive theories have lead to variational and other formulations that are generally more complex, often highly nonlinear, and requiring new tools for their successful resolution. Likewise, there have been significant advances of a computational nature, coupled to the development of new algorithms for solving such problems in discrete form.

It is clear, therefore, that research in the broad area of inelastic media offers contemporary investigators a range of challenges which are most fruitfully addressed through a combination of theoretical, experimental and computational avenues. Furthermore, the field is truly multidisciplinary in nature, drawing on the expertise of specialists in materials science, various branches of engineering, mathematics, and physics, and benefiting from integrative approaches to the solution of problems.

The objective of the IUTAM Symposium on Theoretical, Modelling and Computational Aspects of Inelastic Media, held in Cape Town over the period 14–18 January 2008, was to provide a forum in which experts engaged in a spectrum of activities under the theme of inelastic media could discuss recent developments, and also identify key open problems.

The main success of the Symposium lays in its ability to bring together researchers with backgrounds in mathematics, engineering or physics, whether experimentalists, theoreticians, or computational experts, and with a common interest in inelastic media. The Symposium furthermore provided a forum for the exchange of ideas among researchers working at all scales: the micro or meso scales, and the macroscopic and structural levels. Overall it was a fertile arena for cross-pollination of ideas.

This level of diversity will be apparent from the contents of this volume, which comprises papers based on the talks given by most participants. Topics covered include

- material characterization, the calibration of material models by simulation of experiments, and solution of inverse problem for parameter identification
- development, analysis and implementation of solution algorithms in quasistatic and dynamic plasticity
- discontinuous Galerkin formulations
- multiscale modeling, analysis and computation, and micro- and mesoscale modeling, including gradient theories and simulation of micorostructure

- damage and fracture
- shakedown and limit analysis
- experimental investigations at the structural level

The Symposium was attended by 59 delegates, including 8 students or postdoctoral researchers, from 15 countries. Altogether 46 lectures were presented in the scientific programme.

The Symposium was dedicated to the memory of JB Martin, a major figure in theoretical and applied mechanics in South Africa and internationally.

It is a pleasure to thank those organizations which provided financial support, and which through their generosity made it possible to support a number of participants, and to ensure the scientific and social success of the event. Organisations which supported the symposium in this way were the University of Cape Town, the National Research Foundation (South Africa), the South African Department of Science and Technology, the Centre for High Performance Computing (South Africa), and Springer.

The event was a significant one for South Africa as it gave researchers and students in the region the opportunity to interact at a high level with international experts in the field. It has given a boost to activities in mechanics, broadly speaking, in the country.

The support and encouragement of IUTAM, through the General Assembly and its Executive, are acknowledged with thanks. The South African Association for Theoretical and Applied Mechanics lent strong support to the Symposium, and this support is acknowledged with thanks.

Members of the International Scientific Committee contributed greatly to the quality and success of the event through their proposals for participants, and their general guidance on organizational and other matters. Andrew McBride and Olivia Goodhind served on the Local Organising Committee, while Heidi Tait and Meg Winter of the Centre for Professional Development at the University of Cape Town were responsible for much of the administration. All ensured through their work that the organization ran smoothly. The group of eight graduate student assistants provided enthusiastic support to delegates, whether during technical sessions, providing general support, or in responding to queries about local conditions. These contributions are all acknowledged with thanks.

The Symposium would not have been the success that it was without the lively participation of the delegates, who ensured that the technical sessions were stimulating and enjoyable, and who collectively created a memorable atmosphere. It is a pleasure to thank all delegates for their participation.

It is a pleasure to thank reviewers for their constructive comments which in many cases led to improvements in the papers making up this volume. And finally, thanks are due to Jolanda Karada of Karada Publishing Services for her professional preparation of the volume prior to its submission to Springer.

Daya Reddy *June 2008*
Chairman, Scientific Committee
University of Cape Town

List of Committee Members and Sponsors

Scientific Committee

R. de Borst (Eindhoven University of Technology)
G. Maier (Politecnico di Milano) – IUTAM Representative
C. Miehe (Universität Stuttgart)
A. Needleman (Brown University)
M. Ortiz (California Institute of Technology)
D.R.J. Owen (University of Swansea)
B.D. Reddy (University of Cape Town) – Chairman

Sponsors

IUTAM
University of Cape Town
National Research Foundation
South African Department of Science and Technology
Centre for High Performance Computing (South Africa)
Springer

List of Participants

Professor Amit Acharya, Carnegie Mellon University, Civil & Environmental Engineering, PA 15213, Pittsburgh, USA
acharyaamita@cmu.edu

Professor Hans-Dieter Alber, Technische Universität Darmstadt, FB Mathematik, Schlossgartenstrasse 7, D-64289 Darmstadt, Germany
alber@mathematik.tu-darmstadt.de

Professor Francisco Armero, University of California at Berkeley, Department of Civil & Environmental Engineering, 713 Davis Hall, Berkeley, CA 94720-1710, USA
armero@ce.berkeley.edu

Professor Ahmed Benallal, Directeur de recherche au CNRS, Ecole Normal Supérieure de Cachan, 61 Av du Président Wilson, F-94235 Cachan Cedex, France
benallal@lmt.ens-cachan.fr

Professor Amine Benzerga, Texas A&M University, Department of Aerospace Engineering, 736C H.R. Bright Building, 3141 Tamu, College Station, TX 77843-3141, USA
benzerga@aero.tamu.edu

Professor Gabriella Bolzon, Politecnico di Milano, Dipartimento di Ingegneria Strutturale, Piazza Leonardo da Vinci 32, I-20133 Milano, Italy,
bolzon@stru.polimi.it

Mr Marius Botha, Element Six, 1 Debid Road, Nuffield, 1559, Springs, South Africa
marius.botha@e6.com

Professor Salvatore Caddemi, Università degli Studi di Catania, Dipartimento di Ingegneria Civile ed Ambientale, Viale Andrea Doria 6, I-95125 Catania, Italy
caddemi@diseg.unipa.it

Professor Carsten Carstensen, Humboldt-Universität zu Berlin, Institut für Mathematik, Unter den Linden 6, D-10099 Berlin, Germany
cc@math.hu-berlin.de

Dr Steeve Chung Kim Yuen, University of Cape Town, BISRU, Department of Mechanical Engineering, Private Bag X3, 7701 Rondebosch, South Africa
chnste010@uct.ac.za

Dr Paul Cleary, CSIRO, CSIRO Mathematical and Information Services, 77 Normanby Road, Clayton, VIC 3169, Australia
paul.cleary@cmis.csiro.au

Mr Trevor Cloete, University of Cape Town, BISRU, Department of Mechanical Engineering, Private Bag X3, 7701, Rondebosch, South Africa
trevor.cloete@uct.ac.za

Dr François Ebobisse, University of Cape Town, Department of Mathematics and Applied Mathematics, Private Bag X3, 7701 Rondebosch, South Africa
francois.ebobisseBille@uct.ac.za

Professor Krishna Garikipati, University of Michigan at Ann Arbor, Mechanical Engineering & Applied Physics, 3003B EECS, Ann Arbor, MI 48109, USA
krishna@umich.edu

Professor Marc Geers, Eindhoven University of Technology, Department of Mechanical Engineering P O Box 513, WH4.135, 5600 MB Eindhoven, The Netherlands
m.g.d.geers@tue.nl

Mr Reuben Govender, University of Cape Town, BISRU, Mechanical Engineering, Private Bag X3 7701 Rondebosch, South Africa
grnreu001@uct.ac.za

Professor Morton E Gurtin, Carnegie Mellon University, 5807 Pembroke Place, Pittsburgh, PA 15213, USA
mg0c@andrew.cmu.edu

Professor Klaus Hackl, Ruhr Universität Bochum, Lehrstuhk für Allgemeine Mechanik, D-44780 Bochum, Germany
hackl@am.bi.rub.de

Professor Andrzej Hanyga, University of Bergen, Institute for Solid Earth Physics, Allegaten 41, N-5007 Bergen, Norway
andrzej.hanyga@geo.uib.no

List of Participants

Professor Mohammed Hjiaj, INSA de Rennes, LGCGM, Geomechanics & Structures, 20 Avenue des Buttes de Coesmos, F-35043 Rennes Cedex, France
mohammed.hjiaj@insa-rennes.fr

Mr Ernesto Ismail, University of Cape Town, BISRU, Mechanical Engineering, Private Bag X3, 7701 Rondebosch, South Africa
ernesto.ismail@uct.ac.za

Dr Dora Karagiozova, Institute of Mechanics, Bulgarian Academy of Sciences, Acad. G. Bonchev Street, Block 4, Sofia 1113, Bulgaria
dorakar@yahoo.com

Dr Samuel Kwofie, Nkrumah – University of Science & Technology, Department of Materials Engineering, College of Engineering KNUST, Kumasi, Ghana
drskwofie@yahoo.com

Dr Genevieve Langdon, University of Cape Town, BISRU, Mechanical Engineering, Private Bag X3, 7701 Rondebosch, South Africa
genevieve.langdon@uct.ac.za

Professor Adrian J Lew, Stanford University, Department of Mechanical Engineering, Durand 207, 496 Lomita Mall, Stanford, CA 94305-4040, USA
lewa@stanford.edu

Professor Javier Llorca, Catedratico de Universidad, Universidad Politecnica de Madrid, Departamento de Ciencia de Materiales, ETS de Ingenieros de Caminos Canales Y Puertos, Ciudad Universitaria, E-28040 Madrid, Spain
jllorca@mater.upm.es

Dr S Löhnert, Universität Hannover, Institut für Baumechanik und Numerische Mechanik, Appelstrasse 9a, D-30167 Hannover, Germany
loehnert@ibnm.uni-hannover.de

Professor Giulio Maier, Politecnico di Milano, Dipartimento di Ingegneria Strutturale, Piazza Leonardo da Vinci, 32, I-20133 Milano, Italy
giulio.maier@polimi.it

Professor Francesco Marotti de Sciarra, Università di Napoli Federico II, Dipartimento di Ingegneria Strutturale, via Claudio 21, I-80125 Naples, Italy
marotti@unina.it

Professor Hermann G Matthies, Institute of Scientific Computing, Technical University of Braunschweig, Hans-Sommer-Strasse 65 (Computing Center), D-38092 Braunschweig, Germany h.matthies@tu-bs.de

Mr Andrew McBride, University of Cape Town, Centre for Research in Computational and Applied Mechanics, 7701 Rondebosch, South Africa
andrew.mcbride@uct.ac.za

Professor Christian Miehe, Universität Stuttgart, Institut für Mechanik (Bauwesen), Pfaffenwaldring 7, D-70550 Stuttgart, Germany
miehe@mechbau.uni-stuttgart

Professor Alexander Mielke, WIAS, Mohrenstrasse 39, D-10117 Berlin, Germany
mielke@wias-berlin.de

Professor Nicolas Moës, Institut GeM-UMR CNRS 6183, Ecole Centrale de Nantes, 1 Rue de la Noe, F-44321 Nantes Cedex 3, France
nicolas.moës@ec-nantes.fr

Dr Rouven M. Mohr, University of Kaiserslautern, Building 44, Room 436, P O Box 3049, D-67653 Kaiserslautern, Germany
rmohr@rhrk.uni-kl.de

Ms Ekaterina Muravleva, Institution Lomonosov, Moscow State University, Department of Mechanics and Mathematics, Leninskie Gory 1, 119992, Moscow, Russia
catmurav@gmail.com

Professor Alan Needleman, Brown University, Division of Engineering, 182 Hope Street, Providence, RI 02912, USA
alan_needleman@brown.edu

Dr Patrizio Neff, Technische Universität Darmstadt, FB Mathematik, Schlossgartenstrasse 7, D-64289 Darmstadt, Germany
neff@mathematik.tu-darmstadt.de

Professor Giorgio Novati, Politecnico di Milano, Dipartimento di Ingegneria Strutturale, Piazza Leonardo da Vinci, 32, I-20133 Milano, Italy
giorgio.novati@polimi.it

Professor Gerald Nurick, University of Cape Town, BISRU, Mechanical Engineering, 7701 Rondebosch, South Africa
gerald.nurick@uct.ac.za

Dr Stacey Oerder, University of Cape Town, BISRU, Mechanical Engineering, Private Bag X3, 7701 Rondebosch, South Africa

List of Participants

stacy.oerder@uct.ac.za

Professor Graeme Oliver, Cape Peninsula University of Technology, Department of Mechanical Engineering, P O Box 1906, Bellville 7535, South Africa
oliverg@cput.ac.za

Professor D R J Owen, University of Wales Swansea, Department of Civil Engineering, Singleton Park, Swansea SA2 8PP, United Kingdom
d.r.j.owen@swansea.ac.uk

Professor Djordje Perić, University of Wales Swansea, Department of Civil and Computational Engineering, Singleton Park, Swansea SA2 8PP, United Kingdom
d.peric@swansea.ac.uk

Professor Alan R S Ponter, University of Leicester, Department of Engineering, University Road, Leicester LE1 7RH, United Kingdom
asp@le.ac.uk

Professor Daya Reddy, University of Cape Town, Centre for Research in Computational and Applied Mechanics, Private Bag X3, Rondebosch 7701, South Africa
daya.reddy@uct.ac.za

Professor Stefanie Reese, Institute of Solid Mechanics, Technical University of Braunschweig, Schleinitzstr. 20, D-38106 Braunschweig, Germany
s.reese@tu-bs.de

Professor Suresh Shrivastava, McGill University, Department of Civil Engineering and Applied Mechanics, 817 Sherbrooke Street West, Montreal, Quebec H3A 2K6, Canada
suresh.shrivastava@mcgill.ca

Professor Winston Soboyejo, Princeton University, Mechanical and Aerospace Engineering, Engineering Quad, Princeton, NJ 08544, USA
soboyejo@princeton.edu

Professor R Svendsen, University of Dortmund, Department of Mechanical Engineering, Maschinenbau, Room 106, Leonhard-Euler-Str. 5, D-44221 Dortmund, Germany
bob.svendsen@mech.mb.uni-dortmund.de

Professor Viggo Tvergaard, Technical University of Denmark, Dept of Mechanical Engineering, Nils Koppels Alle, Bldg 404, Room 134, DK-2800 Kgs. Lyngby, Denmark
vit@mek.dtu.dk

Professor Dieter Weichert, Institute of General Mechanics, Aachen University of Technology, Templergraben 64, D-52062 Aachen, Germany
weichert@iam.rwth-aachen.de

Professor Garth N Wells, University of Cambridge, Department of Engineering, Trumpington Street, Cambridge CB2 1PZ, United Kingdom
gnw20@cam.ac.uk

Prof Dr Christian Wieners, Universität Karlsruhe, Fakultät für Mathematik, Lehrstuhl für Wissenschaftliches Rechnen, Englerstraee 2, D-76128 Karlsruhe, Germany
wieners@math.uni-karlsruhe.de

Professor Barbara Wohlmuth, Universität Stuttgart, Institut für Angewandte Analysis & numerische Simulation, Pfaffenwaldring 57, D-70569, Stuttgart, Germany
Barbara.wohlmuth@mthematik.uni-stuttgart.de

Professor Peter Wriggers, Universität Hannover, Institut für Baumechanik und Numerische Mechanik, Appelstrasse 9a, D-30167 Hannover, Germany
wriggers@ibnm.uni-hannover.de

Multiscale Modeling and Simulation of Microstructure

On Computational Procedures for Multi-Scale Finite Element Analysis of Inelastic Solids

D. Perić, D.D. Somer, E.A. de Souza Neto and W.G. Dettmer

Civil and Computational Engineering Centre, School of Engineering,
University of Wales Swansea, Singleton Park, Swansea SA2 8PP, UK
E-mail: {d.peric, d.d.somer, e.desouzaneto, w.g.dettmer}@swansea.ac.uk

Abstract. This work is concerned with issues related to computational procedures for a family of multi-scale constitutive models, based on the volume averaging of stress and strain (or deformation gradient) tensors over a representative volume element (RVE). The computational model relies on a variational framework for multi-scale analysis of solids, which leads to a compact direct numerical procedure within fully coupled two-scale displacement based finite element environment. A particular attention is given to the techniques for efficient computational implementation of multi-scale modelling framework, and, in this context, some recently developed computational procedures are discussed. A numerical example is presented in order to illustrate the scope and benefits of the developed strategy.

Key words: multi-scale analysis, computational homogenization, finite elements, inelastic solids, heterogeneous materials.

1 Introduction

Since the basic principles for the micro-macro modelling of heterogeneous materials were introduced (see Suquet [10]), this technique has proved to be a very effective way to deal with arbitrary physically non-linear and time dependent material behaviour at micro-level. During the last decade or so various approaches and techniques for the micro-macro modelling and simulation of heterogeneous materials have been proposed. Among these we highlight the contributions by Suquet and co-workers [9], Feyel and Chaboche [3], Miehe and co-workers [8], Kouznetsova et al. [5], Ladevèze et al. [6], Terada et al. [11] and Zohdi and Wriggers [12].

This paper discusses aspects of computational implementation of multi-scale constitutive models, based on the volume averaging of stress and strain (or deformation gradient) tensors over a representative volume element (RVE). The multi-scale models are implemented within a fully coupled displacement based finite element framework.

Kinematical constraints, typical of the present class of models, are imposed on the RVE boundary by master-slave type dependencies, through the fluctuating part of boundary displacements. The resulting constitutive relationship is fully linearized and formulated in terms of these dependencies, so that asymptotically quadratic convergence of the macro equilibrium problem is achieved.

The computational efficiency of the overall procedure can be further improved by introducing an independently discretized interface that is based on a finite element type interpolation. Furthermore a local sub-stepping procedure is introduced, which significantly improves the robustness of the basic scheme and also allows for larger load steps to be taken during the numerical solution of the macro scale problem, thus improving the efficiency of the overall multi-scale methodology.

Numerical example is presented to illustrate the scope and benefits of the developed computational strategy.

2 Homogenization-Based Multiscale Constitutive Theory

This section provides a brief summary of the multi-scale constitutive theory. as a basis for the computational developments presented in this work. Consider a (macroscopic) continuum occupying a domain $\bar{\Omega}$ with boundary $\partial\bar{\Omega}$ in its reference configuration. The starting point of the present family of multi-scale theories is the assumption that at each point $\bar{x} \in \bar{\Omega}$ of the macro continuum the deformation gradient $\bar{F}(\bar{x})$ is the volume average of its microscopic counterpart, F, over the domain Ω of the RVE associated to point $\bar{x}$:

$$\bar{F}(\bar{x}) = \frac{1}{V_0} \int_\Omega F(x) \, dV = I + \frac{1}{V_0} \int_\Omega \nabla u(x) \, dV, \qquad (1)$$

where u denotes the displacement field of the RVE (the *microscopic* displacement field), ∇ is material gradient operator and V_0 is the volume of the RVE in its reference configuration.

The volume averaging of the micro deformation gradient sets a constraint upon the possible deformation fields of the RVE. By splitting the microscopic displacement field, without loss of generality, as a sum

$$u = u^* + \tilde{u}, \qquad (2)$$

of a *homogeneous gradient displacement*

$$u^*(x) = (\bar{F} - I)\,x \qquad \forall\, x \in \Omega, \qquad (3)$$

and a *displacement fluctuation* $\tilde{u}(Y)$, this constraint is formally expressed as

$$\tilde{u} \in \tilde{\mathcal{K}}^*; \quad \tilde{\mathcal{K}}^* \equiv \left\{ v, \text{sufficiently regular} \,\middle|\, \int_{\partial\Omega} v \otimes n \, dA = \mathbf{0} \right\}, \qquad (4)$$

where $\tilde{\mathcal{K}}^*$ is the *minimally constrained space of kinematically admissible microscopic displacement fluctuations* [1] of the RVE and n denotes the outward unit normal to the boundary $\partial\Omega$ of the RVE. Different classes of constitutive model can then be obtained with the choice of different kinematic constraints to be imposed on a given RVE. This is done by postulating suitable spaces

$$\mathcal{V} \subset \tilde{\mathcal{K}}^* \qquad (5)$$

of kinematically admissible displacement fluctuations which, in turn, coincide with the corresponding space of variations of kinematically admissible displacements (virtual displacements) of the RVE [1].

Analogously to the homogenized deformation gradient, its work conjugate, the macroscopic first Piola–Kirchoff stress tensor, $\bar{P}$, at a point x of the macro continuum is defined as the volume average of the micro stress field P of the RVE associated with $\bar{x}$:

$$\bar{P}(\bar{x}) = \frac{1}{V_0} \int_\Omega P(x)\, dV = \frac{1}{V_0}\left(\int_{\partial\Omega} t \otimes x\, dA - \int_\Omega b \otimes x\, dV\right), \qquad (6)$$

where t and b denote, respectively, the reference boundary surface traction and body force fields of the RVE.

As a consequence of the Hill–Mandel Principle of Macro-homogeneity (see e.g. [4, 7]), both t and b are purely reactive to the actual choice (5) of kinematical constraints and, hence, cannot be prescribed independently. They belong to the functional space orthogonal to $\mathcal{V}$, whose elements do not produce virtual work. Then, the RVE equilibrium problem, expressed in terms of the Principle of Virtual Work, consists in finding, for a given macroscopic displacement gradient $\bar{F}$, a displacement fluctuation field $\tilde{u} \in \mathcal{V}$ such that:

$$\int_\Omega P : \nabla\eta\, dV = \int_\Omega \mathcal{G}([\bar{F} + \nabla\tilde{u}]^t) : \nabla\eta\, dV = 0 \quad \forall \eta \in \mathcal{V}, \qquad (7)$$

where $\mathcal{G}$ denotes the constitutive response functional of the RVE material which, in general, varies from point to point of Ω.

Three classes of RVE kinematical constraints which satisfy (4) are commonly employed within the present multi-scale framework:

- *Minimal kinematical constraint*, i.e., we choose

$$\mathcal{V} = \tilde{\mathcal{K}}^*. \qquad (8)$$

In this case, the (reactive) RVE boundary surface traction t, orthogonal to $\mathcal{V}$, is uniform:

$$t = \bar{P}\, n. \qquad (9)$$

Hence, this model is also referred to as the *Uniform Boundary Traction Model*.
- *Periodic RVE boundary fluctuations*. This is the assumption usually employed in the modelling of media with periodic microstructure. The RVE geometry

here cannot be arbitrary as it must satisfy the periodicity constraint. In two-dimensional models, quadrilateral or hexagonal RVEs, with equally sized opposing sides are typically used. The periodicity kinematical constraint is defined by the choice

$$\mathcal{V} = \mathcal{V}^{per} \equiv \left\{ \tilde{u} \in \tilde{\mathcal{K}}^* \mid \tilde{u}(x^+) = \tilde{u}(x^-) \ \forall \ \text{pairs} \ \{x^+, x^-\} \in \partial \Omega \right\}, \quad (10)$$

where $\{x^+, x^-\}$ are pairs of points defined by a one-to-one correspondence lying on opposing sides of the RVE polygonal boundary. The orthogonality between the (reactive) boundary traction t and $\mathcal{V}^{per}$ implies that t in the present case is *anti-periodic* on $\partial \Omega$, i.e., $t(u^+) = -t(u^-)$.

- *Linear RVE boundary displacements.* This case corresponds to the choice

$$\mathcal{V} = \mathcal{V}^{lin} \equiv \left\{ \tilde{u} \in \tilde{\mathcal{K}}^* \mid \tilde{u}(x) = 0 \ \forall \ x \in \partial \Omega \right\}. \quad (11)$$

For all three classes of models, the body force orthogonal to $\mathcal{V}$ is zero.

3 Finite Element Approximation

This section summarizes the general computational implementation of the multi-scale constitutive theory outlined in the above within a non-linear finite element framework.

3.1 Discretized Multi-Scale Constitutive Model

The material response within the RVE in the present context is generally modelled by means of conventional internal variable-based constitutive theories. Hence, a numerical approximation of the type

$$\bar{P}^{n+1} = \hat{P}(\bar{F}^{n+1}, \bar{\xi}^n). \quad (12)$$

resulting from time-discretization, is introduced at the micro-scale. We note that the expression (12) represents an *incremental constitutive relation* such that, for a typical time interval $[t^n, t^{n+1}]$, with known set $\bar{\xi}^n$ of internal variables at t^n, the stress $\bar{P}^{n+1}$ at t_{n+1} can be viewed as a function of the deformation gradient $\bar{F}^{n+1}$. This expression, together with (7), leads to the *RVE incremental equilibrium problem*, which consists in finding, for a given macroscopic deformation gradient $\bar{F}^{n+1}$ and known field ξ^n of microscopic internal variables, a kinematically admissible RVE displacement fluctuation field $\tilde{u}^{n+1} \in \mathcal{V}$ such that

$$\int_\Omega \mathbf{P}^{n+1}(\bar{\mathbf{F}}^{n+1} + \nabla \tilde{\mathbf{u}}^{n+1}, \boldsymbol{\xi}^n) : \nabla \boldsymbol{\eta}\, dV = 0 \quad \forall \boldsymbol{\eta} \in \mathcal{V}. \tag{13}$$

With the solution $\tilde{\mathbf{u}}^{n+1}$ at hand, the macroscopic stress can be computed according to (6) as

$$\bar{\mathbf{P}}^{n+1} = \frac{1}{V_0} \int_\Omega \mathbf{P}^{n+1}(\bar{\mathbf{F}}^{n+1} + \nabla \tilde{\mathbf{u}}^{n+1}, \boldsymbol{\xi}^n)\, dV. \tag{14}$$

The fully (time- and space-) discrete multi-scale constitutive model is obtained by introducing a conventional finite element approximation to (13,14). That is, the infinite-dimensional functional space $\mathcal{V}$ is replaced with a finite-dimensional counterpart, $^h\mathcal{V}$, spanned by the finite element shape functions of a mesh h, and the domain Ω is replaced with an approximated counterpart $^h\Omega$ comprising an assembly of finite element domains. The corresponding fully discrete version of (13) consists in finding a vector $\tilde{\mathbf{u}}^{n+1} \in {}^h\mathcal{V}$ of global nodal displacements fluctuations such that

$$\int_{^h\Omega} \boldsymbol{\eta}^T \left[\mathbf{G}^T\, \mathbf{P}^{n+1}(\bar{\mathbf{F}}^{n+1} + \mathbf{G}\,\tilde{\mathbf{u}}^{n+1}, \boldsymbol{\xi}^n) \right] dV = 0 \quad \forall \boldsymbol{\eta} \in {}^h\mathcal{V}, \tag{15}$$

where $\mathbf{G}$ denotes the global discrete gradient matrix containing the appropriate shape function derivatives, $\mathbf{P}^{n+1}$ is the incremental constitutive functional at the RVE level that delivers the array of the First Piola–Kirchhoff stress components, $\bar{\mathbf{F}}^{n+1}$ is the array of macroscopic deformation gradient components and $\boldsymbol{\eta}$ is the vector of global nodal virtual displacements.

Problem (15) together with the finite element-discrete version of (14) defines the fully discretized multi-scale constitutive model which, in turn, can also be symbolically expressed as in (12). The resulting functional $\hat{\mathbf{P}}$ in the present case represents the operations of solving (15) with the subsequent use of (14). The macro set $\boldsymbol{\xi}^n$ here represents the *field* $\boldsymbol{\xi}^n$ over $^h\Omega$.

3.2 Discretized Kinematical Constraints

Problem (15) may differ from finite element versions of conventional solid mechanics problems only in the construction of the relevant finite-dimensional space $^h\mathcal{V}$. For the linear boundary displacements constraint, the solution of the finite element RVE problem follows the same route as conventional solid mechanics problems, as the construction of space $^h\mathcal{V}$ is trivial and, according to (11), requires only that all nodal degrees of freedom of the RVE boundary be fixed as zero. For the minimally constrained and periodic boundary displacements kinematical assumptions, however, the spaces $^h\mathcal{V}$ are discrete versions of (4) and (10), respectively. The kinematical constraints in such cases are non-conventional but can nevertheless be imposed in a straightforward manner, as summarized in the following.

Firstly, note that the constraints embedded in space definitions (4) and (10) – and also (11) – involve only the boundary of the RVE. The constraints themselves are *linear dependencies* among the RVE boundary degrees of freedom. Hence, an arbitrary vector $\mathbf{v} \in {}^h\mathcal{V}$ can be conveniently arranged as

$$\mathbf{v} = \begin{bmatrix} \mathbf{v}_i \\ \mathbf{v}_m \\ \mathbf{v}_d \end{bmatrix} \qquad (16)$$

where the linear dependence constraint requires that

$$\mathbf{v}_d = \boldsymbol{\alpha}\, \mathbf{v}_m, \qquad (17)$$

and where $\mathbf{v}_i$ contains the degrees of freedom of the *interior* nodes, $\mathbf{v}_m$ contains the *free* (or 'master') degrees of freedom of the boundary, $\mathbf{v}_d$ contains the *dependent* boundary degrees of freedom, and $\boldsymbol{\alpha}$ is a matrix of constraint coefficients expressing the linear dependencies that characterize the model in question. For a periodic fluctuations model with pairing boundary nodes located at pairing points x^+ and x^- lying on opposite sides of $\partial^h \Omega$, $\boldsymbol{\alpha}$ is an identity matrix. For the minimally constrained case, $\boldsymbol{\alpha}$ is obtained from the discrete version of the integral constraint of definition (4):

$$\mathbf{C} \begin{bmatrix} \mathbf{v}_m \\ \mathbf{v}_d \end{bmatrix} = \mathbf{0}, \qquad (18)$$

where $\mathbf{C}$ is the global matrix obtained by a standard assembly of elemental matrices which, in two dimensions, for an element e with p nodes on the intersection $\Gamma^{(e)}$ between its boundary and the boundary of the RVE, reads

$$\mathbf{C}^{(e)} = \begin{bmatrix} \int_{\Gamma^{(e)}} N_1^{(e)} n_1\, dA & 0 & \cdots & \int_{\Gamma^{(e)}} N_p^{(e)} n_1\, dA & 0 \\ 0 & \int_{\Gamma^{(e)}} N_1^{(e)} n_2\, dA & \cdots & 0 & \int_{\Gamma^{(e)}} N_p^{(e)} n_2\, dA \\ \int_{\Gamma^{(e)}} N_1^{(e)} n_2\, dA & \int_{\Gamma^{(e)}} N_1^{(e)} n_1\, dA & \cdots & \int_{\Gamma^{(e)}} N_p^{(e)} n_2\, dA & \int_{\Gamma^{(e)}} N_p^{(e)} n_1\, dA \end{bmatrix}, \qquad (19)$$

Here we have assumed that the nodes of element e lying on $\Gamma^{(e)}$ are locally numbered 1 to p sequentially, $N_i^{(e)}$ are the associated local shape functions and n_1 and n_2 are the global Cartesian components of the outward unit normal to the boundary of the discretized RVE.

3.3 Newton–Raphson Solution Procedure

The final reduced set of non-linear algebraic finite element equations to be solved is obtained by introducing representation (16)–(17) for $\boldsymbol{\eta}$ and $\tilde{\mathbf{u}}^{n+1}$ in (15). After straightforward matrix manipulations, this gives

$$\begin{bmatrix} \mathbf{g}_i \\ \mathbf{g}_m + \boldsymbol{\alpha}^T \mathbf{g}_d \end{bmatrix} = \begin{bmatrix} \mathbf{0} \\ \mathbf{0} \end{bmatrix}, \qquad (20)$$

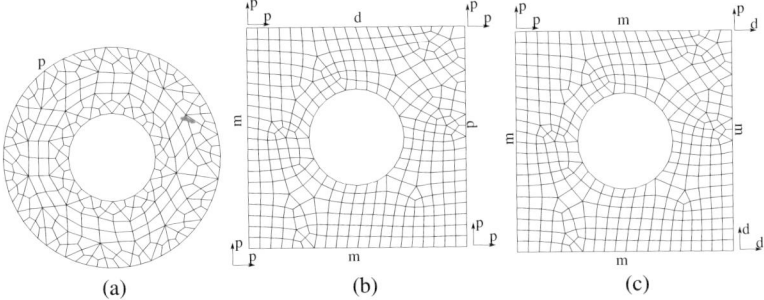

Fig. 1 Boundary nodes for (a) linear b.c., (b) periodic b.c., and (c) uniform traction b.c.

where $\mathbf{g}_i$, $\mathbf{g}_m$ and $\mathbf{g}_d$ are the components of the global vector

$$\mathbf{g} \equiv \int_{{}^h\bar{\Omega}} \mathbf{G}^T \bar{\mathbf{P}}^{n+1}(\bar{\mathbf{F}}^{n+1} + \mathbf{G}\tilde{\mathbf{u}}^{n+1}, \bar{\boldsymbol{\xi}}^n) \, dV, \qquad (21)$$

associated, respectively, with the internal, master and dependent degrees of freedom of the RVE.

The Newton–Raphson iterative scheme is used to solve (20) resulting in a robust and efficient numerical scheme characterised by a quadratic rate of asymptotic convergence.

4 Enhancements of the Basic Scheme

4.1 Interface Discretization

The interaction surface can be redefined by introducing an independently discretized interface that is based on a finite element type interpolation similar to the procedure described in [2]. This type of interface discretisation separates the RVE discretization from the interface constraints, and can be used to reduce the time cost substantially, at a negligible loss of accuracy. The interface is imposed by master-slave type dependencies that are derived from a finite element type interpolation on the boundary, as shown in Figure 2.

4.2 Sub-Stepping

A commonly encountered problem in multi-scale simulation of complex non linear problems is the local failure of the solution procedure at one or more RVEs, which

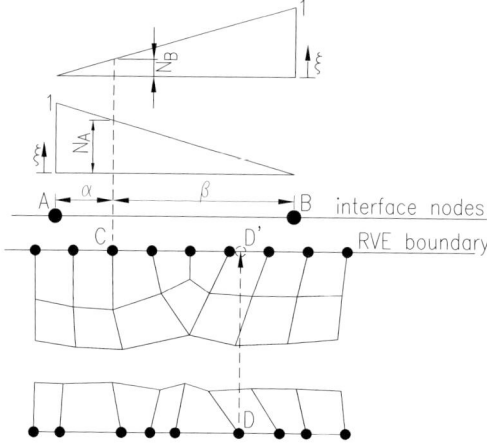

Fig. 2 Finite Element interpolation of the RVE boundary.

enforces smaller time steps for the overall problem. We propose local sub-stepping as a means of acquiring a reliable estimate, so that the overall problem can be solved using larger time steps, as well as preserving the asymptotically quadratic convergence. Typically, a generally non physical intermediate step is introduced recursively until convergence is achieved. The resulting displacement field is then used as an estimate for the originally failed step.

5 Numerical Example

The purpose of this numerical example is to demonstrate the overall performance of the two-scale computational algorithm in the solution of macroscopic initial boundary value problems characterized by a multi-scale constitutive description of the underlying material. In particular the improvements of the basic scheme achieved with the enhancements of the basic procedure described in Section 4 are illustrated.

A plate measuring 20×36 cm containing a 5 cm radius hole in its centre is stretched longitudinally by 0.5%. Because of symmetry, only a quarter of the plate is considered shown in Figure 4(a). The plate is assumed to be made of a porous elasto-plastic material with computational multi-scale constitutive law described by a unit square RVE with 15% void placed in the centre. Plane stress condition is assumed and the RVE matrix material is modelled by the von Mises type elastoplastic law with Young's modulus $E = 70$ GPa, Poisson's ratio $\nu = 0.2$, initial yield stress $\sigma_0 = 243$ MPa and linear hardening modulus $H = 0.2$ GPa. Under the large strains assumption a Hencky hyperelasticity-based multiplicative extension of the infinitesimal constitutive law is adopted. Two types of RVE kinematical constraints

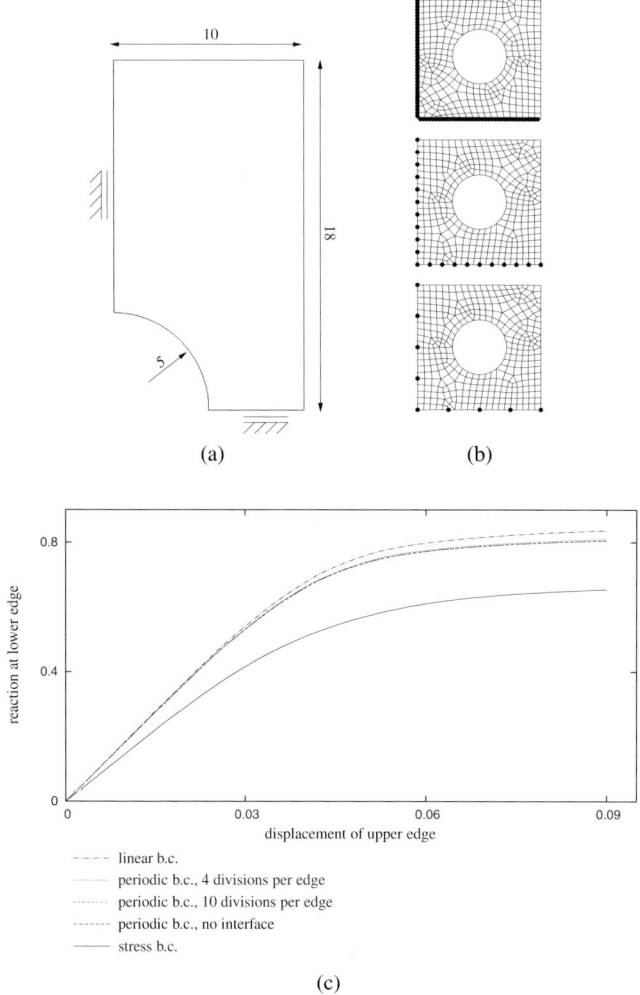

Fig. 3 Comparison of three interface discretisation levels for periodic boundary condition: (a) macro geometry and boundary conditions; (b) RVE FE mesh and interface nodes: (top) without the interface, (middle) interface with 10 and, (bottom) with 4 divisions per side; (c) reaction/displacement diagram.

are considered: periodic boundary fluctuations and minimum kinematical constraint (uniform boundary traction).

The loading program consists in applying incrementally a uniform vertical (stretching) displacement to the nodes of the top edge of the mesh until a total axial straining of 0.5% is achieved.

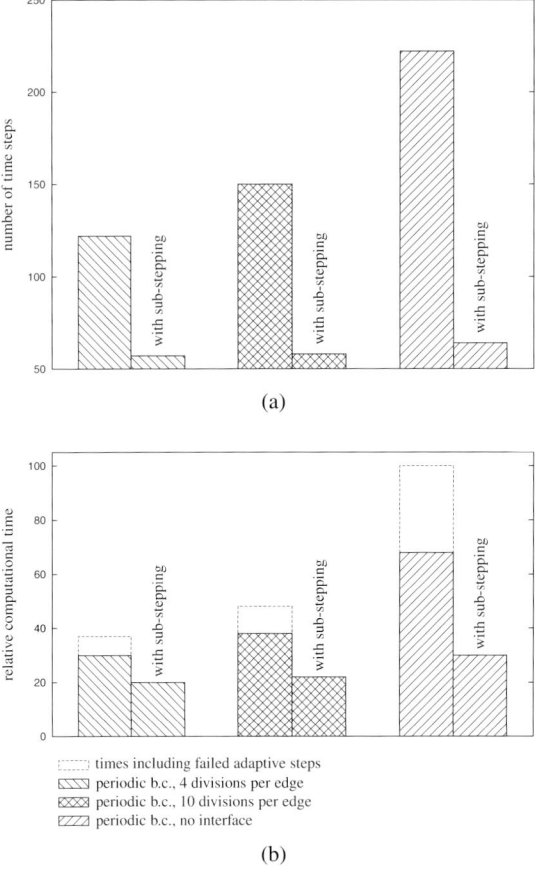

Fig. 4 Comparison of three interface discretisation levels for periodic boundary condition: (a) number of time steps; (b) relative computation time required for solution.

The target number of time steps is set at 50, and the problem is solved for various interface discretization. The solution has been obtained by performing simulations both with and without incremental sub-stepping.

Numerical results are shown in Figures 3 and 4, and include a comparison between different schemes. As depicted in Figure 4(b), up to 5 times improvement has been achieved in computational time with negligible variation of the results (see Figure 3(c)) when reduced interface discretisation was employed together with sub-stepping.

6 Conclusion

A compact and efficient computational framework has been developed for the homogenisation based multiscale finite element analysis of solids, characterised by a direct numerical treatment of kinematical constraints. The computational framework relies on an elegant variational formulation, which, in a natural way, introduces a hierarchy of boundary conditions at the micro-scale, and allows for direct treatment of micro-to-macro transitions. It is illustrated that the efficiency and robustness of the basic multi-scale computational procedure can be significantly improved by introducing independently discretised interfaces and local sub-stepping. It should be noted that the present analysis has been performed for a relatively small two-dimensional problem, and we expect that the computational gains for three-dimensional problems will be much more substantial. This work is currently in progress and the findings will be reported in the near future.

References

1. de Souza Neto EA, Feijóo RA, (2006) Variational Foundations of Multi-Scale Constitutive Models of Solid: Small and Large Strain Kinematical Formulation. National Laboratory for Scientific Computing (LNCC), Brazil, Internal Research and Development Report No. 16/2006.
2. Dettmer W, Perić D (2006) A computational framework for fluid-structure interaction: Finite element formulation and applications. *Comput Methods Appl Mech Engng* 195:5754–5779.
3. Feyel F and Chaboche J-L (2000) FE^2 multiscale approach for modelling the elastoviscoplastic behaviour of long fibre SiC/Ti compiste materials. *Comput Methods Appl Mech Engng* 183.309–330.
4. Hill R (1972) On constitutive macro-variables for heterogeneous solids at finite strain. *Proc Roy Soc London* 326:131–147.
5. Kouznetsova VG, Brekelmans WAM, Baaijens FPT (2001) An approach to micro-macro modelling of heterogeneous materials. *Comput Mech* 27:37–48.
6. Ladevèze P, Loiseau O, Dureisseix D (2001) A micro-macro and parallel computational strategy for highly heterogeneous structures. *Int J Numer Meth Engng* 52:121–138.
7. Mandel J (1971) *Plasticité Classique et Viscoplasticité*, CISM Courses and Lectures No. 97, Springer-Verlag, Udine, Italy.
8. Miehe C, Koch A (2002) Computational micro-to-macro transitions of discretized microstructures undergoing small strains. *Arch Appl Mech* 72:300–317.
9. Michel JC, Moulinec H, Suquet P (1998) Effective properties of composite materials with periodic microstructure: a computational approach. *Comput Methods Appl Mech Engng* 172:109–143.
10. Suquet PM (1987) Elements of homogenization for inelastic solid mechanics. In: Sanchez-Palencia E and Zaoui A (Eds), *Homogenization Techniques for Composite Media*. Springer-Verlag, Berlin.
11. Terada K, Saiki I, Matsui K, Yamakawa Y (2003) Two-scale kinematics and linearization for simultaneous two-scale analysis of periodic heterogeneous solids at finite strain, *Comput Methods Appl Mech Engng* 192:3531–3563.
12. Zohdi TI, Wriggers P (2005) *Introduction to Computational Micromechanics*. Springer, Berlin

Material Characterization Based on Micro-Structural Computations and Homogenization

P. Wriggers[1] and M. Hain[2]

[1]*Institute for Mechanics and Computational Mechanics, Leibniz Universität Hannover, 30167 Hannover, Germany*
E-mail: wriggers@ibnm.uni-hannover.de
[2]*Daimler Chrysler AG, 70546 Stuttgart, Germany*
E-mail: hain@daimler.de

Abstract. Based on a micro-structural finite-element model using computer-tomography scans at micro meter length-scale, damage due to frost within hardened cement paste (HCP) is evaluated. In order to establish microscopic constitutive equations comparison with experimental data at macro-level are performed together with parameter identification. Subsequently, damage due to frost is simulated numerically: the water filled pores of HCP increase in volume during a freezing process which yields an inelastic material behavior. Numerical simulations at micro-structural level are performed for different moistures and temperatures. Upscaling then leads to an effective correlation between moisture, temperature and the inelastic material behaviour. Finally, thermo-mechanically coupling is introduced on the macro-scale and an effective constitutive equation for HCP is developed using the abovementioned temperature-moisture-damage correlation.

Key words: finite elements, damage, frost, multi-scale, cement paste.

1 Micro-Structure of HCP

The finite element model of the micro-structure is based on CT-scans of the hardened cement paste, see Figure 1(a). It includes different material phases: the unhydrated clinker phases, the hydration products as well as the pores. These phases produce in general an inelastic repsonse, see Figure 1(b). Hence elastic and inelastic response at micro-scale level will influence the overall damage behaviour due to frost.

1.1 Effective Elastic Behaviour

The elastic constants for these phases can be found in the literature. The unhydrated clinker phases include reaction products such as C3S, C2S, C3A and C4AF. The Young's modulus of these quantities varies from 125,000 N/mm^2 to 145,000 N/mm^2 [1]. Due to the small volume part of only 2% a mean value of $E = 132,700$ N/mm^2

B.D. Reddy (ed.), IUTAM Symposium on Theoretical, Modelling and Computational Aspects of Inelastic Media, 15–26.
© *Springer Science+Business Media B.V. 2008*

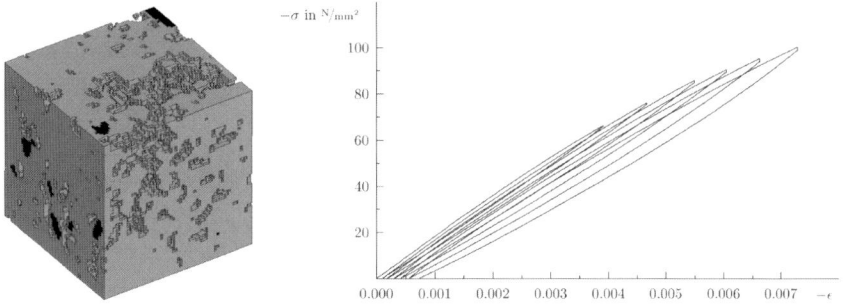

Fig. 1 (a) Micro-structure and (b) stress-strain curve of cement paste.

will be selected for the numerical simulations together with a Poisson ratio of $\nu = 0.30$. The hydration products are summarized by the CSH phase. Its Young's modulus varies from $20,000\pm2,000$ N/mm^2 [1] to $21,700\pm2,200$ N/mm^2 [3]. In the following homogenization a value of $24,000$ N/mm^2 is selected together with a Poisson ratio of 0.24, see also [2].

Homogenization of the cement paste yields probability densities for the effective material parameters, see [5]. Thus randomly distributed elastic material parameters characterize the effective material model. The randomness is described by distribution functions, see [13], and can be introduced at Gauss point level within a finite element discretization. Hence individual elastic properties E_{ig} and ν_{ig} are computed at every integration point $\bullet_{ig}$ by random numbers $rnd_{ig\,1,2} \in [0;1]$, based on the mean values $\bullet^{\mathrm{med}}$ and the standard deviation $\bullet^{\mathrm{std}}$

$$E_{ig} = E^{\mathrm{med}} + \begin{cases} E^{\mathrm{std}}\sqrt{-2\ln rnd_{ig\,1}}\sin(2\pi\, rnd_{ig\,2}); & ig \text{ even}, \\ E^{\mathrm{std}}\sqrt{-2\ln rnd_{ig\,1}}\cos(2\pi\, rnd_{ig\,2}); & ig \text{ odd}. \end{cases} \quad (1)$$

This procedure, used in the same way for ν_{ig}, approximates the Gauss distribution of the effective material properties of cement paste, see [6]. Since the elastic material properties are randomly distributed an inhomogeneous response of a probe under uniform loading can be observed, as depicted in Figure 2. These stress deviations have an influence on the inelastic response.

1.2 Inelastic Micro-Structural Model

A material equation is needed to describe the constitutive behaviour on microstructural level. Many different constitutive models can be found for brittle materials, see e.g. [7, 8]. Since also ductile phases are included in the microstructure a visco-plastic equations, see Perzyna [14], will be combined with an isotropic damage model, see [11]. The constitutive model describes the main mechanical proper-

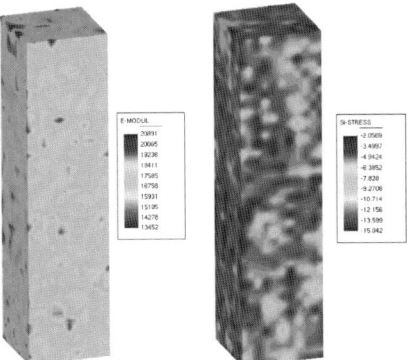

Fig. 2 Randomly distributed Young's modulus and principal stress σ_I within a probe of $40 \times 40 \times 160$ mm under uniform loading.

ties but cannot consider cracks of finite size, furthermore the damage depends here only on the mechanical loading.

The visco-plastic model can be derived from the maximum dissipation postulate which yields a constraint optimizaton problem for the dissipation $\mathcal{D}$

$$-\mathcal{D} \to \min \quad \text{under the constraints} \quad f \leq 0 \quad \text{and} \quad S^u \leq 0. \tag{2}$$

The constraints relate to the yield surface f and to the damage surface S^u. The yield surface is defined in stress space while the damage surface is formulated in strain space. In order to solve the above problem a combined penalty-Lagrange method is used which leads to

$$\mathcal{P}: \quad -\boldsymbol{\sigma}:\dot{\boldsymbol{\epsilon}}^{\text{pl}} - Y\dot{D}^u + \frac{1}{\eta}\phi(f) + \dot{\chi}S^u \to \text{stat}, \tag{3}$$

The penalty parameter $\frac{1}{\eta}$ is associated with the viscosity of the material, $\phi(f)$ defines the penalty functional and χ the Lagrange multiplier. A partial differentiation of $\mathcal{P}$ with respect to the stresses $\boldsymbol{\sigma}$ results in the evolution equation

$$\dot{\boldsymbol{\epsilon}}^{\text{pl}} = \frac{1}{\eta}\phi^+ \frac{\partial f}{\partial \boldsymbol{\sigma}}, \tag{4}$$

in which ϕ^+ denotes the differentiation of the penalty function $\phi(f)$. It is given by

$$\phi^+ = \begin{cases} 0 & ; f \leq 0, \\ f^k & ; f > 0, \end{cases}$$

where $\phi(f)$ was defined as the $(k+1)$-th power of the yield surface f:

$$\phi(f) = \frac{1}{k+1} f^{k+1} \quad \text{for} \quad f > 0.$$

By setting $k = 1$ it is possible to describe nonlinear viscous behaviour. The classical J_2 von Mises model is chosen for the yield surface f which then depends upon the stress deviator dev σ

$$f := \|\text{dev } \sigma\| - \sqrt{\frac{2}{3}} k_f \leq 0, \tag{5}$$

k_f is a material parameter defining the yield limit.

The partial differentiation of $\mathcal{P}$ with respect to the elastic rate of energy Y yields a second evolution equation. It is independent on the strains since the penalty function $\phi(f)$ was defined in stress space

$$\dot{D}^u = \dot{\chi} \frac{\partial S^u}{\partial Y}. \tag{6}$$

The damage surface S^u is defined by an exponential term, see [4],

$$S^u = 1 - \exp\left[-\left(\frac{\epsilon^{\text{eq}} - a}{b}\right)^c\right] - D^u \quad \text{with} \quad \epsilon^{\text{eq}} = \frac{1}{2} \epsilon^{\text{el}} : \mathbb{C}_0 : \epsilon^{\text{el}}, \tag{7}$$

in which a, b, and c are material parameters. Implicit time integration using an Euler backward scheme leads to an algorithmic expression for the evolution of the internal variables

$$\epsilon^{\text{pl}}_{n+1} = \epsilon^{\text{pl}}_n + \frac{\Delta t}{\eta} \phi^+ \frac{\partial f}{\partial \sigma}, \quad D^u_{n+1} = D^u_n + \Delta \chi \frac{\partial S^u}{\partial Y}. \tag{8}$$

The proposed effective material model for the micro-structure is then linearized consistently in order to obtain an efficient numerical scheme for subsequent finite element simulations.

Experimental investigations of cement paste were performed using a probe with dimensions of $40 \times 40 \times 160$ mm. The effective constitutive model includes material parameters $\kappa = (k_f, \eta, \Delta t, b, a)^T$ and c which cannot be found in the literature and can also not be obtained by special tests nor by classical numerical homogenization. Hence a parameter identification has to be performed to obtain the constitutive parameters which describe the mechanical damage.

For the numerical identification of the inelastic material parameters the experimental tested probe will be discretized using finite elements. At the lower side all displacements are fixed and the specimen is loaded at the upper side by a prescribed displacement. Due to the nonlinearities involved the identification is quite time consuming and thus was performed in parallel on a client-server system using six servers.

The identified parameters $\kappa^\star$ were computed as $k_f \approx 48$, $\eta \approx 8{,}240$, $\Delta t \approx 0.0265$, $a \approx 1{,}000$, $b \approx 2{,}920$, while c was pre-selected as $c = 3$ based on [4]. Experimental and numerical results are compared for the identified parameters in Figure 3. The accuracy is satisfactory.

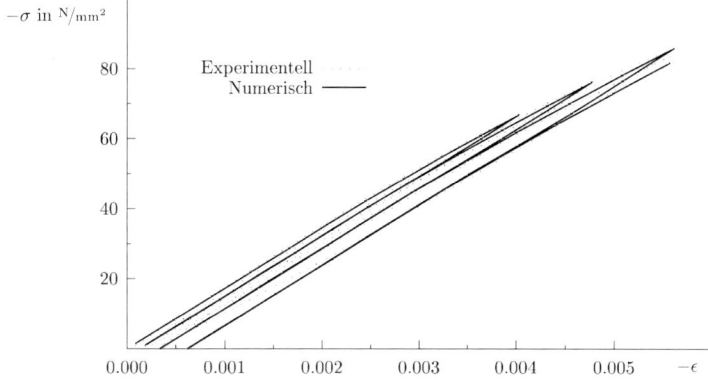

Fig. 3 Identified stress-strain curve for cement paste.

2 Micro-Structural damage Due to Frost

Numerical simulations which model damage due to frost at micro-structural level need a model for the process of freezing of water which describes the volume change of about 9 Vol.%. The freezing process is temperature dependent hence the temperature field has to be determined within the micro-structure by a coupled thermo-mechanical analysis. Within this analysis it is assumed that the temperature gradient is negligible within the micro-structure due to its smallness. Thus the temperature can be controlled within the micro-structure by an external parameter which depends on the actual load step and will be called prescribed temperature in the following. Hence thermo-mechanical coupling is not necessary and all simulations can be performed based on mechanical analysis.

An absorption of water within thermal load cycles can be observed in experiments which leads to a continuous increase of the water contents. This effect is not considered here, instead it will be assumed that the micro-structure is in a steady state which was determined in [10] numerically. Figure 4 depicts some cut planes of microstructures at different level of water contents w_h. As expected, the water (here in black) first covers the boundary of a pore and after that the inner part of the pore. The unhydrated phases are depicted by black dots while the hydration products are shown in grey in Figure 4. The pores are white.

A volume increase of waterfilled pores occurs once the prescribed temperature is below a given freezing temperature Θ_f which depends on the actual radius of the pore. The pore radii vary from 0.5 μm to 2.0 μm in the micro-structures stemming from the CT scans. This is related to the resolution of the scans which are not able to resolve smaller pores which can reach a radius down to 10 nm. Thus the finite element simulation neglects the influence of smaller pores. A relation between the freezing temperature Θ_f of water in micro pores and the pore radius is provided by

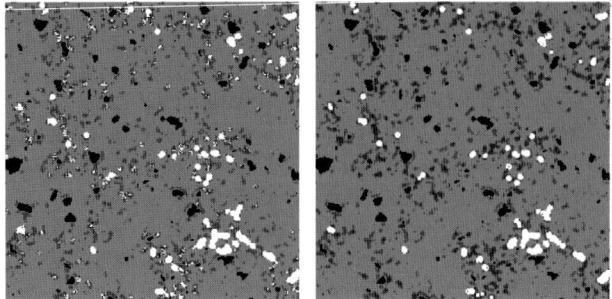

Fig. 4 Water filled cement paste of (256 × 256 μm) for water contents of $w_h = 0.70$ and $w_h = 0.80$.

Table 1 Freezing temperatures of water in micro pores.

r_{pore} in μm	Θ_f in ^{0}C
0.5	−0.18058
1.0	−0.09021
1.5	−0.06013
2.0	−0.04509

$$r_{\text{pore}} = \frac{3.3}{\ln\left(\frac{273.15 \text{ K}}{\Theta_f}\right)} + 10 \quad \Rightarrow \quad \Theta_f = \exp\left[\ln 273.15 \text{ K} - \frac{3.3}{r_{\text{pore}} - 10}\right], \quad (9)$$

see [12]. Here Θ_f is measured in Kelvin (273.15 K = $^\circ$C) and the pore radius r_{pore} in Angström (1 Å = 10^{-4} μm). The relation considers a lower freezing temperature of water in smaller pores due to the influence of an increaseing surface tension. The freezing temperatures of micro pores with different sizes are tabulated in Table 1. The real cement paste includes, as mentioned above, also pores of nanosize with very low freezing limits which are not resolved by the CT-scans. Hence the resolution of the CT-Scans limits the accuracy of the numerical frost simulation.

A thermal strain ϵ^Θ is defined to describe the volume increase within the freezing process

$$\epsilon^\Theta := \frac{\alpha_t}{e} \exp\left[1 - \exp\left(\frac{\Theta - \Theta_f}{0.008}\right)\right] \mathbf{1}, \quad (10)$$

in which e is the Euler number, $\alpha_t = \sqrt[3]{1.09} - 1$ describes a thermal strain coefficient and Θ is the prescribed temperature of the micro structure. The latter is controlled by

$$\Theta := a_0 + a_1 t + a_2 t^2 \quad (11)$$

where t defines the actual time step of the mechanical analysis and a_i are parameters. The water filled pores are described by a linear elastic material including thermal strains. A possible visco-plastic response is neglected since there are at the moment no material models for ice at micro-structural level available.

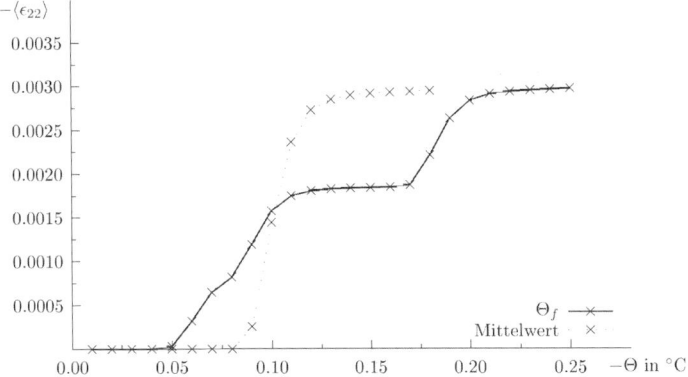

Fig. 5 Artificial thermal strain.

The elastic properties of ice vary from $E = 9{,}170\ldots 9{,}940$ N/mm^2, $\nu = 0.31\ldots 0.36$, see [9], to $E = 9{,}000$ N/mm^2, $\nu = 0.33$, see [15]. For the numerical simulation the mean values $E = 9{,}280$ N/mm^2 and $\nu = 0.333$ are selected.

The developed material model is then linked to the waterfilled pores and the prescribed temperature will be decreased within the finite element simulation. The expansion of ice, as an outcome of the freezing process, leads to temperature strains. Figure 5 depicts the resulting prescribed temperature strain within a micro-structure with 64^3 finite elements and a water contents of $w_h = 0.80$. Within the simulation a radii dependent freezing temperature Θ_f was used for all pores based on the mean value $\Theta_f = -0.097°$C. It is clear that the different pore radii have a strong influence and thus cannot be neglected within a numerical simulation.

A reliable numerical simulation of frost needs a statistical analysis of different water filled micro-structures and thus different finite element discretizations. These simulations are carried out for different degrees of water filling. The statistical analysis employed is based on 64 different micro-structures for each degree of water filling of the size $64 \times 64 \times 64$ μm. The total solution time of such analysis is about 5 h when using a CG-solver on a (P4 with 3 GHz).

A number of $n = 64$ effective damage-temperature relations $\langle d(\Theta)_i \rangle$ are determined for a filling degree of $w_h = 0.80$ under the assumption of an isotropic total damage of the micro-structure. Here a volumetric averaging of the damage $d(\Theta)$ makes sense for moderate damage, otherwise the assumption of isotropy might not be fulfilled. This leads to

$$\langle d(\Theta) \rangle = \frac{1}{|\Omega|} \int_\Omega d(\Theta) \, d\Omega \, . \tag{12}$$

As can be observed in Figure 6 most simulations yield an effective damage of $\langle d(\Theta) \rangle \approx 0.05$. However there are some which depict much more damage up to

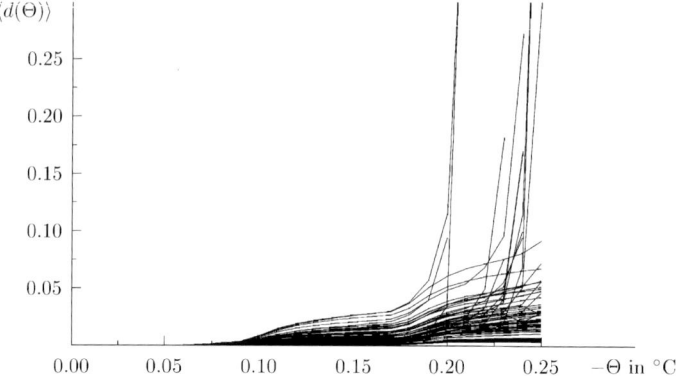

Fig. 6 Effective damage-temperature relation for ($w_h = 0.80$).

$\langle d(\Theta) \rangle \approx 0.30$. This result is in good agreement with experiments in which damage due to frost is only observed in very small areas of the probe.

The assumption of a polynomial damage-temperature relation can now be used to approximate mean value and standard deviation of the effective damage-temperature relations by

$$d(\Theta)^{\text{med}} := \frac{1}{n} \sum_{i=1}^{n} \langle d(\Theta)_i \rangle , \quad d(\Theta)^{\text{std}} := \sqrt{\frac{1}{n-1} \sum_{i=1}^{n} [\langle d(\Theta)_i \rangle - d(\Theta)^{\text{med}}]^2} . \tag{13}$$

Based on the least square method it is possible to determine the polynomials

$$D(\Theta)^{\text{med}} := \sum_{i=0}^{i \leqslant 8} c_i^{\text{med}} \Theta^i , \quad D(\Theta)^{\text{std}} := \sum_{i=0}^{i \leqslant 6} c_i^{\text{std}} \Theta^i , \tag{14}$$

which approximate $d(\Theta)^{\text{med}}$ and $d(\Theta)^{\text{std}}$. Additonally frost simulations were carried out for $n = 64$ different micro-structures with filling degrees of $w_h = 0.75$ and $w_h = 0.70$. The computed effective damage-temperature relations are shown in Figure 7. There is almost no difference between the results for $w_h = 0.75$ and $w_h = 0.70$. This is correlated to the free pore space which allows almost a strain free expansion of the ice during freezing. Based on the high degree of the polynomial there occur small nonphysical oscillations for $\Theta < -0.05$ °C, hence the polynomials are only used within the range $-0.05°C \leqslant \Theta \leqslant -0.25°C$ where damage occurs.

The qualitative results are now applied to characterize the frost damage at macro-level.

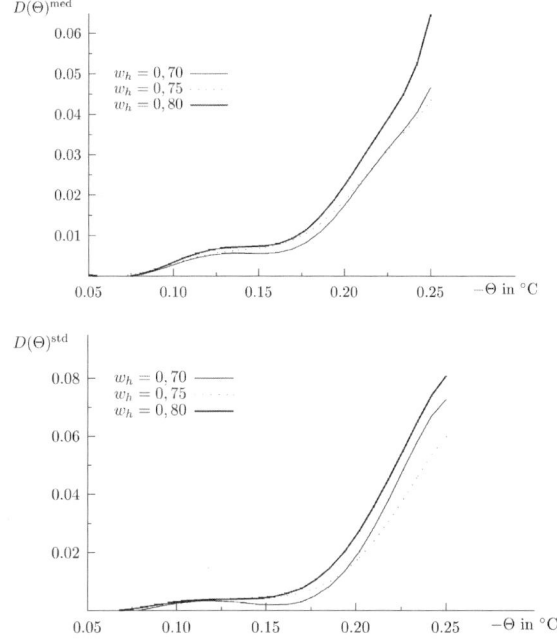

Fig. 7 Mean value and standard deviation of $D(\Theta)$ for different degrees of water filling.

3 Effective Inelastic Behaviour with Frost Damage

Upscaling of the results at micro-level, see Section 1.2, leads to a thermo-mechanical coupled constitutive model at macro-level. There the temperature is computed by assuming nonstationary heat conduction which then provides temperatures at each Gauss point. Additionally the isotropic damage depends on the mechanical stress as well as on the moisture and the temperature. This leads to an additive split of the damage into a thermal D^Θ and a mechanical D^u part

$$D := D^u + D^\Theta. \tag{15}$$

The mechanical part was already considered in Section 1.2. For the thermal part the damage surface S^Θ is defined which describes the temperature-moisture-damage relation stemming from the micro-structural computations

$$S^\Theta - D^\Theta \leqslant 0. \tag{16}$$

For $S^\Theta \leqslant 0$ no further damage due to frost occurs otherwise the damage increases. The damage surface S^Θ is evaluated at each integration point $\bullet_{ig}$ depending on random numbers rnd_1, rnd_2 and the polynomials $D(\Theta, w_h)^{\text{med}}$ and $D(\Theta, w_h)^{\text{std}}$

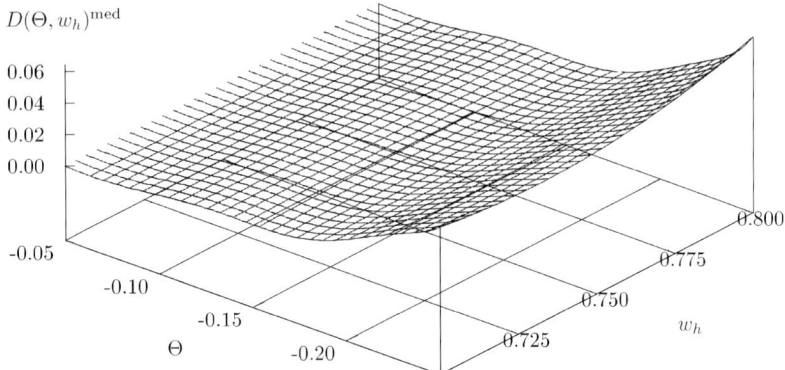

Fig. 8 Mean value of the effective temperature-moisture-damage relation.

$$S_{ig}^{\Theta} := D(\Theta, w_h)^{\text{med}} + \begin{cases} D(\Theta, w_h)^{\text{std}} \sqrt{-2\ln rnd_1} \sin(2\pi\, rnd_2); & ig \text{ even,} \\ D(\Theta, w_h)^{\text{std}} \sqrt{-2\ln rnd_1} \cos(2\pi\, rnd_2); & ig \text{ odd.} \end{cases} \quad (17)$$

The polynomials interpolate the mean value and the standard deviation of the microstructural results for damage due to freezing. The mean value $D(\Theta, w_h)^{\text{med}}$ is defined by

$$D(\Theta, w_h)^{\text{med}} := \sum_{i=0}^{i \leq 8} p_i(w_h)^{\text{med}} \Theta^i, \quad (18)$$

where $p_i(w_h)^{\text{med}}$ is a polynomial of second order

$$p_i(w_h)^{\text{med}} := a_{i\,0}^{\text{med}} + a_{i\,1}^{\text{med}} w_h + a_{i\,2}^{\text{med}} w_h^2. \quad (19)$$

The coefficients of this polynomial are determined from the effective temperature-damage relation. A three-dimensional plot can be found in Figure 8 for the mean value. The standard deviation has a similar form.

3.1 Numerical Simulation of Frost Damage

Nonstationary heat conduction is now used within a themo-mechanical coupled numerical simulation at macro level to compute the damage of cement paste due to frost. For that the additional parameters related to the thermal behaviour of cement paste are used: density $\rho = 2{,}120$ kg/m^3, specific heat capacity $c = 1{,}000$ J/kg K, heat conduction $k = 1.4$ W/m K and thermal expansion $\alpha_t = 10^{-5}$ 1/K. The variation of the material and damage variables, as discussed above, are incorporated at Gauss point level within the finite element discretization.

Fig. 9 Development of damage due to frost in a $40 \times 40 \times 160$ mm specimen at times $t = 1$, $t = 50$ and $t = 100$.

A specimen of size $40 \times 40 \times 160$ mm is now discretized using $10 \times 10 \times 40$ finite elements. The specimen is fixed at its lower side. A constant thermal load of $\Theta = -0.25°C$ is prescribed at the upper part. Mechanical loading is not considered. The computed damage is depicted in Figure 9 at different time steps $t \in [0; 100]$ for a moisture of $w_h = 0.80$. The damage due to frost increases during the cooling process as expected. Since the effective material parameters vary randomly also the damage due to frost is not equally distributed. Damage starts directly at the upper surface. This is physically not completely true, here the damage is just beneath the surface due to frost suction. However the latter effect was not included in the present model and needs additional investigations at micro-structural level.

4 Conclusion

In this work a thermo-mechanical analysis of water filled cement paste undergoing freezing is considered. Based on micro-structural investigations on μm scale constitutive relations are obtained and homogenized which then can be applied on the macro-scale. Furthermore the random distribution of the micro-structure is taken into account leading to constitutive parameters on macro-scale which are described by mean values and standard deviation. For the analysis tools like homogenization, parameter identification and finite element simulations are neccessary to describe the constitutive behaviour correctly within an upscaling process. By repeated upscaling a larger range of length scales can be investigated leading to the description of mortar or concrete. Another possibility of this type of analysis could be goal oriented design of micro-structures in order to obtain specific effective properties at macro-scale and thus to be able to enhance the performance of a material.

References

1. P. Acker. Micromechanical analysis of creep and shrinkage mechanisms. In *Creep, Shrinkage and Durability Mechanics of Concrete and Other Quasi-Brittle Materials*, F.-J. Ulm, Z. P. Bazant, and F. H. Wittmann (Eds.), pp. 15–25, Elsevier, 2001.
2. O. Bernard, F. J. Ulm, and E. Lemarchand. A multiscale micromechanics-hydration model for the early age elastic properties of cement-based materials. *Cement and Concrete Research*, 33:1293–1309, 2003.
3. G. Constantinides and F. J. Ulm. The effect of two types of CSH on the elasticity of cement-based materials: Results from nanoindentation and micromechanical modeling. *Cement and Concrete Research*, 34:67–80, 2004.
4. D. Tikhomirov. *Theorie und Finite-Element-Methode für die Schädigungsbeschreibung in Beton und Stahlbeton*. Dissertation, Universität Hannover, 2000.
5. M. Hain and P. Wriggers. On the numerical homogenization of hardened cement paste. *Computational Mechanics*, On-line first, 2007.
6. J. Hartung. *Statistik: Lehr- und Handbuch der angewandten Statistik*. Oldenbourg Verlag, 1993.
7. G. Hofstetter and H. A. Mang. *Computational Mechanics of Reinforced Concrete Structures*. Vieweg, Berlin, 1995.
8. R. H. J. Peerlings. *Enhanced Damage Modelling for Fracture and Fatigue*. Dissertation, Technische Universiteit Eindhoven, 1999.
9. J. Pohe. *Ein Beitrag zur Stoffgesetzentwicklung für polykristallines Eis*. Dissertation, Ruhr Universität Bochum, 1993.
10. M. Koster, J. Hannawald, and W. Brameshuber. Simulation of water permeability and water vapor diffusion through hardened cement paste. *Computational Mechanics*, 37:163–172, 2006.
11. J. Lemaitre. *A Course on Damage Mechanics*. Springer, Berlin/New York, 1996.
12. N. Stockhausen. *Die Dilatation hochporöser Festkörper bei Wasseraufnahme und Eisbildung*. Dissertation, Technische Universität München, 1981.
13. R. Pukl, M. Jansta, J. Cervenka, M. Vorechovsky, D. Novak, and R. Rusina. Spatial variability of material properties in nonlinear computer simulation. In *Computational Modelling of Concrete Structures*, G. Meschke, R. de Borst, H. Mang, and N. Bicanic (Eds.), pp. 891–896, Taylor and Francis, London, 2006.
14. J. C. Simo and T. J. R. Hughes. *Computational Inelasticity*. Springer, New York/Berlin, 1998.
15. G. W. Scherer. Crystallization in pores. *Cement and Concrete Research*, 29:1347–1358, 1999.

Relaxed Potentials and Evolution Equations for Inelastic Microstructures

Klaus Hackl and Dennis M. Kochmann

Institute of Mechanics, Ruhr-University Bochum, D-44780 Bochum, Germany
E-mail: {klaus.hackl, dennis.kochmann}@rub.de

Abstract. We consider microstructures which are not inherent to the material but occur as a result of deformation or other physical processes. Examples are martensitic twin-structures or dislocation walls in single crystals and microcrack-fields in solids. An interesting feature of all those microstructures is, that they tend to form similar spatial patterns, which hints at a universal underlying mechanism. For purely elastic materials this mechanism has been identified as minimisation of global energy. For non-quasiconvex potentials the minimisers are not anymore continuous deformation fields, but small-scale fluctuations related to probability distributions of deformation gradients, so-called Young measures. These small scale fluctuations correspond exactly to the observed microstructures of the material. The particular features of those, like orientation or volume fractions, can now be calculated via so-called relaxed potentials. We develop a variational framework which allows to extend these concepts to inelastic materials. Central to this framework will be a Lagrange functional consisting of the sum of elastic power and dissipation due to change of the internal state of the material. We will obtain time-evolution equations for the probability-distributions mentioned above. In order to demonstrate the capabilities of the formalism we will show an application to crystal plasticity.

Key words: inelasticity, relaxation, microstructures, continuum mechanics.

1 Introduction

We investigate inelastic materials described by so-called internal or history-variables. Examples include elastoplastic but also damaged materials or those undergoing phase-transformations. By considering associated potentials in a time-incremental setting, it is possible to model the onset of the formation of microstructures but not their subsequent evolution [1–5]. Recently, some new approaches have been developed to find an energy-based formulation for the evolution of inelastic materials, see [6, 7]. This one does not require any derivatives and thus is especially suited for the treatment of microstructure. Here, we will pursue a related but different approach based on the use of Young measures. An application of a similar concept to shape-memory-alloys can be found in [8–10].

2 Minimum Principles

In an isothermal setting the state of a general inelastic material will be defined by its deformation gradient $F = \nabla \phi$ and a collection of internal variables: K. Denoting the specific Helmholtz free energy by $\Psi(F, K)$ we introduce thermodynamically conjugate stresses by

$$P = \frac{\partial \Psi}{\partial F}, \qquad Q = -\frac{\partial \Psi}{\partial K}. \tag{1}$$

The evolution of K is then governed either by a so-called inelastic potential $J(K, Q)$ or its Legendre-transform, the dissipation functional

$$\Delta(K, \dot{K}) = \sup \left\{ \dot{K} : Q - J(K, Q) \,\big|\, Q \right\}. \tag{2}$$

The evolution equations are then given in the two equivalent forms

$$\dot{K} \in \frac{\partial J}{\partial Q}, \qquad Q \in \frac{\partial \Delta}{\partial \dot{K}}. \tag{3}$$

The entire evolution problem can now be described in terms of two minimum principles, where we follow ideas presented in [5, 11, 12]. Considering the Gibbs free energy of the entire body

$$\mathcal{I}(t, \phi, K) = \int_\Omega \Psi(\nabla \phi, K)\, \mathrm{d}V - \ell(t, \phi) \tag{4}$$

the deformation is given by the principle of minimum potential energy

$$\phi = \arg\min \left\{ \mathcal{I}(t, \phi, K) \,\big|\, \phi = \phi_0 \text{ on } \Gamma_\varphi \right\}. \tag{5}$$

Here Ω is the material body, Γ_φ a subset of its boundary and $\ell(t, \phi)$ the potential of external forces. On the other hand introducing the Lagrange functional

$$\mathcal{L}(\phi, K, \dot{K}) = \frac{d}{dt} \Psi(\nabla \phi, K) + \Delta(K, \dot{K}) \tag{6}$$

we can write the evolution equation (3) in the form

$$\dot{K} = \arg\min \left\{ \mathcal{L}(\phi, K, \dot{K}) \,\big|\, \dot{K} \right\}. \tag{7}$$

For a thorough investigation of this principle and its relation to the principle of maximum dissipation, see [13]. For rate-independent materials the principle (7) enables us to account for instantaneous change of the value of K, because it can be integrated to yield the balance law

$$\Psi(\nabla \phi, K_1) - \Psi(\nabla \phi, K_0) = -D(K_0, K_1), \tag{8}$$

where

$$D(\boldsymbol{K}_0, \boldsymbol{K}_1) = \inf \left\{ \int_0^1 \Delta(\boldsymbol{K}(s), \dot{\boldsymbol{K}}(s)) \, \mathrm{d}s \, \Big| \, \boldsymbol{K}(0) = \boldsymbol{K}_0, \boldsymbol{K}(1) = \boldsymbol{K}_1 \right\} \quad (9)$$

is called dissipation-distance [12]. When applied to a finite time-increment $[t_n, t_{n+1}]$ equation (8) gives rise to an approximate formulation, where $\boldsymbol{\phi}_{n+1}$ and $\boldsymbol{K}_{n+1}$ at time t_{n+1} are determined for given loading at time t_{n+1} and value of the internal variables $\boldsymbol{K}_n$ at time t_n via the following principle [11, 12]

$$\{\boldsymbol{\phi}_{n+1}, \boldsymbol{K}_{n+1}\} =$$
$$\arg\min \left\{ \int_\Omega \{\Psi(\nabla\boldsymbol{\phi}, \boldsymbol{K}) + D(\boldsymbol{K}_n, \boldsymbol{K})\} \, \mathrm{d}V - \ell(t_{n+1}, \boldsymbol{\phi}) \, \Big| \, \boldsymbol{\phi}, \boldsymbol{K} \right\}. \quad (10)$$

3 Young Measures

Carrying out the minimisation with respect to $\boldsymbol{K}$ in (10) beforehand gives the so-called condensed energy

$$\Psi_{\boldsymbol{K}_n}^{\mathrm{cond}}(\boldsymbol{F}) = \inf \left\{ \Psi(\boldsymbol{F}, \boldsymbol{K}) + D(\boldsymbol{K}_n, \boldsymbol{K}) \, \Big| \, \boldsymbol{K} \right\} \quad (11)$$

which has been used in the literature to calculate the onset of microstructures [1–5]. This approach, however, is not suitable to describe the evolution of microstructures, because then the internal variables already exhibit a microstructure at the beginning of the time-increment as a result of a relaxation process in the preceding time-increment, or they are microstructured through the whole course of continuous evolution. Hence, they have to be given in the form of so-called Young–measures. We are going to give some concepts now, how such a formulation might be derived.

Young measures are probability distributions $\lambda_F > 0$ given for example for the deformation gradient, i.e. on $\mathrm{GL}(d)$, and dependent on the material point. Thus they have the following properties:

$$\int \lambda_F \, \mathrm{d}\boldsymbol{F} = 1, \quad \int \lambda_{\bar{F}} \, \bar{\boldsymbol{F}} \, \mathrm{d}\bar{\boldsymbol{F}} = \boldsymbol{F}. \quad (12)$$

Moreover in the case of the deformation gradient, the probability distribution has to be compatible, i.e. realisable by a deformation field $\boldsymbol{\phi}$. This means that

$$\frac{1}{\Omega_{\mathrm{rep}}} \int_{\Omega_{\mathrm{rep}}} \Psi(\nabla\boldsymbol{\phi}) \, \mathrm{d}V = \int_{\mathrm{GL}(d)} \lambda_{\bar{F}} \Psi(\bar{\boldsymbol{F}}) \, \mathrm{d}\bar{\boldsymbol{F}} \quad (13)$$

has to hold for all quasiconvex potentials Ψ. In this case we call $\lambda_F \in \mathrm{GYM}$ a gradient Young measure.

It is now, at least in principle, possible to define a relaxed energy and dissipation functional via cross-quasiconvexication as

$$\Psi^{\text{rel}}(F, \lambda_K) = \inf \left\{ \int \Lambda_{\bar{F},\bar{K}} \Psi(\bar{F}, \bar{K}) \, d\bar{K} \, d\bar{F} \, \Big| \, \Lambda_{\bar{F},\bar{K}} ; \int \Lambda_{\bar{F},\bar{K}} \, d\bar{K} \, d\bar{F} = 1, \right.$$
$$\left. \int \Lambda_{\bar{F},\bar{K}} \, d\bar{K} \in \text{GYM}, \int \Lambda_{\bar{F},K} \, d\bar{F} = \lambda_K, \int \Lambda_{\bar{F},\bar{K}} \bar{F} \, d\bar{K} \, d\bar{F} = F \right\}, \quad (14)$$

$$\Delta^*(\dot{\lambda}_K) = \inf \left\{ \int \Lambda_{K_0, K_1} D(K_0, K_1) \, dK_0 \, dK_1 \, \Big| \, \Lambda_{K_0, K_1} ; \right.$$
$$\left. \int \Lambda_{K_0, K_1} \, dK_0 \, dK_1 = 1, \int \Lambda_{K_0, K_1} \, dK_0 = \dot{\lambda}_K, \int \Lambda_{K_0, K_1} \, dK_1 = -\dot{\lambda}_K \right\}. \quad (15)$$

Related concepts have already been introduced in somewhat different settings in [3]. With these definitions we recover the original principles (5) and (7), with the only difference that the internal variables K have been replaced by the Young measures λ_K.

4 Approximation via Lamination

In general the expressions (14) and (15) will be very hard to compute. One possible approximation is via so-called lamination. Applications of this procedure to the time-incremental problem can be found in [1]. For brevity we will restrict ourselves here to first-order laminates. Everything stated below can be extended to general laminates in an essentially straightforward manner, but the details of this may become very cumbersome which would exceed the scope of this paper.

A laminate of first order is characterised by N volume fractions λ_i separated by parallel planes with normal vector b, see Figure 1. To every volume fraction there corresponds a value K_i of the internal variable. Moreover in every volume fraction we have the deformation gradient

$$F_i = F(1 + a_i \otimes b). \quad (16)$$

The deformation gradients differ only by tensors of rank one, meaning that there exists a corresponding deformation field. We impose the volume average $\sum_{i=1}^{N} \lambda_i F_i = F$ which is equivalent to $\sum_{i=1}^{N} \lambda_i a_i = 0$. We consider the normal vector b as ingrained into the material, because changing it would require a change of the internal variable and thus lead to dissipation. The amplitudes a_i on the other hand can be changed purely elastically. This suggests to define a relaxed energy as

$$\Psi^{\text{rel}}(F, \lambda, K, b) = \inf \left\{ \sum_{i=1}^{N} \Psi(F_i, K_i) \, \Big| \, a_i ; \sum_{i=1}^{N} \lambda_i a_i = 0 \right\}, \quad (17)$$

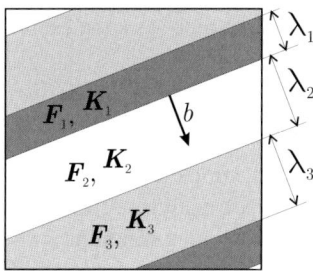

Fig. 1 First-order laminate for $N = 3$ with normal vector $\boldsymbol{b}$.

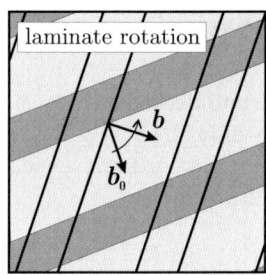 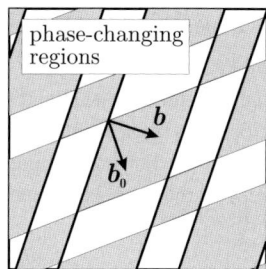

Fig. 2 Rotation of the original laminate with old normal vector $\boldsymbol{b}_0$ to the new normal vector $\boldsymbol{b}$. The right graphic highlights in white those regions changing their phase membership upon rotation and hence causing dissipation.

where we introduced the abbreviations $\lambda = \{\lambda_1, \ldots, \lambda_N\}$ and $\boldsymbol{K} = \{\boldsymbol{K}_1, \ldots, \boldsymbol{K}_N\}$.

If we further assume that the lamination respects the ordering $\{1, \ldots, N\}$ and that the normal vector remains fixed, the relaxation of the dissipation is given by

$$\Delta^*(\lambda, \boldsymbol{K}, \dot{\lambda}, \dot{\boldsymbol{K}}) = \sum_{i=1}^{N} \lambda_i \, \Delta(\boldsymbol{K}_i, \dot{\boldsymbol{K}}_i) + \inf\Bigg\{ \sum_{i,j=1}^{N} \Delta\lambda_{ij}\, D(\boldsymbol{K}_i, \boldsymbol{K}_j) \,\Big|\, \Delta\lambda_{ij}; $$
$$\sum_{i=1}^{N} \Delta\lambda_{ij} = \dot{\lambda}_j, \sum_{j=1}^{N} \Delta\lambda_{ij} = \dot{\lambda}_i, \Delta\lambda_{ij} = 0 \text{ for } |(i-j) \bmod N| \neq 1 \Bigg\}. \quad (18)$$

Now once again from (7) we obtain evolution equations for λ and $\boldsymbol{K}$ for fixed $\boldsymbol{b}$. As can be seen from Figure 2, however, the change of $\boldsymbol{b}$ is not continuous but associated with a fixed amount of dissipation given by

$$D_{\mathrm{b}}(\lambda, \boldsymbol{K}) = \sum_{i,j=1}^{N} \lambda_i \lambda_j D(\boldsymbol{K}_i, \boldsymbol{K}_j). \quad (19)$$

A jump in orientation will now take place as soon as it becomes energetically favorable. This gives

$$\inf\left\{\Psi^{\text{rel}}(F, \lambda, K, b) - \Psi^{\text{rel}}(F, \lambda, K, b_0) \mid b; |b| = 1\right\} + D_b(\lambda, K) \leq 0 \quad (20)$$

for given b_0, λ, K. Equation (20) completes the description of the inelastic evolution of a first-order laminate. A deeper analysis of the concept laid out above will be published in [14]. Because the concept is still quite formal in nature we will specify it now to the case of crystal plasticity.

5 Crystal Plasticity with a Single Slip-System

We are going to demonstrate the general scheme introduced above by applying it to the crystal plasticity model introduced originally in [11]. Let $\kappa > 0$ be a hardening modulus and $\mu > 0$ denote a Lamé parameter. Then, we assume an energy density for an incompressible neo-Hookean material in the form

$$\Psi(F_e, p) = \tfrac{1}{2}\mu \operatorname{tr} F_e^T F_e + \kappa p^4, \quad \det F = 1. \quad (21)$$

Multiplicative split of the deformation gradient into an elastic part F_e and an irreversible, plastic part F_p yields the decomposition $F = F_e F_p$. The single active slip-system is characterised by unit vectors s and m ($|s| = |m| = 1$, $s \cdot m = 0$). Consider the following flow rule with slip-rate $\dot{\gamma}$,

$$\left(\dot{F}_p F_p^{-1}, \dot{p}\right) = (\dot{\gamma}\, s \otimes m, |\dot{\gamma}|) \quad (22)$$

with initial conditions $\gamma(0) = 0$ and $p(0) = 0$. By time-integration we infer

$$F_p^{-1} = \mathbf{1} - \gamma s \otimes m. \quad (23)$$

As dissipation functional we will simply assume

$$\Delta(\dot{\gamma}) = r|\dot{\gamma}|, \quad (24)$$

with a positive constant r, see [11].

Due to a non-convex condensed energy (see [11]), microstructures arise as energy minimisers. Let us assume a first-order laminate microstructure with N phases having interfaces with unit normal b. We define the deformation gradient in phase i according to equation (16). To ensure incompressibility of each laminate phase, we must enforce that $\det F_i = 1$ or $a_i \cdot b = 0$. Taking this constraint into account, the relaxed energy defined in (14) takes the form

$$\Psi^{\text{rel}}(F, \lambda, \gamma, p, b)$$

$$= \kappa \sum_i^N \lambda_i p_i^4 + \frac{\mu}{2} \left[\frac{1}{\sum_i^N \frac{\lambda_i}{b_i \cdot b}} \left(\sum_j^N \sum_k^N \frac{\lambda_j \lambda_k b_j \cdot C b_k}{b_j \cdot b \, b_k \cdot b} - \frac{1}{b \cdot C^{-1} b} \right) \right.$$

$$\left. + \sum_i^N \lambda_i \left(\frac{b_i \cdot b}{b \cdot C^{-1} b} - \frac{b_i \cdot C b_i}{b_i \cdot b} \right) + \text{tr}\, C + \sum_i^N \lambda_i \left(\gamma_i^2 s \cdot C s - 2 \gamma_i s \cdot C m \right) \right],$$
(25)

where
$$b_i = b - \gamma_i (b \cdot m\, s + b \cdot s\, m) + \gamma_i^2 b \cdot s\, s, \tag{26}$$

$C = F^\mathsf{T} F$ is the right Cauchy–Green tensor and $\gamma = \{\gamma_1, \ldots, \gamma_N\}$, $p = \{p_1, \ldots, p_N\}$.

For simplicity, let us reduce the present model to a two-phase laminate ($N = 2$) and define the volume fraction of phase 2 as λ such that, taking into account (24), the dissipation potential defined in (18) may be written in the form

$$\Delta^*(\lambda, \gamma_i, \dot{\lambda}, \dot{\gamma}_i) = r \left(\left| \dot{\lambda}(\gamma_1 - \gamma_2) \right| + (1 - \lambda) \left| \dot{\gamma}_1 \right| + \lambda \left| \dot{\gamma}_2 \right| \right) \tag{27}$$

and the Lagrange functional corresponding to (6) now becomes

$$\mathcal{L}(F, \lambda, \gamma_i, \dot{\lambda}, \dot{\gamma}_i, b) = \frac{d}{dt} \Psi^{\text{rel}}(F, \lambda, \gamma_i, p_i, b) + \Delta(\lambda, \gamma_i, \dot{\lambda}, \dot{\gamma}_i). \tag{28}$$

Via the principle given in (7) we now arrive at evolution equations for λ and γ_i from the above Lagrange functional as

$$-r |\gamma_1 - \gamma_2| \operatorname{sign} \dot{\lambda} \in -q = \frac{\partial \Psi^{\text{rel}}}{\partial \lambda}, \tag{29}$$

$$-r(1 - \lambda) \operatorname{sign} \dot{\gamma}_1 \in \frac{\partial \Psi^{\text{rel}}}{\partial \gamma_1} + \frac{\partial \Psi^{\text{rel}}}{\partial p_1} \operatorname{sign} \dot{\gamma}_1, \tag{30}$$

$$-r \lambda \operatorname{sign} \dot{\gamma}_2 \in \frac{\partial \Psi^{\text{rel}}}{\partial \gamma_2} + \frac{\partial \Psi^{\text{rel}}}{\partial p_2} \operatorname{sign} \dot{\gamma}_2. \tag{31}$$

With our goal of computing the evolution of plastic microstructures in mind, we need to find an incremental formulation to be solved numerically, using finite deformation increments $[F_n, F_{n+1}]$. As discussed in Section 4, we can hence avoid the solution of a global minimization problem. Before outlining the numerical scheme, we need to discuss two important steps: First, a change of λ results in mixing of the formerly pure phases in a small part of Ω. We propose to obtain the updated p_i values by taking the energetic average. For $\lambda_{n+1} = \lambda_n + \Delta \lambda$ and e. g. $\Delta \lambda > 0$ we have (analogously for $\Delta \lambda < 0$)

$$(\lambda_n + \Delta \lambda) p_{2,n+1}^4 = \lambda p_{2,n}^4 + \Delta \lambda p_{1,n}^4, \quad p_{1,n+1} = p_{1,n}. \tag{32}$$

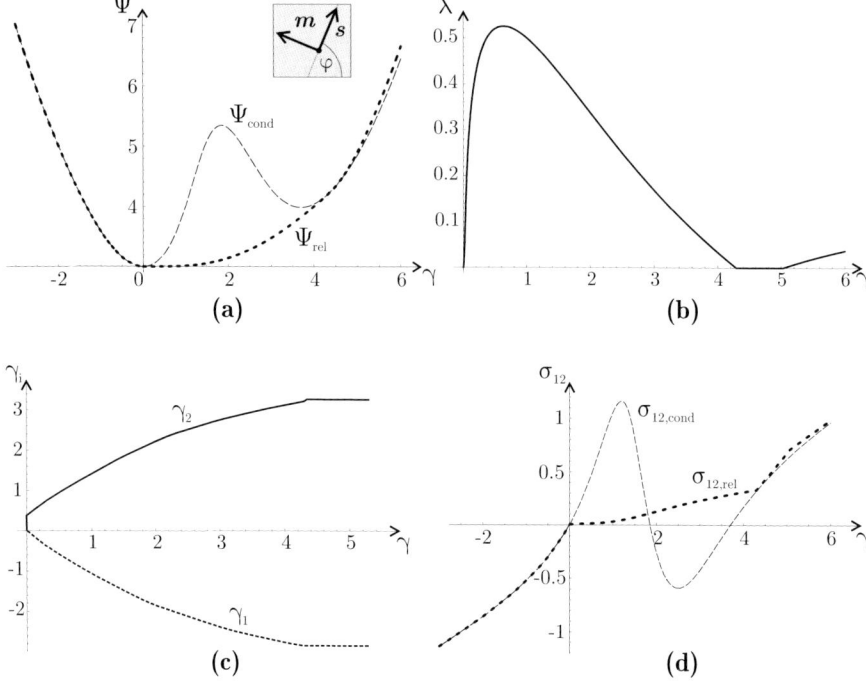

Fig. 3 Plane-strain simple shear test: (a) comparison of condensed and relaxed energy (energy computed via incremental loading), (b) origin and evolution of the volume fraction of phase 2, (c) evolution of the plastic slips γ_i within both laminate phases, (d) comparison of condensed and relaxed Kirchhoff shear stress.

Second, changes of the orientation vector $\boldsymbol{b}$ will be treated using the criterion given in (20). Upon each increment, we investigate if there exists $(\boldsymbol{b}, \gamma_2)$ such that for some value $\epsilon \ll 1$ we have

$$\sup \left\{ q(\lambda = \epsilon, \gamma_{1,n}, \gamma_2, p_{1,n}, \boldsymbol{b}) - r \left| \gamma_{1,n} - \gamma_2 \right| \mid \boldsymbol{b}, \gamma_2 \, ; \, |\boldsymbol{b}| = 1 \right\} \geq 0, \qquad (33)$$

and a laminate forms with this vector $\boldsymbol{b}$. Otherwise, no microstructure originates. Once the laminate has formed, we check with each further increment whether or not a rotation of the laminate is energetically possible by means of (20), i.e. by finding $\boldsymbol{b}_{n+1}$ from

$$\Psi^{\text{rel}}(\boldsymbol{b}_{n+1}) - \Psi^{\text{rel}}(\boldsymbol{b}_n) + 2r\lambda_n(1 - \lambda_n)\left|\gamma_{1,n} - \gamma_{2,n}\right| \leq 0, \qquad (34)$$

where the right-hand side represents the dissipation given by (19).

Now, our numerical scheme takes the form outlined in Algorithm 1. This algorithm computes the microstructure evolution (i.e. plastic slips γ_i and volume fraction λ) by incrementally minimising the functional (28). Since we are using the relaxed energy and dissipation functional this constitutes a well-posed problem and we can resort to solving the stationarity conditions (29), (30), and (31). Each time-step starts with the current state as initial condition and solves the stationarity conditions in order to update the internal variables. For initially homogeneous material (i.e. no laminate present) the interface normal $\boldsymbol{b}$ as well as the slip for the originating second laminate phase are determined via maximisation of the driving force according to condition (33). Once a laminate has been formed the evolution of the variables $\lambda, \gamma_1, \gamma_2$ is calculated by a staggered scheme. In a first step a time-discretised version of (29) is solved for the increment $\Delta\lambda$ for fixed γ_1 and γ_2. Afterwards p_1 and p_2 are updated via (32). Then, in a second step, (30) and (31) are solved for the increments $\Delta\gamma_1$ and $\Delta\gamma_2$ for fixed λ. Afterwards p_1 and p_2 are updated again. Finally the updated values of $\lambda, \gamma_1, \gamma_2, p_1$ and p_2 are transfered to the next time-step.

Algorithm 1. *Incremental evolution formulation:*

(a) incremental load update: $\boldsymbol{F}_{n+1} = \boldsymbol{F}_n + \Delta \boldsymbol{F}$
(b) find λ_{n+1} (assume $\gamma_i = const. = \gamma_{i,n}$):

- *for the initially uniform single-crystal ($\lambda = 0$) find $\boldsymbol{b}$ and γ_2 from*

$$\max_{\boldsymbol{b},\overline{\gamma_2}} \left\{ q(\lambda = \epsilon, \gamma_{1,n}, p_{1,n}, \overline{\gamma_2}, \overline{\boldsymbol{b}}) - r \left|\gamma_{1,n} - \overline{\gamma_2}\right| \right\} \overset{?}{\geq} 0$$

$$\Rightarrow \lambda_{n+1} = \epsilon, \gamma_{2,n+1} = \overline{\gamma_2}, \boldsymbol{b}_{n+1} = \overline{\boldsymbol{b}}$$

- *for an existing laminate microstructure ($\lambda > 0$) solve:*

$$q(\lambda_n + \Delta\lambda, \gamma_{i,n}, p_{i,n}) \operatorname{sign} \Delta\lambda = r \left|\gamma_{1,n} - \gamma_{2,n}\right| \Rightarrow \lambda_{n+1} = \lambda_n + \Delta\lambda$$

- *check for laminate rotation by finding $\boldsymbol{b}$ such that:*

$$\min_{\overline{\boldsymbol{b}}} \Psi^{rel}(\overline{\boldsymbol{b}}) - \Psi^{rel}(\boldsymbol{b}_n) + 2r\lambda_n(1-\lambda_n)\left|\gamma_{1,n} - \gamma_{2,n}\right| \overset{?}{\leq} 0 \Rightarrow \boldsymbol{b}_{n+1} = \overline{\boldsymbol{b}}$$

- *update $p_{i,n}$ according to (32)*

(c) find $\gamma_{i,n+1}$ (assume $\lambda = const. = \lambda_{i,n+1}$) by solving:

$$\begin{bmatrix} \dfrac{\partial \Psi^{rel}}{\partial \gamma_1} + \dfrac{\partial \Psi^{rel}}{\partial p_1} \operatorname{sign} \Delta\gamma_1 \bigg|_{\substack{\gamma_{i,n}+\Delta\gamma_i \\ p_{i,n}+|\Delta\gamma_i|}} = -r(1-\lambda_{n+1})\operatorname{sign}\Delta\gamma_1 \\ \dfrac{\partial \Psi^{rel}}{\partial \gamma_2} + \dfrac{\partial \Psi^{rel}}{\partial p_2} \operatorname{sign} \Delta\gamma_2 \bigg|_{\substack{\gamma_{i,n}+\Delta\gamma_i \\ p_{i,n}+|\Delta\gamma_i|}} = -r\lambda_{n+1}\operatorname{sign}\Delta\gamma_2 \end{bmatrix}$$

$$\Rightarrow \gamma_{i,n+1} = \gamma_{i,n} + \Delta\gamma_i, \quad p_{i,n+1} = p_{i,n} + |\Delta\gamma_i|$$

6 Numerical Results

The numerical scheme outlined above can be applied to arbitrary deformations, so long as the deformation remains volume-preserving to account for incompressible material. The first example treats the microstructure evolution during a plane-strain simple shear test parametrised by the macroscopic deformation gradient

$$F = \begin{pmatrix} 1 & \gamma & 0 \\ 0 & 1 & 0 \\ 0 & 0 & 1 \end{pmatrix}. \tag{37}$$

Computations were performed with $\mu = 2$, $\kappa = 0.01$, $r = 0.001$, $\epsilon = 0.001$ and with constant increments $\Delta\gamma = 0.001$ up to the extremal loads $\gamma_{max} = 6$ and $\gamma_{min} = -3$. The slip-system was oriented under an angle of $\varphi = 150°$ (see Figure 3(a) for the definition of φ). Because of the non-aligned slip-system the material stability of the homogeneous deformation is lost and microstructures arise for $\gamma > 0$. Due to the convexity of Ψ^{cond} for $\gamma < 0$, no microstructures form with negative strain γ. Figure 3 demonstrates the evolution of the laminate microstructure by illustrating the paths of energy, volume fractions, plastic slips and shear stress upon straining.

In the second example we investigate the microstructure evolution for a plane-strain tension-compression test with the macroscopic deformation gradient

$$F = \begin{pmatrix} e^{\delta} & 0 & 0 \\ 0 & e^{-\delta} & 0 \\ 0 & 0 & 1 \end{pmatrix}. \tag{38}$$

Computations were carried out with $\mu = 2$, $\kappa = 0$ (no hardening), $r = 0.001$, $\epsilon = 0.001$ and with constant increments $\Delta\delta = 0.001$ up to the extremal loads $\delta_{max} = 2$ and $\delta_{min} = -0.5$. The slip-system was oriented under an angle of $\varphi = 75°$ (see Figure 3). Again, due to the loss of rank-one convexity, the homogeneous deformation state is not stable and decomposes into micro-deformations. Figure 4 shows the microstructure evolution, displaying the path of energy, volume fractions, and stress components upon straining.

In both examples, the body behaves elastically first, until a second phase with finite, non-zero slip γ_2 originates from the uniform ground state, until finally phase 1 becomes plastic, too. Once, the laminate has formed with a distinct orientation vector b, we do not observe laminate rotations due to the large amount of dissipation necessary for a rotation. Rotation commonly only occurs when the body is in an almost uniform state (i.e. $\lambda \approx 0$ or $\lambda \approx 1$).

Finally, let us investigate the behavior of the model presented above during cyclic loading. Figure 5 illustrates the numerical results of a cyclic test in plane-strain simple shear as in (37). Computations were carried out with $\mu = 2$, $\kappa = 0.01$, $r = 0.001$, $\epsilon = 0.001$ and with constant increments $\Delta\gamma = 0.001$ up to the maximal load $\gamma = \pm 0.1$. The slip-system was oriented under an angle of $\varphi = 150°$ (see Figure 3).

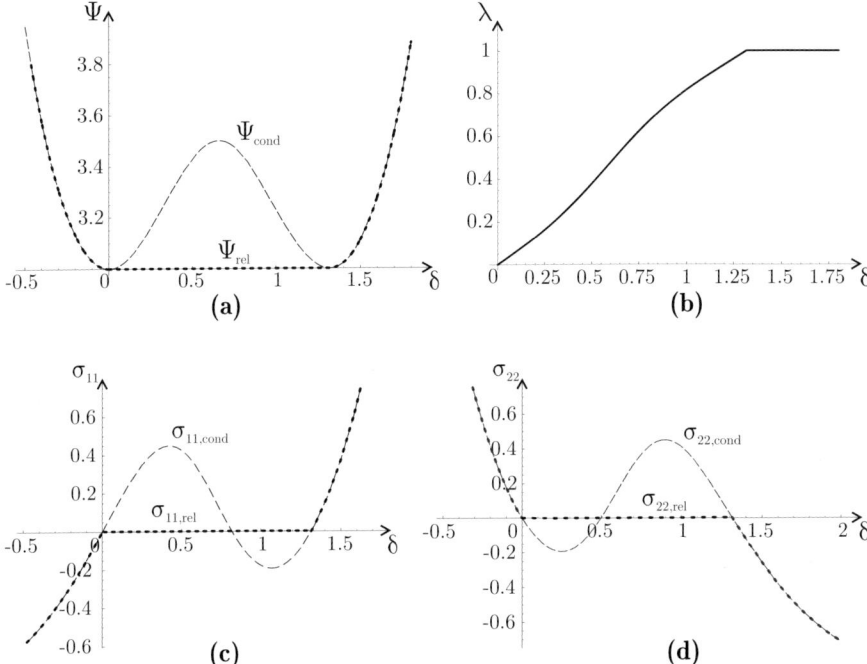

Fig. 4 Plane-strain tension-compression test: (a) comparison of condensed and relaxed energy (energy computed via incremental loading), (b) origin and evolution of the volume fraction of phase 2, (c) and (d) comparison of condensed and relaxed Kirchhoff stress components.

It is crucial to discuss the laminate's capability to rotate, i.e. to change the orientation of b. If we only admit small rotations of the fully established laminate (i.e. find b_{n+1} locally in the vicinity of b_n, cyclic loading yields the results presented in Figures 5(a) through (c): During the first two loading cycles, laminate microstructures originate and evolve. Jumps in the curves in Figure 5 result from small changes of the normal vector b. Due to hardening, all curves do not follow the same reversible paths. After the second cycle (with constant maximum cyclic deformation), the microstructure is frozen ($\lambda \approx 0$). If, however, we admit large rotations and find b_{n+1} globally, then the stress-strain curve exhibits the behavior shown in Figure 5(d): The microstructure is not frozen and still evolves (with frequently changing b), hence producing the illustrated shifting stress-strain hystereses.

7 Conclusion

We have shown how the evolution of microstructures can be described efficiently by employing relaxation of nonconvex potentials. The approach is universally ap-

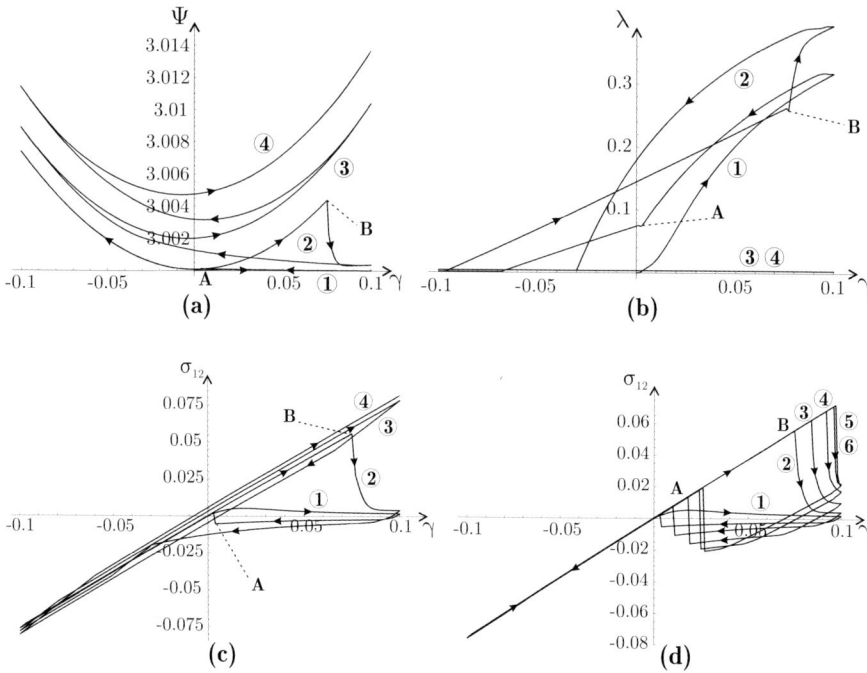

Fig. 5 Cyclic test under simple shear: evolution of (a) the relaxed energy, (b) the volume fraction of phase 2, (c) the Kirchhoff shear stress, if we admit only small rotations of the normal vector *b*. If we admit large rotations, the stress-strain curve takes the form as in (d). Numbers in circles denote the loading cycle. In points **A** and **B** the normal vector *b* rotates in cases (a) through (c). In case (d) *b* rotates at the jumps in the stress-strain curve in the regions denoted by **A** and **B**.

plicable to any kind of material exhibiting the formation of microstructures. Relaxation leads by definition to well-posed problems with regular solutions. We specified the general procedure introduced to first-order laminates and applied it to the time-continuous evolution of microstructures in elastoplastic single crystals. Generalizations in several directions are desirable and will be the topic of subsequent research.

References

1. Bartels S, Carstensen C, Hackl K, Hoppe U (2004) *Comp Meth Appl Meth Eng* 193:5143–5175.
2. Conti S, Theil F (2005) *Arch Rat Mech Anal* 178:125–148.
3. Mielke A (2004) *Comp Meth Appl Meth Eng* 193:5095–5127.
4. Lambrecht M, Miehe C, Dettmar J (2003) *Int J Solids Struct* 40:1369–1391.
5. Ortiz M, Repetto EA (1999) *J Mech Phys Solids* 47:397–462.
6. Mielke A, Ortiz M (2007) *ESAIM Control Optim Calc Var*, online since December 21.

7. Conti S, Ortiz M (2008) *J Mech Phys Solids* 56:1885–1904.
8. Hackl K, Schmidt-Baldassari M, Zhang W (2003) *Mat Sci Eng A* 378:503–506.
9. Hackl K, Heinen R (2008) *Continuum Mech Thermodyn* 19:499–510.
10. Hackl K (2006) Relaxed potentials and evolution equations. In: Gumbsch P (Ed) *Proceedings Third International Conference on Multiscale Materials Modeling*. Fraunhofer IRB Verlag.
11. Carstensen C, Hackl K, Mielke A (2002) *Proc R Soc London A* 458:299–317.
12. Mielke A (2002) Finite elastoplasticity, Lie groups and geodesics on SL(d). In: Newton P, Weinstein A, Holmes P (Eds), *Geometry, Dynamics, and Mechanics*. Springer, Berlin.
13. Hackl K, Fischer FD (2008) *Proc Roy Soc London A* 464:117–132.
14. Hackl K, Mielke A (2008) manuscript, in preparation.

Towards Effective Simulation of Effective Elastoplastic Evolution

Carsten Carstensen and Robert Huth

Humboldt Universität zu Berlin, Unter den Linden 6, Berlin, Germany
E-mail: {cc, huth}@mathematik.hu-berlin.de

Abstract. This paper summarises the general strategy for time evolving finite elastoplasticity and outlines encountered computational challenges in form of numerical benchmarks. Each time-step of some natural implicit time-discretisation is eventually recast into a possibly non-convex minimisation problem. Finite plasticity seems to imply the lack of lower semicontinuity of the energy functional and so leads to enforced fine strain oscillations called microstructures with required generalised solution concepts. The adaptive spacial discretisation is possible for convexified formulations from the relaxation finite element method (RFEM). For single-slip finite plasticity, one requires to relax numerically with laminates or semiconvexity notions.

Key words: finite elastoplasticity, non-convex minimisation, quasiconvexity, numerical relaxation, FEM.

1 Introduction

The outcome of RFEM is the macroscopic behaviour of the highly nonlinear microscopic material also called effective behaviour and models the macroscopic energy and the macroscopic stress fields. The numerical simulation is equally important and difficult in many situations and the model example of our choice is the single-slip model. The numerical relaxation is performed via successive layers of fine microstructures and leads to approximations of the quasiconvex hull.

The numerical simulation of elastoplastic evolutions experiences severe difficulty in the interplay of adaptive timespace discretisation and numerical relaxation. The overall algorithm is depicted in the subsequent box.

Time stepping: ∀ time steps
Adaptivity: ∀ level ℓ
Macroscopic FEM : $\forall T \in \mathcal{T}_\ell$ compute deformation and internal variables
Numerical Relaxation: compute energy and stress

The construction of effective algorithms and the erroranalysis in space as well as in time is even more challenging due to the unknown relaxation error. Since the numerical relaxation is in the deepest loop, time consumption is also a crucial factor.

The remainder of this paper is organised as follows. Section 2 recalls the generalised rate independent material and the notion of quasiconvexity. Its failure leads to nonexistence of solutions and the observation of microstructures in finite plasticity. Section 3 introduces the single-slip elastoplasticity without a closed form relaxation and so motivates the necessity of numerical relaxation. Section 4 is devoted to the introduction of the relaxation finite element Method (RFEM) and outlines the benchmark of computational microstructures [9] with closed form relaxation. Section 5 outlines adaptive mesh-refining algorithms. In Section 6 we list numerical relaxation schemes and shortly discuss their advantages and shortcomings. The hysteresis benchmark of Section 7 outlines the incremental problem with closed form condensed relaxation and spatial error control but without accumulated time discretisation errors.

2 Rate-Independent Materials

2.1 Standard Generalised Materials

Let $\varphi(\cdot, t)$ represent the deformation of a material body $\mathcal{B}$ from a reference configuration $\Omega_0 \subset \mathbb{R}^n$ to the current configuration $\varphi(\Omega_0, t) = \Omega_t$ and let $z(\cdot, t) : \Omega_0 \to \mathbb{R}^m$ denote internal variable like hardening or softening at the time t. Given the free Helmholtz energy $W(F, z)$ and the dissipation potential $\Delta(z, \dot{z})$, in terms of the deformation gradient F and the internal variable z plus its rate $\dot{z}$, we consider the outer energy L from applied forces and define [16, 21]

$$\text{Gibb's energy} \quad \mathcal{E}(t, \varphi, z) = \int_\Omega W(D\varphi(x, t), z(x, t))\, dx - L(t, \varphi(\cdot, t)),$$

$$\text{Dissipation distance} \quad \mathcal{D}(z_0, z_1) = \inf_{\substack{z \in C^1([0,1]; \mathbb{R}^m) \\ z(0)=z_0, z(1)=z_1}} \int_0^1 \Delta(z(s), \dot{z}(s))\,ds.$$

The dissipation distance is then the amount of energy which must at least be dissipated in a smooth transition from state z_0 to state z_1.

2.2 Continuous Formulation

The unknown deformation $\varphi : \Omega_0 \times \mathbb{R}_+ \to \mathbb{R}^n$ and internal variable $z : \Omega_0 \times \mathbb{R}_+ \to \mathbb{R}^m$ satisfy the following set of inequalities for all $0 \le t \le T$ and $0 \le t_1 \le t_2 \le T$ [21]

$$\mathcal{E}(t,\varphi(t),z(t)) \leq E(t,\hat{\varphi},\hat{z}) + \int_\Omega \mathcal{D}(z(t),\hat{z})\,dx \quad \text{for all } (\hat{\varphi},\hat{z}) \in V,$$

$$\mathcal{E}(t_1,\varphi(t_1),z(t_1)) + \text{Diss}(z;t_0,t_1) \leq \mathcal{E}(t_0,\varphi(t_0),z(t_0)) - \int_{t_0}^{t_1} L_t(s,\varphi(s))\,ds,$$

$$\text{with}\quad \text{Diss}(z;t_0,t_1) = \sup_{\substack{N\in\mathbb{N} \\ t_0\leq\tau_0<\ldots<\tau_N\leq t_1}} \sum_{j=1}^N \int_\Omega \mathcal{D}(z(\tau_{j-1}),z(\tau_j))\,dx.$$

2.3 Incremental Formulation

The time discretisation of the continuous problem results in the incremental problem, where for each time step t_j, (φ_j, z_j) approximates $(\varphi(\cdot,t_j), z(\cdot,t_j))$ and minimises the functional $E(t_j,\varphi,z) + \int_\Omega \mathcal{D}(z_{j-1},z)\,dx$.

The partial minimisation with respect to z can be solved separately for each material point and gives rise to the condensed energy [8, 22]

$$W_{\text{cond}}(z_{j-1}; D\varphi) := \min_z \left(W(D\varphi, z) + \mathcal{D}(z_{j-1}, z) \right),$$

$$E_{\text{cond}}(\varphi) := \int_\Omega W_{\text{cond}}(z_{j-1}; D\varphi)\,dx - L(\varphi).$$

The incremental problem is equivalently recast into the minimisation of E_{cond} among all admissible deformations. In practice, the condensed energy density W_{cond} has to be computed by analytical manipulations or some extra inner loop.

In the calculus of variations, the existence of minimisers of E_{cond} follows with its direct method, in the situation where W_{cond} is quasiconvex. In a typical finite plasticity problem, however, this is not the case and enforced microstructures are observed.

2.4 Generalised Notions of Convexity

The state-of-the-art calculus of variations [12] for the minimisation of the energy

$$E(v) = \int_\Omega W(x, v(x), Dv(x))\,dx \quad \text{for all } v \in V \tag{M}$$

is concerned with semiconvexity of $W(x, v(x), \cdot)$. Besides growth and continuity conditions on the energy E, a sufficient condition for the existence of a minimiser of E is sequential lower weak semicontinuity equivalent to the quasiconvexity of W.

One calls function $W : \mathbb{R}^{m\times n} \to \mathbb{R}$ quasiconvex if for some open subset $\omega \subset \mathbb{R}^n$ and all $E \in \mathbb{R}^{m\times n}$;

$$W(F) = \inf_{\varphi \in C_c^\infty(\omega,\mathbb{R}^m)} \frac{1}{|\omega|} \int_\omega W(F + D\varphi)\,dx$$

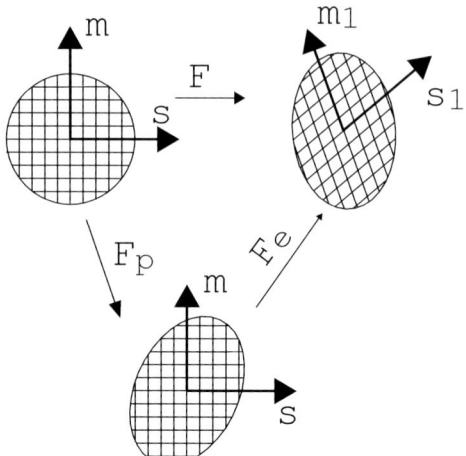

Fig. 1 Split of plastic and elastic deformation.

holds. The quasiconvex envelope W^{qc} is given

$$W^{qc}(F) = \sup \left\{ W^*(F) : W^* \leq W \text{ and } W^* \text{is quasiconvex} \right\}.$$

It is clear that W^{qc} is also quasiconvex. The quasiconvex energy

$$E^{qc}(v) = \int_\Omega W^{qc}(x, v(x), Dv(x)) \, dx \quad \text{for all } v \in V \tag{Q}$$

models relevant macroscopic properties like the displacement- or stress-field. However the notion of quasiconvexity is a difficult concept. A related concept is the notion of rank-one convexity: A function W is called rank-one convex, if for all matrices F and all rank-one matrices $a \otimes b$ the function $f : \mathbb{R} \to \mathbb{R}$ with $f(t) = W(F + ta \otimes b)$ is convex. The conjecture that all rank-one convex functions are quasiconvex has been an open problem for decades, before Sverak [23] found a counterexample based on the function $\varphi(A) = xyz$ for all 3×2 matrix $A^T = (\text{diag}(x, y), [z, z])$. Even these days it remains an open question whether or not rank-one convexity is equivalent to quasiconvexity on $\mathbb{R}^{2\times 2}$.

Another related semiconvexity notion is polyconvexity: A function $W(A)$ is polyconvex if it is a convex function of a vector of all minors of A. Similar to the quasiconvex envelope one defines the convex, polyconvex and the rank-one convex envelope W^{**}, W^{pc}, W^{rc} for which it is known that $W^{**} \leq W^{pc} \leq W^{qc} \leq W^{rc}$.

3 Single-Slip Finite Plasticity

In this example the resulting energy is not quasiconvex and the formation of microstructures are expected. The relaxation can not be done analytically and therefore numerical relaxation schemes have to be employed. The local deformation F is supposed to consist of an elastic deformation superimposing a plastic deformation,

$$F = D\varphi = F_e F_p .$$

Based on the given slip directions s and $m \in \mathbb{R}^n$ with $|m| = 1 = |s|$, the plastic deformation

$$F_p = I + \gamma s \otimes m$$

depends only one the real shear parameter γ. With material parameters h, μ, the critical shear stress τ_{cr} and a neo Hookean energy U, the free energy reads [8]

$$W(F, z) = U(\det F_e) + \frac{\mu}{2}\text{tr}(F_e^T F_e) + \frac{h}{2}p^2 .$$

The dissipation potential Δ and dissipation distance $\mathcal{D}$ are defined as

$$\Delta = \begin{cases} \tau_{cr}|\dot\gamma| & \text{if } |\dot\gamma| + \dot p \le 0, \\ \infty & \text{otherwise,} \end{cases}$$

$$\mathcal{D}(\gamma_0, \gamma_1) = \begin{cases} \tau_{cr}|\gamma_1 - \gamma_0| & \text{if } |\gamma_1 - \gamma_0| \le p_0 - p_1, \\ \infty & \text{otherwise.} \end{cases}$$

The resulting condensed energy E_{cond} from Subsection 2.3 utilises [8, 22]

$$W_{\text{cond}}(F) = U(F) + \frac{\mu}{2}(|F|^2 - 2) - \frac{1}{2}\frac{(\max(0, \mu|\mathbf{Cs} \cdot \mathbf{m}| - \tau_{cr}))^2}{\mu \mathbf{Cs} \cdot \mathbf{s} + h} .$$

Since W_{cond} is not quasiconvex its unknown quasiconvex envelope has to be computed as a benchmark for numerical relaxation algorithms. which is unknown. For a special case of U and $\tau_{cr} = 0$ the quasiconvex envelope is known [11].

4 Relaxation Finite Element Method (RFEM)

The piecewise constant strain can not develop any microstructures on each element and hence oscillations are mesh dependent as seen in Figure 2. To improve the element ansatz functions further, the RFEM allows an arbitrary ansatz function on each element subject to the elementwise affine boundary conditions. The resulting RFEM minimises the energy

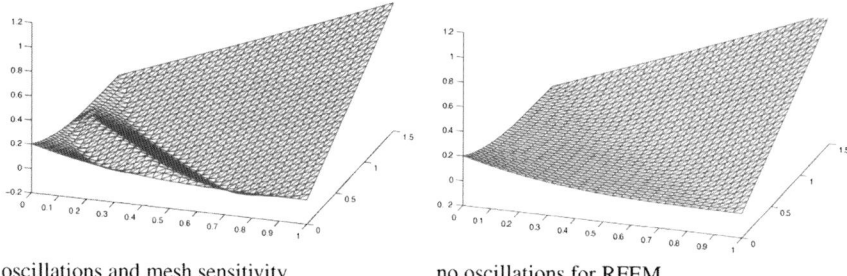

oscillations and mesh sensitivity   no oscillations for RFEM

Fig. 2 Microstructures for the 2 well benchmark [9].

$$E_\ell(v_\ell) = \sum_{T \in \mathcal{T}_\ell} \left(\inf_{w \in C_c^\infty(T, \mathbb{R}^m)} \int_T W(x, v_\ell(x), Dv_\ell(x) + Dw(x)) dx \right) - L(v_\ell).$$

The infimum on each triangle is the relaxation and allows the direct simulation of the macroscopic displacement field. For linear FE functions the infimum is the integral of the quasiconvex envelope of W,

$$E_\ell(v_\ell) = \sum_{T \in \mathcal{T}_\ell} \left(\int_T W^{qc}(x, v_\ell(x), Dv_\ell(x)) dx \right) - L(v_\ell).$$

The numerical outcome for the computational microstructures benchmark is shown in Figure 2 [9].

This model is a typical example which is in its original formulation not quasiconvex and therefore oscillations and meshdependance of numerical solutions are observed. The relaxation in closed form for this example yields meshindependent solutions.

In this example we are looking for $\varphi : \mathbb{R}^2 \to \mathbb{R}$ and define the energy density

$$W(F) = |F - F_1|^2 |F - F_2|^2$$

for two different wells $F_1 = -F_2 = -(3, 2)/\sqrt{13} \in \mathbb{R}^2$. The energy reads

$$E(v) = \int_\Omega \left(W(Dv) + |v - f|^2 \right) dx \quad \text{for all } v \in V,$$

for a function given $f(x, y) = -3t^5/128 - t^3/3 \in C^1(\Omega); t = (3(x-1)+2y)/\sqrt{13}$.

The quasiconvex envelope of W coinsides with the convex envelope $W^{**}(F) = W^{qc}(F) = \left((|F|^2 - 1)_+\right)^2 + 4|F|^2 - \left((3, 2) \cdot F)^2/13\right)$ [9]. The original problem faces severe difficulties in its numerical treatment and mesh dependent microstructures are observed.

Elastoplastic Evolution

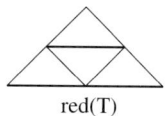
red(T)
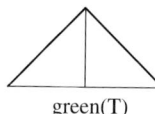
green(T)
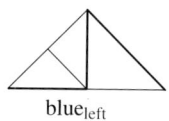
blue$_{\text{left}}$
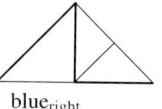
blue$_{\text{right}}$

Fig. 3 Refinement rules for the red-green-blue strategy.

5 Adaptive Finite Element Method (AFEM)

An adaptive mesh refining algorithm consists of a successive loop of

$$\text{solve} \Rightarrow \text{estimate} \Rightarrow \text{mark} \Rightarrow \text{refine} \ .$$

The refinement indicator $\eta_E = h_E^{1/p'} \|[\sigma_\ell]\nu_E\|_{L^{p'}(E)}$ from subroutine *estimate* is monitored in *mark* for possible refinement of the edge E. Figure 3 shows admissible refinements of a triangle up to rotation. The error estimator through the sum over all η_E suffers from a reliability/efficiency gap [9]. This does not prevent a convergence proof for the AFEM in [4]. Moreover, optimal complexity is visible in numerical experiments for the microstructures benchmark [9].

6 Numerical relaxation

Since direct approaches to approximate the quasiconvex envelope of a given function $W : \mathbb{R}^{n \times n} \to \mathbb{R}$, are of the same complexity as findig the solution of the original problem, numerical relaxation algorithms fall back to the notion of polyconvexity and rank-one convexity, which gives upper and lower bounds for the quasiconvex envelope and may even coinside with the latter.

6.1 Polyconvexification

A reliable and efficient computation of the polyconvex envelope

$$W^{pc}(F) = \inf \left\{ \sum_{\ell=1}^{\tau+1} \lambda_\ell W(A_\ell) | A_\ell \in \mathbb{R}^{n \times n}, \lambda_\ell \geq 0, \sum_{\ell=1}^{\tau+1} \lambda_\ell T(A_\ell) = T(F) \right\}$$

in terms of a vector of all minors $T(A) \in \mathbb{R}^\tau$ of matrices $A \in \mathbb{R}^{n \times n}$ is studied in [1, 14].

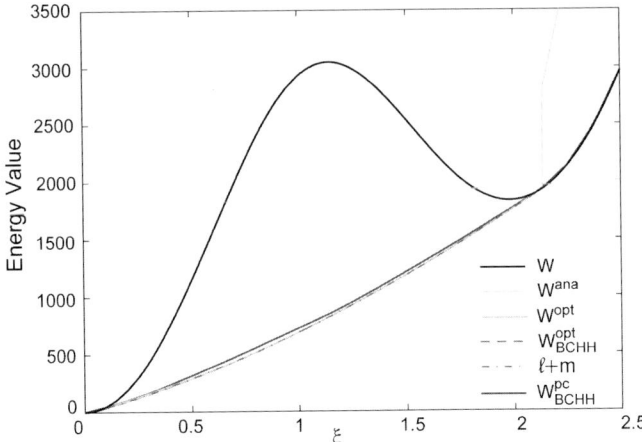

Fig. 4 Comparison of various relaxation methods along some rank-one line [5].

6.2 Lamination

The approximation of the rank-one convex envelope by successive lamination [2, 7, 12, 13, 15, 17–19, 22] results in an upper bound for the rank-one convex envelope and quasiconvex envelope.

On the other hand, in finite lamination, one approximates W^{rc} by a laminate of second order, which are in practice sufficient approximations. The lamination is parametrised and a difficult nonlinear optimisation problem arises [3].

A combined numerical and analytical relaxation exploits the special structure of the single-slip plasticity analytically in [5]. This leads to initial guesses for a local minimisation algorithm to compute layers within layers and a dramatic reduction in the numerical effort.

6.3 Numerical Relaxation Benchmark

Figure 4 displays a comparison of several numerical relaxation methods for the single-slip plasticity benchmark. Therein W^{opt}_{BCHH} denotes the outcome of finite lamination and W^{pc}_{BCHH} of polyconvexification as an upper and lower bound for the quasiconvex envelope [3]. The new lamination solution W^{opt} as an upper bound and the approximated lower bound $\ell + m$ are taken from [5]. W^{ana} is an analytical relaxation obtained for a simplification of the underlying model [5].

The overall impression of the numerical outcome of Figure 4 displays roughly the same energy. It is less clear whether or not they coinside.

7 Shape-Memory Alloys Benchmark

A rate independent hysteresis of phase transitions in shape memory alloys is modelled by a mixture $\chi^{(1)} = 1 - \chi^{(2)}$ of two materials, with given material constants $\kappa, W_{0,1}, W_{0,2} \in \mathbb{R}$ and $F_1, F_2 \in \mathbb{R}^2$, and resulting free energy

$$E(t, \chi) = \inf_{v \in V} \int_\Omega \sum_{i=1}^{2} \chi^{(j)} W_j(\varepsilon(v)) \, dx - L(t, v),$$

$$W_j(F) = \frac{1}{2} \left(F - E_j, \mathbb{C} \, (F - E_j) \right)_{\mathbb{R}^{n \times n}} + W_{0,j},$$

and the dissipation for re-arrangement of a phase mixture $\zeta \to \chi$,

$$\mathcal{D}(\chi, \zeta) = \int_\Omega \kappa |\chi - \zeta| \, dx.$$

The incremental problem formulation allows a condensed form, which requires a quasiconvexification [6, 20]. The effective model reads: for fixed time t_j compute the minimiser $\varphi_j \in V$ of

$$E_{\chi_{n-1}}(v) = \int_\Omega \left(W_2(\varepsilon(v)) + 2\gamma H(\chi_{n-1}^{(1)}, \ell(\varepsilon(v))) \right) dx - L(t_j, v)$$

and thereafter compute the update $\chi_j^{(1)} = M(\chi_{j-1}^{(1)}, \ell(\varepsilon(u_j)))$ with the definition $\gamma = \gamma(E_1, E_2, \mathbb{C})$, [10, 20] and

$$\ell(E) = \frac{1}{2\gamma} \left(W_2(F) - W_1(E) \right) + \frac{1}{2},$$

$$H(r, s) = \begin{cases} \frac{\kappa}{2\gamma} r & \text{if } s \leq -\frac{\kappa}{2\gamma}, \\ \frac{\kappa}{2\gamma} r - \frac{1}{2}(s + \frac{\kappa}{2\gamma})^2 & \text{if } -\frac{\kappa}{2\gamma} \leq s \leq r - \frac{\kappa}{2\gamma}, \\ \frac{1}{2} r^2 - rs & \text{if } r - \frac{\kappa}{2\gamma} \leq s \leq r + \frac{\kappa}{2\gamma}, \\ -\frac{\kappa}{2\gamma} r - \frac{1}{2}(s - \frac{\kappa}{2\gamma})^2 & \text{if } r + \frac{\kappa}{2\gamma} \leq s \leq 1 + \frac{\kappa}{2\gamma}, \\ \frac{\kappa}{2\gamma}(1-r) + \frac{1}{2} - s & \text{if } s \geq 1 + \frac{\kappa}{2\gamma}, \end{cases}$$

$$M(r, s) = \begin{cases} 0 & \text{if } s \leq -\frac{\kappa}{2\gamma}, \\ s + \frac{\kappa}{2\gamma} & \text{if } -\frac{\kappa}{2\gamma} \leq s \leq r - \frac{\kappa}{2\gamma}, \\ r & \text{if } r - \frac{\kappa}{2\gamma} \leq s \leq r + \frac{\kappa}{2\gamma}, \\ s - \frac{\kappa}{2\gamma} & \text{if } r + \frac{\kappa}{2\gamma} \leq s \leq 1 + \frac{\kappa}{2\gamma}, \\ 1 & \text{if } s \geq 1 + \frac{\kappa}{2\gamma}. \end{cases}$$

The algorithms are derived in [10] where spatial a priori and a posteriori error analysis is developed for one time step only. It remained as an open problem to control the error accumulated over various time steps.

Acknowledgements

The authors thank S. Conti, A. Mielke and A. Orlando for stimulating discussions and collaborations and the DFG for support via the FG797.

References

1. S. Bartels (2003) Reliable and efficient approximation of polyconvex envelopes. *SIAM J. Numer. Anal.* **43**(1), 363–385.
2. S. Bartels (2003) Linear convergence in the approximation of rank.one convex envelopes. *ESAIM M2AN* **38**(5), 811–820.
3. S. Bartels, C. Carstensen, K. Hackl and U. Hoppe (2003) Effective relaxation for microstructure simulations: Algorithms and applications. *Comput. Meth. Appl. Mech. Engng.* **193**, 5143–5175.
4. C. Carstensen (2007) Convergence of an adaptive FEM for a class of degenerate convex minimization problems. *IMA Journal of Numerical Analysis*, published online.
5. C. Carstensen, S. Conti and A. Orlando (2008) Mixed analytical-numerical relaxation in finite single-slip crystal plasticity (submitted).
6. S. Govindjee, A. Mielke and G.J. Hall. (2002) The free-energy of mixing for n-variant martensitic phase transformations using quasi-convex analysis. *Mech. Physics Solids* **50**, 1879–1922.
7. K. Hackl and U. Hoppe (2003) On the calculation of microstructures for inelastic materials using relaxed energies. in *Proceedings of IUTAM Symposium on Computational Mechanics of Solids at Large Strains*, C. Miehe (Ed.), pp. 77–86, Kluwer Academic Publishers, Dordrecht.
8. C. Carstensen, K. Hackl and A. Mielke (2002) Non-convex potentials and microstructures in finite-strain plasticity. *Proc. Royal Soc. London, Ser. A* **458**, 299–317.
9. C. Carstensen and K. Jochimsen (2003) Adaptive finite element methods for miocrostructures? Numerical experiments for a 2-well benchmark. *Computing* **71**, 175–204.
10. C. Carstensen and P. Plechac (2001) Numerical analysis of a relaxed variational model of hysteresis in two-phase solids. *M2AN* **35**, 865–878.
11. S. Conti (2006) Relaxation of single-slip single-crystal plasticity with linear hardening. In *Multiscale Materials Modeling*, Fraunhofer IRB, pp. 30–35.
12. B. Dacorogna (2007) *Direct Methods in the Calculus of Variations.* Applied Mathematical Sciences Vol. 78, Springer, New York.
13. G. Dolzmann (1999) Numerical computation of rank one convex envelopes. *SIAM J. Numer. Anal.* **26**, 1621–1635.
14. G. Dolzmann (2003) *Variational Methods for Crystalline Microstructure – Analysis and Computation.* Lecture Notes in Mathematics, Vol. 1803, Springer, Berlin/Heidelberg.
15. M. Kruzik, A. Mielke and T. Roubicek (2005) Modelling of microstructure and its evolution in shape-memory-alloy single-crystals, in particular in CuAlNi. *Meccanica* **40**, 389–418.
16. B. Halphen and Q.S. Nguyen (1975) Sur les materiaux standards generalizes. *J. Mech.* **14**, 39–63.
17. C. Miehe (2002) Strain-driven homogenization of inelastic microstructures and composites based on an incremental variational formulation. *Int. J. Numer. Meth. Engrg.* **55**(11), 1285–1322.
18. C. Miehe and M. Lambrecht (2003) A two-scale finite element relaxation analysis of shear bands in non-convex inelastic solids: Small-strain theory for standard dissipative materials. *Comput. Methods Appl. Mech. Engrg.* **192**, 473–508.
19. C. Miehe, J. Schotte and M. Lambrecht (2002) Homogenization of inelastic solid materials at finite strains based on incremental minimization principles. *J. Mech. Phys. Solids* **50**, 2123–2167.

20. A. Mielke and F. Theil (1999) A mathematical Model for rate-independant phase transformations with hysteresis. In *Proceedings Workshop of Continuum Mechanics in Analysis and Engineering*, Aachen, pp. 117–129.
21. A. Mielke (2003) Energetic formulation of multiplicative elasto-plasticity using dissipation distances. *Cont. Mech. Thermodynamics* **15**, 351–382.
22. M. Ortiz and E.A. Repetto (1999) Non-convex energy minimization and dislocation structures in ductile single crystals. *J. Mech. Phys. Solids* **47**, 397–462.
23. V. Sverak (1992) Rank-one convexity does not imply quasiconvexity. *Proc. Roy. Soc. Edinb.* **120A**, 185–189.

Numerical Approximation Techniques for Rate-Independent Inelasticity*

Alexander Mielke

Weierstraß-Institut für Angewandte Analysis und Stochastik, Mohrenstr. 39, 10117 Berlin and Institut für Mathematik, Humboldt Universität zu Berlin, Rudower Chaussee 25, 12489 Berlin, Germany
E-mail: mielke@wias-berlin.de

Abstract. Some recent advances in the numerical analysis of rate-independent material models are surveyed. A general concept of convergence of numerical approximations is discussed und the basis of Γ-convergence. It provides convergence of subsequences to true solutions under minimal regularity assumptions but gives no rates of convergence. Applications to elastoplasticity and damage are discussed.

Key words: rate-independent systems, energetic solutions, time-incremental minimization problems, Γ-convergence, damage.

1 Introduction

Incremental minimization techniques play a crucial role in the modeling of many inelastic effects. In particular, for rate-independent material models they are closely link to the solutions of the so-called energetic formulation, see [24–26]. We present recent advances in the field of space-time discretizations of such models. Using techniques from Γ-convergence of functionals, which were established in [21], we are able to establish numerical convergence results for quite general systems, including models with evolution of microstructure in terms of Young measures. Here we present the results of [16, 20] in a form that shows its easy applicability in many cases of rate-independent inelastic or hysteretic material behavior.

As an easy application of the theory we show that we obtain a simple proof of the result in [12], which states that the space-time discretization for linearized elastoplasticity with hardening converge to the solution of the space-time continuous problem. This paper seems to be the first one addressing the subtle issue of the proving this convergence without assuming any additional temporal or spatial smoothness of the solutions, as is commonly done, see e.g. [1, 11] and the references therein.

---

* Partially supported by DFG via the Research Center MATHEON "Mathematics for Key Technologies" (Project C18).

B.D. Reddy (ed.), IUTAM Symposium on Theoretical, Modelling and Computational Aspects of Inelastic Media, 53–63.
© *Springer Science+Business Media B.V. 2008*

Our work is in the same spirit in using a very weak solution concept and in obtaining convergence under general conditions. In fact, we are dealing with the rather general concept of *energetic solutions*, which allows solutions to have jumps with respect to time and whose spatial regularity is only determined by the fact that they have finite energy. As is common in the nonconvex rate-independent setting, we cannot expect uniqueness of solutions and as a consequence we will only be able to show that suitable subsequences of the numerical approximations converge. Moreover, we are not able to derive convergence rates in terms of the discretization parameters.

In the last section we will address some computational results in a damage model introduced in [8] and further developed in [3,9,17,22]. A similar numerical approach to a model for shape-memory alloys is discussed in [2,20].

2 Energetic Rate-Independent Systems

Fully rate independent models for processes describing material models occur as limits when the loading rate slows down to 0. This makes the model simpler by omitting all effects due to interior relaxation processes. However, the resulting rate-independent mathematical models are somehow degenerate. In particular, in many cases solutions for a given initial datum are no longer unique and may have jumps in time. Nevertheless, as a subclass of the generalized standard materials [7, 10], such models are widely used in engineering, in particular in the isothermal case. Mathematical analysis of such processes can be based on the notion of energetic solutions introduced in [18, 23], and there is now a variety of applications in finite-strain elastoplasticity, shape-memory alloys, ferroelectric and ferromagnetic materials, in delamination, and damage, see [14] and the references therein. In fracture and crack propagation the same concept is used but often called *irreversible quasistatic evolution*, see [4–6].

Here we remain mostly in the abstract setting and refer to [16] for more elaborations on numerical approximations and to [13] for a numerical convergence result involving gradient Young measures.

We consider situations where the state of the body $\Omega \subset \mathbb{R}^d$ can be described by the displacement $u : \Omega \to \mathbb{R}^d$ and an internal variable $z : \Omega \to Z \subset \mathbb{R}^m$. Here z may be a collection of internal variables, either scalars (like in damage), vectors (like magnetization or polarization) or tensors (like the plastic deformation). The pair $q = (u, z)$ is called the state of the system and it is assumed to lie in the Banach space $\mathcal{Q} = \mathcal{U} \times \mathcal{Z}$, where $\mathcal{U}$ is the set of admissible displacements which is specified via Dirichlet boundary conditions on $\Gamma_{\text{Dir}} \subset \partial\Omega$.

The properties of the body are described via an energy storage functional $\mathcal{E}(t,q) \in \mathbb{R}_\infty := \,]{-}\infty, \infty]$ and a dissipation potential $\mathcal{R}(z, \dot{z}) \in [0, \infty]$. In most cases one can assume that these functional are given via integration over the body as

$$\mathcal{E}(t, u, z) = \int_{\Omega} W(x, e(u)(x), z(x)) + \kappa |\nabla z(x)|^r \, dx - \langle \ell(t), u \rangle$$

where $\langle \ell(t), u \rangle = \int_{\Omega} f_{\text{vol}}(t, x) \cdot u(x) \, dx + \int_{\partial \Omega \setminus \Gamma_{\text{Dir}}} f_{\text{surf}}(t, x) \cdot u(x) \, da$,

$$\mathcal{R}(z, v) = \int_{\Omega} R(x, z(x), v(x)) \, dx,$$

where the linearized strain is $e(u) = \frac{1}{2}(\nabla u + \nabla u^{\mathsf{T}})$.

The rate-independent evolution can be written as the system given via the elastic equilibrium and the balance of the internal forces, also called Biot's law, flow rule, or switching condition:

$$\begin{aligned}
\text{elastic equilibrium} \quad & 0 = D_u \mathcal{E}(t, u(t), z(t)), \\
\text{flow rule} \quad & 0 \in \partial_{\dot{z}} \mathcal{R}(z(t), \dot{z}(t)) + D_z \mathcal{E}(t, u(t), z(t)),
\end{aligned} \quad (1)$$

where $\partial \mathcal{R}(z, v)$ denotes the set-valued subdifferential of the convex and 1-homogeneous function $v \mapsto \mathcal{R}(z, v)$.

However, in many situations it is not possible to show that (1) has solutions. Hence, we will use the energetic solutions introduced in [14,18,23]. For this we need the dissipation distance $\mathcal{D}(z_0, z_1) \in [0, \infty]$, which denotes the minimal energy that is dissipated along a smooth path when changing the internal state from z_0 into z_1:

$$\mathcal{D}(z_0, z_1) := \inf \{ \text{Diss}_{\mathcal{D}}(\tilde{z}, [0, 1]) \mid \tilde{z}(0) = z_0, \tilde{z}(1) = z_1 \}, \quad (2)$$

where $\text{Diss}_{\mathcal{D}}(\tilde{z}, [t_0, t_1]) = \int_{t_0}^{t_1} \mathcal{R}(\tilde{z}(s), \dot{\tilde{z}}(s)) \, ds$. Note that $\mathcal{R}$ has the physical dimension of a power whereas $\mathcal{D}$ has the dimension of energy. We will call the triple $(\mathcal{Q}, \mathcal{E}, \mathcal{D})$ an *energetic system*.

Definition 1. *The process $q : [0, T] \to \mathcal{Q}$ is called an* energetic solution *of the energetic systems $(\mathcal{Q}, \mathcal{E}, \mathcal{D})$, if for all $t \in [0, T]$, we have stability (S) and energy balance (E):*

(S) $\quad \mathcal{E}(t, q(t)) \leq \mathcal{E}(t, \tilde{q}) + \mathcal{D}(q(t), \tilde{q})$ *for all* $\tilde{q} \in \mathcal{Q}$.

(E) $\quad \mathcal{E}(t, q(t)) + \text{Diss}_{\mathcal{D}}(q, [0, t]) = \mathcal{E}(0, q(0)) + \int_0^t \partial_s \mathcal{E}(s, q(s)) \, ds$.

We continue to write $\mathcal{D}(q, \tilde{q})$ for $\mathcal{D}(z, \tilde{z})$ and $\text{Diss}_{\mathcal{D}}(q, [0, t])$ for $\text{Diss}_{\mathcal{D}}(z, [0, t])$, whenever it is clear that $q = (u, z)$ and $\tilde{q} = (\tilde{u}, \tilde{z})$.

It is interesting to note that the subdifferential form (1) and the energetic form (S) & (E) are in fact extremal principle in the sense of [25,26], particularly the definition (2) of the dissipation distance.

It is discussed in [14, 19] under which conditions on $\mathcal{E}$ and $\mathcal{D}$ the notion of energetic solutions is equivalent to the solutions of (1). The point is that for general, nonconvex functionals $\mathcal{E}(t, \cdot)$ one cannot expect to find solutions of (1) while there exist energetic solutions under quite general situations. The typical assumptions for an existence theory are the following. Assume that $\mathcal{Q} = \mathcal{U} \times \mathcal{Z}$ is a reflexive Banach space, e.g. $\mathcal{U}$ is a closed subspace of $W^{1,p}(\Omega; \mathbb{R}^d)$ and $\mathcal{Z} = W^{1,r}(\Omega; \mathbb{R}^m)$. We introduce the sets $S(t)$ of stable states at time t via

$$S(t) = \{q \in Q \mid \mathcal{E}(t, q) < \infty \text{ and } \mathcal{E}(t, q) \leq \mathcal{E}(t, \tilde{q}) + \mathcal{D}(q, \tilde{q}) \text{ for all } \tilde{q} \in Q \}.$$

If $\mathcal{E}$ and $\mathcal{D}$ satisfy the following conditions (3)–(7), then for each initial condition $q_0 \in S(0)$ an energetic solution exists (see [14] for a survey):

$$\text{for all } z_1, z_2, z_3 \in Z \text{ we have}$$
$$\text{positivity:} \quad \mathcal{D}(z_1, z_2) = 0 \iff z_1 = z_2, \tag{3}$$
$$\text{triangle inequality: } \mathcal{D}(z_1, z_3) \leq \mathcal{D}(z_1, z_2) + \mathcal{D}(z_2, z_3);$$

$$\mathcal{D} : Z \times Z \to [0, \infty] \text{ is weakly lower semi-continuous;} \tag{4}$$

$$\mathcal{E} : [0, T] \times Q \text{ is weakly lower semi-continuous and coercive;} \tag{5}$$

$$\text{there exist constants } c_0^{\mathcal{E}}, c_1^{\mathcal{E}} \text{ such that}$$
$$\mathcal{E}(0, q) < \infty \text{ implies } \mathcal{E}(\cdot, q) \in C^1([0, T]) \text{ with} \tag{6}$$
$$|\partial_t \mathcal{E}(t, q)| \leq c_1^{\mathcal{E}} \big(\mathcal{E}(t, q) + c_0^{\mathcal{E}}\big);$$

$$\text{for each sequence } (t_n, q_n)_{n \in \mathbb{N}} \text{ with } (t_n, q_n) \rightharpoonup (t_*, q_*),$$
$$q_n \in S(t_n), \text{ and } \sup_{n \in \mathbb{N}} \mathcal{E}(t_n, q_n) < \infty \text{ we have} \tag{7}$$
$$\text{(a) } q_* \in S(t_*) \quad \text{and (b) } \partial_t \mathcal{E}(t_*, q_n) \to \partial_t \mathcal{E}(t_*, q_*).$$

Here conditions (3) and (4) are standard assumptions on the dissipation distance; the triangle inequality follows easily from definition (2). Conditions (5) and (6) relate only to the energy functional. The first one is the standard condition in the calculus of variations, while the second one is called an *energetic control* of the power of the external forces. This condition is crucial to obtain uniform a priori bounds.

Condition (7) may be called a *compatibility condition* as it relates, via the stable sets $S(t_j)$, the properties of $\mathcal{E}$ and $\mathcal{D}$ in an intrinsic manner. While part (b) is often easy to establish (time $t = t_*$ is fixed in the power $\partial_t \mathcal{E}(t, q)$), part (a) is the most delicate point. One way to establish this condition is the *joint recovery condition*, namely

$$\text{for all } q_*, \tilde{q} \in Q, \ (t_n, q_n)_{n \in \mathbb{N}} \text{ with } (t_n, q_n) \rightharpoonup (t_*, q_*), \ q_n \in S(t_n),$$
$$\text{and } \sup_{n \in \mathbb{N}} \mathcal{E}(t_n, q_n) < \infty \text{ there exists } (\tilde{q}_n)_{n \in \mathbb{N}} \text{ with } \tilde{q}_n \rightharpoonup \tilde{q} \text{ such that} \tag{8}$$
$$\limsup_{n \to \infty} \mathcal{E}(t_n, \tilde{q}_n) + \mathcal{D}(q_n, \tilde{q}_n) - \mathcal{E}(t_n, q_n) \leq \mathcal{E}(t_*, \tilde{q}) + \mathcal{D}(q_*, \tilde{q}) - \mathcal{E}(t_*, q_*).$$

Proposition 1. *Conditions* (5) *and* (8) *imply* (7a).

Proof. By (5) we have $\mathcal{E}(t_*, q_*) \leq \liminf_{n \to \infty} \mathcal{E}(t_n, q_n) \leq \sup_{n \in \mathbb{N}} \mathcal{E}(t_n, q_n) < \infty$, where the last inequality is assumed in (7). Next, for $\tilde{q} \in Q$ arbitrary, choose $\tilde{q}_n \in Q$ as in (8). By definition $q_n \in S(t_n)$ says that $\mathcal{E}(t_n, \tilde{q}_n) + \mathcal{D}(q_n, \tilde{q}_n) - \mathcal{E}(t_n, q_n) \geq 0$. Taking the limsup and using (8) gives

$$0 \leq \limsup_{n \to \infty} \mathcal{E}(t_n, \tilde{q}_n) + \mathcal{D}(q_n, \tilde{q}_n) - \mathcal{E}(t_n, q_n) \leq \mathcal{E}(t_*, \tilde{q}) + \mathcal{D}(q_*, \tilde{q}) - \mathcal{E}(t_*, q_*),$$

which is the desired stability of q_*, since $\widetilde{q}$ was arbitrary. □

3 Space-Time Discretization

We consider two positive parameters τ and h, were $\tau > 0$ represents the fineness of a time discretization by a partition (not necessarily equidistant) of the time interval $[0, T]$. We assume that partitions $\Pi^\tau = \{0 = t_0^\tau < t_1^\tau < \cdots < t_{N_\tau-1}^\tau < t_{N_\tau}^\tau = T\}$ are given such that

$$\text{fineness}(\Pi^\tau) := \max\{ t_j^\tau - t_{j-1}^\tau \mid j = 1, \ldots, N^\tau \} \leq \tau. \tag{9}$$

The parameter $h > 0$ denotes a discretization of the state space $\mathcal{Q}$ by subsets $\mathcal{Q}_h$ again having the structure $\mathcal{Q}_h := \mathcal{U}_h \times \mathcal{Z}_h$. We assume that each $\mathcal{Q}_h$ is closed and the family $(\mathcal{Q}_h)_{h>0}$ is dense, namely

$$\begin{aligned} &\text{for each } (t, q) \in [0, T] \times \mathcal{Q} \text{ there exist } (q_h)_{h>0} \text{ such that} \\ &q_h \in \mathcal{Q}_h, \ q_h \to q, \text{ and } \mathcal{E}(t, q_h) \to \mathcal{E}(t, q). \end{aligned} \tag{10}$$

Hence each space-time discretization is denoted by a pair (τ, h) and we now define the approximation via an incremental minimization problem for the partition Π^τ in the discrete space $\mathcal{Q}_h$ as follows. For a given initial value $q_0^h \in \mathcal{Q}_h$ we define $(q_j^{\tau,h})_{j=0,1,\ldots,N_\tau}$ via

$$q_j^{\tau,h} \in \underset{\widetilde{q} \in \mathcal{Q}_h}{\text{Argmin}} \ \mathcal{E}(t_j^\tau, \widetilde{q}) - \mathcal{E}(t_{j-1}^\tau, q_{j-1}^{\tau,h}) + \mathcal{D}(q_{j-1}^{\tau,h}, \widetilde{q}). \tag{11}$$

Existence of these minimizers follows easily if we assume (4) and (5).

Using these time-discrete approximations in $\mathcal{Q}_h$ we define piecewise constant interpolants $\overline{q}_{\tau,h} : [0, T] \to \mathcal{Q}_h$ via

$$\overline{q}_{\tau,h}(t) = q_k^{\tau,h} \text{ for } t \in [t_k, t_{k+1}[\text{ and } k = 0, \ldots, N^\tau - 1 \text{ and } \overline{q}_{\tau,h}(T) = q_{N^\tau}^{\tau,h}.$$

The first result we give may be considered as a weak analog of *stability of a numerical scheme*. In fact, it provides uniform a priori estimates.

Theorem 1. *Let (3)–(6) hold, then the approximations* $\overline{q}_{\tau,h} : [0, T] \to \mathcal{Q}_h$ *exist and satisfy the following conditions:*

discrete stability
$$\overline{q}_{\tau,h}(t) \in \mathcal{S}_h(t) \text{ for all } t \in \Pi^\tau = \{t_j^\tau \mid j = 0, 1, \ldots, N^\tau\}; \tag{12}$$

upper energy estimate (for $0 \leq j < k \leq N^\tau$)
$$\mathcal{E}(t_k^\tau, \overline{q}_{\tau,h}(t_k^\tau)) + \text{Diss}_{\mathcal{D}}(\overline{q}_{\tau,h}, [t_j^\tau, t_k^\tau])$$
$$\leq \mathcal{E}(t_j^\tau, \overline{q}_{\tau,h}(t_j^\tau)) + \int_{t_j^\tau}^{t_k^\tau} \partial_s \mathcal{E}(s, \overline{q}_{\tau,h}(s)) \, ds; \tag{13}$$

a priori estimates for all $t \in [0, T]$
$$\mathcal{E}(t, \overline{q}_{\tau,h}(t)) \leq \exp(c_1^{\mathcal{E}} t)\big(\mathcal{E}(0, q_0^h) + c_0^{\mathcal{E}}\big) - c_0^{\mathcal{E}} \text{ and} \tag{14}$$
$$\text{Diss}_{\mathcal{D}}(\overline{q}_{\tau,h}, [0, t]) \leq \exp(c_1^{\mathcal{E}} t)\big(\mathcal{E}(0, q_0^h) + c_0^{\mathcal{E}}\big). \tag{15}$$

Here the stable sets $\mathcal{S}_h(t)$ are defined in the obvious way

$$\mathcal{S}_h(t) := \{ q_h \in \mathcal{Q}_h \mid \mathcal{E}(t, q_h) < \infty, \ \mathcal{E}(t, q_h) \leq \mathcal{E}(t, \widetilde{q}_h) + \mathcal{D}(q_h \widetilde{q}_h) \text{ for } \widetilde{q}_h \in \mathcal{Q}_h \}.$$

Note that these stable sets may be substantially larger than $\mathcal{S}(t) \cap \mathcal{Q}_h$.

To formulate the main convergence result, we need to adjust the compatibility condition (7) to sequences of spatial approximations:

$$\begin{array}{c} \text{for each sequence } (h_n, t_n, q_n)_{n \in \mathbb{N}} \text{ with } (h_n, t_n, q_n) \rightharpoonup (0, t_*, q_*), \\ q_n \in \mathcal{S}_{h_n}(t_n), \text{ and } \sup_{n \in \mathbb{N}} \mathcal{E}(t_n, q_n) < \infty \text{ we have} \\ \text{(a) } q_* \in \mathcal{S}(t_*) \quad \text{and (b) } \partial_t \mathcal{E}(t_*, q_n) \to \partial_t \mathcal{E}(t_*, q_*). \end{array} \tag{16}$$

As given in Proposition 1 the crucial part (a) can be derived via the correspondingly adjusted *joint recovery condition*, namely

$$\begin{array}{c} \text{for all } q_*, \widetilde{q} \in \mathcal{Q}, \ (h_n, t_n, q_n)_{n \in \mathbb{N}} \\ \text{with } (h_n, t_n, q_n) \rightharpoonup (0, t_*, q_*), \ q_n \in \mathcal{S}_{h_n}(t_n), \text{ and } \sup_{n \in \mathbb{N}} \mathcal{E}(t_n, q_n) < \infty, \\ \text{there exists } (\widetilde{q}_n)_{n \in \mathbb{N}} \text{ with } \mathcal{Q}_{h_n} \ni \widetilde{q}_n \rightharpoonup \widetilde{q} \text{ such that} \\ \limsup_{n \to \infty} \mathcal{E}(t_n, \widetilde{q}_n) + \mathcal{D}(q_n, \widetilde{q}_n) - \mathcal{E}(t_n, q_n) \leq \mathcal{E}(t_*, \widetilde{q}) + \mathcal{D}(q_*, \widetilde{q}) - \mathcal{E}(t_*, q_*). \end{array} \tag{17}$$

Our main result provides the convergence of space-time discretizations. Because of the implicit nature of the incremental minimization problem (11) there is no "stability restriction" on the size of τ in relation to h. Of course, we cannot expect convergence of the full sequence of approximations, since in general the energetic systems $(\mathcal{Q}, \mathcal{E}, \mathcal{D})$ may have several solutions for a given initial value $q_0 \in \mathcal{S}(0)$, and subsequences may converge to different solutions. Nevertheless, any accumulation point of the approximations is an energetic solution for $(\mathcal{Q}, \mathcal{E}, \mathcal{D})$. Thus, there are no spurious solutions and we may call this property *consistency of the numerical scheme*.

Theorem 2. *Assume that $\mathcal{E}$ and $\mathcal{D}$ satisfy (3)–(6). Let $(\Pi^\tau)_\tau$ and $(\mathcal{Q}_h)_h$ be given such that (9), (10), and (16) hold. Let $q_0 \in \mathcal{S}(0)$ be given and choose $q_0^h \in \mathcal{Q}_h$ with*

$q_0^h \to q_0$ and $\mathcal{E}(0, q_0^h) \to \mathcal{E}(0, q_0)$, and construct approximate solutions $\overline{q}_{\tau,h}$: $[0, T] \to \mathcal{Q}_h$. Then, there exists a subsequence $(\tau_n, h_n)_{n \in \mathbb{N}}$ with $(\tau_n, h_n) \to (0, 0)$ for $n \to \infty$ and an energetic solution $q : [0, T] \to \mathcal{Q}$ of $(\mathcal{Q}, \mathcal{E}, \mathcal{D})$ with $q(0) = q_0$ such that, with the shorthand $\overline{q}_n = (\overline{u}_n, \overline{z}_n) := \overline{q}_{\tau_n, h_n}$, for all $t \in [0, T]$ the following holds:

$$\mathcal{E}(t, \overline{q}_n(t)) \to \mathcal{E}(t, q(t)) \text{ for } n \to \infty; \tag{18a}$$

$$\mathrm{Diss}_{\mathcal{D}}(\overline{q}_n; [0, t]) \to \mathrm{Diss}_{\mathcal{D}}(q, [0, t]) \text{ for } n \to \infty; \tag{18b}$$

$$\overline{z}_n(t) \rightharpoonup z(t) \text{ in } \mathcal{Z} \text{ for } n \to \infty, \tag{18c}$$

there exists a subsequence $(n'_l)_{l \in \mathbb{N}}$ such that

$$\overline{u}_{n'_l}(t) \rightharpoonup u(t) \text{ in } \mathcal{U} \text{ for } l \to \infty; \tag{18d}$$

$$\partial_t \mathcal{E}(\cdot, \overline{q}_n(\cdot)) \to \partial_t \mathcal{E}(\cdot, q(\cdot)) \text{ in } L^1(0, T) \text{ for } n \to \infty. \tag{18e}$$

If additionally, $\mathcal{E}(t, \cdot, z) : \mathcal{U} \to \mathbb{R}_\infty$ is strictly convex, then (18d) can be strengthened into $\overline{u}_n(t) \rightharpoonup u(t)$ in $\mathcal{U}$ (without further subsequences).

If there is only one energetic solution q for $(\mathcal{Q}, \mathcal{E}, \mathcal{D})$ with $q(0) = q_0$, then the whole sequence converges, i.e., $\overline{q}_{\tau,h}(t) \rightharpoonup q(t)$ in $\mathcal{Q}$ for $(\tau, h) \to (0, 0)$.

For a proof of this and even much more general results we refer to [16, 20]. In fact, the proof is an adaptation of the proof of theorem 3.4 in [21] which is based on general ideas of Γ-convergence for sequences of rate-independent systems $(\mathcal{Q}, \mathcal{E}_n, \mathcal{D}_n)_{n \in \mathbb{N}}$.

Since the energetic solutions are not unique in general, one may ask the opposite question. Is it possible to obtain each energetic solution of $(\mathcal{Q}, \mathcal{E}, \mathcal{D})$ as limit of a subsequence? It is shown in [15] that this cannot be expected. However, if one uses approximate minimizers in (11), then this is true. Here approximate minimizers means that $q_j^{\tau,h}$ must be such that the functional under "Argmin" is minimized up to an error δ. In [21] it is shown that the above convergence of subsequences still holds if $\delta = o(\tau)$ for $\tau \to 0$.

4 Linearized Plasticity with Hardening

To start with, we want to demonstrate the applicability of our theory in a simple situation, namely in rate-independent linearized elastoplasticity with hardening. In fact, we are thus able to recover that result in [12], where convergence (without rates) of space-time discretization was shown for the first time under *conditions of minimal regularity*, viz. thus that are known from the classical existence theory.

Here $\mathcal{Q} = \mathcal{U} \times \mathcal{Z}$ with Hilbert spaces $\mathcal{U} = \mathrm{H}^1_{\Gamma_{\mathrm{Dir}}}(\Omega; \mathbb{R}^d)$ and $\mathcal{Z} = \mathrm{L}^2(\Omega; Z)$, where Z is $\mathbb{R}^{d \times d}_{0, \mathrm{sym}} = \{A \in \mathbb{R}^{d \times d} \mid A = A^\mathsf{T}, \mathrm{tr}\, A = 0\}$ for kinematic hardening and $Z = \mathbb{R}^{d \times d}_{0, \mathrm{sym}} \times \mathbb{R}$ for isotropic hardening. The energy functional is quadratic and takes the form

$$\mathcal{E}(t,u,z) = \frac{1}{2}\langle\!\langle \mathcal{A}\begin{pmatrix}u\\z\end{pmatrix}, \begin{pmatrix}u\\z\end{pmatrix}\rangle\!\rangle - \langle \ell(t), u\rangle$$

with a bounded, symmetric and positive definite operator $\mathcal{A} : \mathcal{Q} \to \mathcal{Q}^*$. The dissipation distance reads $\mathcal{D}(q_0, q_1) = \Psi(z_1 - z_0)$ with $\Psi^* = \chi_\mathcal{K}$, where the closed convex cone $\mathcal{K} \subset \mathcal{Z}^*$ is called the elastic domain.

It is easy to see that $\mathcal{E}$ and $\mathcal{D}$ satisfy the assumptions (3)–(6). Moreover, part (b) in the compatibility condition (16) is also valid, as the power $\partial_t \mathcal{E}(t, q) = -\langle \dot{\ell}(t), u\rangle$ is linear and, hence, weakly continuous.

It remains to establish part (a) of (16) by using the joint recovery condition (17). For this assume that there exist interpolation operators $B_h : \mathcal{Q} \to \mathcal{Q}_h$ such that

$$B_h q \to q \text{ strongly in } \mathcal{Q} \text{ and } \Psi(B_h q) \to \Psi(q). \tag{19}$$

While the first case is the usual interpolation condition, the second condition states that this has to be consistent with the dissipation potential Ψ. Since in general Ψ is not strongly continuous, this is nontrivial. However, as $\Psi(v) = \int_\Omega \psi(x, v(x))\,\mathrm{d}x$ and $\psi(x, \cdot)$ is convex, it is sufficient to choose B_h such that z_h in $(u_h, z_h) = B_h q$ is piecewise constant taking the average value over the polyhedra in the spatial discretization. We choose $\widetilde{q}_n \in \mathcal{Q}_{h_n}$ in (17) via

$$\widetilde{q}_n = q_n + B_{h_n}(\widetilde{q} - q_*) \text{ giving } \begin{cases} \widetilde{q}_n \to \widetilde{q}, \\ \widetilde{q}_n - q_n \to \widetilde{q} - q_*. \end{cases} \tag{20}$$

Clearly, we have $\mathcal{D}(q_n, \widetilde{q}_n) = \Psi(B_{h_n}(\widetilde{q} - q_*)) \to \Psi(\widetilde{q} - q_*) = \mathcal{D}(q_*, \widetilde{q})$. Moreover, in the energy can use the quadratic nature to profit from cancellation effects:

$$\mathcal{E}(t_n, \widetilde{q}_n) - \mathcal{E}(t_n, q_n) = \langle\!\langle \tfrac{1}{2}\mathcal{A}(\widetilde{q}_n + q_n) - \begin{pmatrix}\ell(t_n)\\0\end{pmatrix}, \widetilde{q}_n - q_n\rangle\!\rangle$$
$$\to \langle\!\langle \tfrac{1}{2}\mathcal{A}(\widetilde{q} + q_*) - \begin{pmatrix}\ell(t_*)\\0\end{pmatrix}, \widetilde{q} - q_*\rangle\!\rangle = \mathcal{E}(t_*, \widetilde{q}) - \mathcal{E}(t_*, q_*).$$

Here (20) guarantees that the first term in $\langle\!\langle \cdot, \cdot\rangle\!\rangle$ converges weakly and the second strongly. It follows that (17) and whence (16) hold, and Theorem 2 provides convergence of the *whole discretization sequence*, since that continuous problem $(\mathcal{Q}, \mathcal{E}, \mathcal{D})$ has a unique solution for each $q(0) \in \mathcal{S}(0)$.

5 A Damage Model

Finally we consider a damage model introduced in [8, 9] and analyzed in [17, 22] using the energetic approach. While the displacement $u \in \mathcal{U} = \mathrm{H}^1_{\Gamma_{\mathrm{Dir}}}(\Omega; \mathbb{R}^d)$ is as above, the internal variable is now a scalar damage variable with $z(t, x) \in Z := [0, 1]$, where $z = 1$ denotes an undamaged material whereas $z = 0$ means that all breakable pieces are broken. However, depending on the model, $z = 0$ may still have some remaining elasticity. As space of internal states we let

$$\mathcal{Z} := \{ z \in W^{1,r}(\Omega) \mid z(x) \in [0,1] \} \Subset C^0(\overline{\Omega}),$$

where we assume $r > d$ to have the indicated embedding. The dissipation distance is chosen in such a way that increase of damage (decrease of z) costs proportional to the increase and the damaged volume. To forbid healing we set the dissipation ∞ for increasing z:

$$\mathcal{D}(z_0, z_1) = \int_\Omega \psi(x, z_1(x) - z_0(x)) \, dx \text{ with } \psi(x, v) = \begin{cases} \delta(x)|v| & \text{for } v \leq 0, \\ \infty & \text{for } v > 0. \end{cases}$$

For simplicity we assume that the linearized elasticity can be assumed giving a quadratic energy functional $\mathcal{E}$:

$$\mathcal{E}(t, u, z) = \int_\Omega \frac{1}{2}(e(u) + e_\mathrm{D}(t)) : \mathbf{C}(z) : (e(u) + e_\mathrm{D}(t)) + \kappa |\nabla z|^r \, dx,$$

where $e_\mathrm{D}(t) = e(u_\mathrm{D}(t))$ and $u_\mathrm{D} \in C^1([0, T], H^1(\Omega; \mathbb{R}^d))$ is given. The elasticity tensor is monotone in z, i.e., $\mathbf{C}'(z) \geq \mathbf{0}$ in the sense of symmetric operators. Moreover, the coercivity

$$e : \mathbf{C}(z) : e \geq (\alpha_0 + \alpha_1 z^\gamma) |e|^2 \text{ for all } z \in [0, 1] \text{ and } e \in \mathbb{R}^{d \times d}_\mathrm{sym}$$

will be basic, where $\alpha_1, \gamma > 0$ and $\alpha_0 \geq 0$. The case $\alpha_0 > 0$ corresponds to incomplete damage like in [6, 17], and α_0 allows for complete damage as studied in [3, 22]. To treat the latter case it is necessary to eliminate the displacement u, since it may not be well-defined because of missing coercivity. This is done via introducing the quadratic functional

$$\mathcal{V}(z, e_\mathrm{D}) = \liminf_{z_n \to z} \min_{u \in \mathcal{U}} \int_\Omega \frac{1}{2}(e(u) + e_\mathrm{D}) : \mathbf{C}(z_n) : (e(u) + e_\mathrm{D}) \, dx,$$

which even allows us to control the equilibrium stresses via $D_{e_\mathrm{D}} \mathcal{V}$.

Again it is easy to check the assumption (3)–(6); as usual the main difficulty lies in part (a) of (16). Assume again that $B_h : \mathcal{Q} \to \mathcal{Q}_h$ is an interpolant to piecewise affine functions on a triangulation $\mathcal{T}_h$ of Ω such that $B_h q = (B_h^\mathcal{U} u, B_h^\mathcal{Z} z) \to q$ in $\mathcal{Q}$ strongly. To employ the joint recovery condition (17) we choose for given $\widetilde{q} \in \mathcal{Q}$ and $q_n \in \mathcal{Q}_{h_n}$ the joint recovery sequence

$$\widehat{q}_n = (B_{h_n}^\mathcal{U} \widetilde{u}, \max\{0, B_{h_n}^\mathcal{Z} \widetilde{z} - \rho_n\}) \in \mathcal{Q}_{h_n} \text{ with } \rho_n = \|\max\{0, B_{h_n}^\mathcal{Z} \widetilde{z} - z_n\}\|_{L^\infty(\Omega)}.$$

Since we only need to check condition (16a) for $\mathcal{D}(z, \widetilde{z}) < \infty$ we may assume $\widetilde{z} \leq z_*$. Now using the embedding $\mathcal{Z} \Subset C^0(\overline{\Omega})$ we find $z_n \to z_*$ in L^∞ and similarly $B_{h_n}^\mathcal{Z} \widetilde{z} \to \widetilde{z} \leq z_*$. Thus, we have $\rho_n \to 0$ and conclude $\widetilde{q}_n \to \widetilde{q}$ in $\mathcal{Q}$ strongly. This in turn implies $\mathcal{E}(t_n, q_n) \to \mathcal{E}(t_*, q_*)$ and $\mathcal{D}(z_n, \widetilde{z}_n) \to \mathcal{D}(z_*, \widetilde{z})$. Using the lower semicontinuity of $\mathcal{E}$, condition (16a) is established.

References

1. J. Alberty and C. Carstensen. Numerical analysis of time-depending primal elastoplasticity with hardening. *SIAM J. Numer. Anal.*, 37, 1271–1294 (electronic), 2000.
2. F. Auricchio, A. Mielke, and U. Stefanelli. A rate-independent model for the isothermal quasi-static evolution of shape-memory materials. M^3AS *Math. Models Meth. Appl. Sci.*, 18(1), 125–164, 2008.
3. G. Bouchitté, A. Mielke, and T. Roubíček. A complete-damage problem at small strains. *Z. Angew. Math. Phys. (ZAMP)*, 2007.
4. G. Dal Maso and C. Zanini. Quasi-static crack growth for a cohesive zone model with prescribed crack path. *Proc. R. Soc. Edinb., Sect. A, Math.*, 137(2), 253–279, 2007.
5. G. Dal Maso, G. Francfort, and R. Toader. Quasistatic crack growth in nonlinear elasticity. *Arch. Rational Mech. Anal.*, 176, 165–225, 2005.
6. G. Francfort and A. Garroni. A variational view of partial brittle damage evolution. *Arch. Rational Mech. Anal.*, 182, 125–152, 2006.
7. M. Frémond. *Non-Smooth Thermomechanics*. Springer, Berlin, 2002.
8. M. Frémond and B. Nedjar. Damage, gradient of damage and principle of virtual power. *Internat. J. Solids Structures*, 33, 1083–1103, 1996.
9. M. Frémond, K. Kuttler, B. Nedjar, and M. Shillor. One-dimensional models of damage. *Adv. Math. Sci. Appl.*, 8, 541–570, 1998.
10. B. Halphen and Q. S. Nguyen. Sur les matériaux standards généralisés. *J. Mécanique*, 14, 39–63, 1975.
11. W. Han and B. D. Reddy. *Plasticity (Mathematical Theory and Numerical Analysis)*, volume 9 of *Interdisciplinary Applied Mathematics*. Springer-Verlag, New York, 1999.
12. W. Han and B. D. Reddy. Convergence of approximations to the primal problem in plasticity under conditions of minimal regularity. *Numer. Math.*, 87(2), 283–315, 2000.
13. M. Kružík, A. Mielke, and T. Roubíček. Modelling of microstructure and its evolution in shape-memory-alloy single-crystals, in particular in CuAlNi. *Meccanica*, 40, 389–418, 2005.
14. A. Mielke. Evolution in rate-independent systems (Ch. 6). In C. Dafermos and E. Feireisl (Eds.), *Handbook of Differential Equations, Evolutionary Equations, vol. 2*, Elsevier, Amsterdam, pp. 461–559, 2005.
15. A. Mielke and F. Rindler. Reverse approximation of energetic solutions to rate-independent processes. *Nonl. Diff. Eqns. Appl. (NoDEA)*, 2007.
16. A. Mielke and T. Roubíček. Numerical approaches to rate-independent processes and applications in inelasticity. *M2AN Math. Model. Numer. Anal.*, 2006. Submitted. WIAS Preprint 1169.
17. A. Mielke and T. Roubíček. Rate-independent damage processes in nonlinear elasticity. *Math. Models Methods Appl. Sci.*, 16, 177–209, 2006.
18. A. Mielke and F. Theil. A mathematical model for rate-independent phase transformations with hysteresis. In H.-D. Alber, R. Balean, and R. Farwig (Eds.), *Proceedings of the Workshop on "Models of Continuum Mechanics in Analysis and Engineering"*, Aachen. Shaker-Verlag, pp. 117–129, 1999.
19. A. Mielke and F. Theil. On rate–independent hysteresis models. *Nonl. Diff. Eqns. Appl. (NoDEA)*, 11, 151–189, 2004.
20. A. Mielke, L. Paoli, and A. Petrov. On the existence and approximation for a 3d model of thermally induced phase transformations in shape-memory alloys, 2008, in preparation.
21. A. Mielke, T. Roubíček, and U. Stefanelli. Γ-limits and relaxations for rate-independent evolutionary problems. *Calc. Var. Part. Diff. Equ.*, 31, 387–416, 2008.
22. A. Mielke, T. Roubíček, and J. Zeman. Complete damage in elastic and viscoelastic media and its energetics. *Comput. Methods Appl. Mech. Engrg.*, 2007. Submitted. WIAS preprint 1285.
23. A. Mielke, F. Theil, and V. I. Levitas. A variational formulation of rate–independent phase transformations using an extremum principle. *Arch. Rational Mech. Anal.*, 162, 137–177, 2002.

24. M. Ortiz and E. Repetto. Nonconvex energy minimization and dislocation structures in ductile single crystals. *J. Mech. Phys. Solids*, 47, 397–462, 1999.
25. B. D. Reddy and J. B. Martin. Algorithms for the solution of internal variable problems in plasticity. *Comput. Methods Appl. Mech. Engrg.*, 93(2), 253–273, 1991.
26. B. D. Reddy, J. B. Martin, and T. B. Griffin. Extremal paths and holonomic constitutive laws in elastoplasticity. *Quart. Appl. Math.*, 45(3), 487–502, 1987.

Damage and Fracture

A Computational Methodology for Modeling Ductile Fracture

Ahmed Amine Benzerga

Department of Aerospace Engineering, Texas A&M University,
College Station, TX 77843-3141, USA
E-mail: benzerga@aero.tamu.edu

Abstract. In recent work, the author and co-workers have introduced and developed a new computational ductile fracture methodology. This approach accounts for certain types of initial and induced anisotropy, and has been further refined to account for coupling between void shape and the effects of anisotropy. The effectiveness of the method in capturing the link between microstructure and fracture properties is explored through a three-dimensional finite element analysis of ductile fracture in notched bars.

Key words: ductile fracture, homogenization, void growth, void coalescence, induced anisotropy, finite element analysis.

1 Introduction

An attractive computational methodology for modeling and simulation of ductile fracture was developed in the 1980s by Tvergaard and Needleman [1–3]. Its success was in part due to employing an elegant micromechanics-based model of plastic behavior applicable to porous materials [4]. A unique feature of Gurson's yield criterion is that it constitutes, for the chosen representative volume element (RVE) a rigorous upper bound, which also happens to lie very close to the exact criterion. Under large plastic deformations, however, the basic structural unit, i.e. the RVE itself, evolves. This microstructural evolution leads to induced anisotropy. To account for this, extensions of the Gurson model were developed in the 1990s to incorporate void shape effects [5, 6] and plastic anisotropy of the matrix material [7, 8]. Both have been shown to affect void growth rates in a drastic way. Micromechanical unit-cell calculations of the type pioneered by Koplik and Needleman [9] have also documented the effect of void shape on void coalescence [10]. This has motivated the development of improved models of void coalescence [7, 10–14]. Based on the above extensions of the Gurson model, Benzerga and co-workers [7, 15, 16] introduced a new ductile fracture computational methodology, which accounts for certain types of initial and induced anisotropy. In particular, they proposed a heuristic combination of void shape and plastic anisotropy effects. In a recent paper, their methodology was refined to account for such coupling in a rigorous manner [17]. To illustrate the advantage of the approach in capturing the link between microstructure and fracture properties, ductile failure in notched bars is analyzed based on three-dimensional finite element calculations.

2 Basic Micromechanisms

To set the stage and emphasize the relevance of a micromechanics based methodology, the basic micromechanisms of ductile fracture are recalled. Most often, ductile fracture occurs due to void nucleation, growth and coalescence. The latter is a precursor to crack initiation in an otherwise crack-free material. In structural alloys, voids originate at second phase particles as illustrated in Figure 1a. Subsequently, void growth is driven by plastic flow and strongly affected by the state of stress. An example of enlarged cavities is shown in Figure 1b. Most cavities are actually elongated in the direction normal to the plane of view and are shown end-on. After sufficient growth, plastic deformation tends to localize in the inter-void ligament thus leading to accelerated (mostly lateral) void growth (see Figure 1c).

Void growth to coalescence (Figures 1b, c) is a mere expression of substantial plastic deformation of the surrounding matrix. Therefore, it is not surprising that the ease (or difficulty) with which the material flows plastically would affect void enlargement. In particular, material texture effects are expected to play a key role. In previous works [7, 18], a systematic attempt was made to quantify the effect of plastic anisotropy on void growth and coalescence. Figure 2c shows the evolution with axial plastic strain, of the strain ratio, R^X, defined as the ratio of transverse plastic strain rates for loading along X. For an isotropic material, $R^X = 1$ for all X. Consequence of the fabrication process, plastic anisotropy effects of the type depicted in Figure 2 are generally neglected in quantitative analyses of ductile fracture. In the following section, some of these effects are theoretically explored by means of simulation.

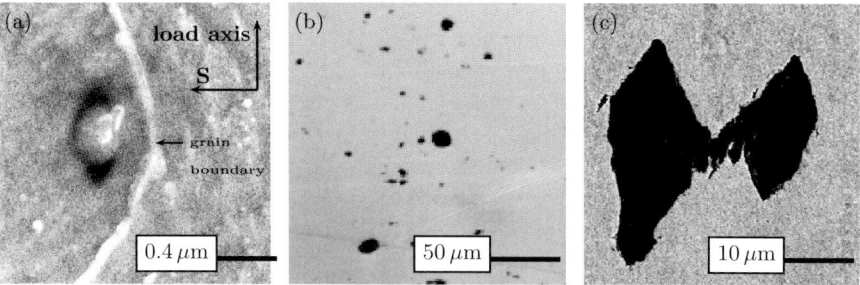

Fig. 1 Micromechanisms of ductile fracture by (a) void nucleation, (b) growth, and (c) coalescence. In (b) loading is normal to plane of view. In (c) void impingement, the final stage of coalescence, is shown. Adapted from [7, 18].

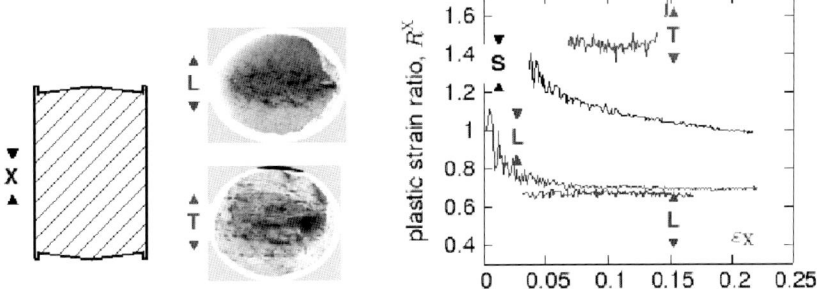

Fig. 2 Monitoring the evolution of plastic anisotropy. (a) Cylinder under compression (arrows pointing inward). (b) Fracture surfaces showing oval cross-sections in tension (arrows pointing outward). (c) Measured strain ratios versus axial plastic strain in typical tension and compression tests. Key for loading directions, L: rolling; T: transverse; S: through-thickness. After [12].

3 FE Micromechanical Calculations

Finite-element micromechanical calculations have proven useful in guiding the development of improved ductile fracture models over the past two decades [8–10]. The calculations are based on the concept of a unit-cell containing a void (Figure 3) as elaborated upon in [1]. FE calculations of that type are presented in this preliminary section to recall the phenomenology of void growth to coalescence and demonstrate the drastic effect of material plastic anisotropy on void growth rates. To that end, invariance of material flow properties about an axis is assumed so that axisymmetric calculations can be used as in [8]. The object-oriented FE code ZeBuLoN [19] is employed. A Lagrangian formulation of the field equations is used as

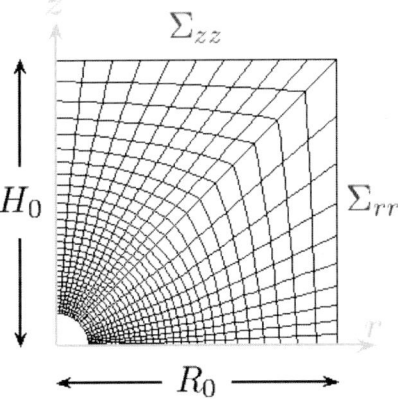

Fig. 3 Finite element mesh used in a typical unit-cell calculation showing some of the notation used.

in [16]. A spherical void is embedded in a matrix, as sketched in Figure 3. The mesh consists of sub-integrated quadratic quadrilateral elements. The matrix is taken to be viscoplastic Norton-like and plastically anisotropic of the Hill type. The material parameters are chosen such that the matrix behaves as a power-law strain-hardening material with exponent $n = 0.1$. The initial void volume fraction is $f_0 = 0.001$. Special boundary conditions are formulated such that the ratio θ of net axial stress, Σ_{zz}, to net lateral stress, Σ_{rr}, remains constant throughout the calculation. Stress triaxiality is measured by the ratio $\mathcal{T}$ of the mean normal stress, Σ_m, to the equivalent stress Σ_e, given by:

$$\Sigma_e = |\Sigma_{zz} - \Sigma_{rr}|, \quad \Sigma_m = \frac{1}{3}(\Sigma_{zz} + 2\Sigma_{rr}), \quad \mathcal{T} = \frac{\Sigma_m}{\Sigma_e} = \frac{1}{3}\frac{2\theta + 1}{|1 - \theta|} \quad (1)$$

A Riks algorithm [20] is used to integrate the nonlinear constitutive equations in order to keep the stress ratio θ, and hence $\mathcal{T}$, constant. Here $\mathcal{T} = 1$. All material parameters are kept fixed except the Hill anisotropy factors that characterize plastic flow of the matrix material. Two sets of Hill factors are used which are representative of an Al alloy and Zr alloy.

The corresponding mesoscale stress versus strain responses are compared in Figure 4 with that of an isotropic solid. In a typical calculation, the stress drop coincides with the onset of void coalescence, which is accompanied by accelerated lateral void growth, see case of Zr alloy in figure. It is this shift to a regime of fast and anisotropic void growth that is known as *onset of void coalescence* [9,13], which is not to be confused with void impingement. The results in Figure 4 clearly show the effect of plastic anisotropy of the matrix on the overall response of the unit cell. It remains to be seen how much of the details of plastic deformation need to be accounted for in analyzing void growth in structural materials.

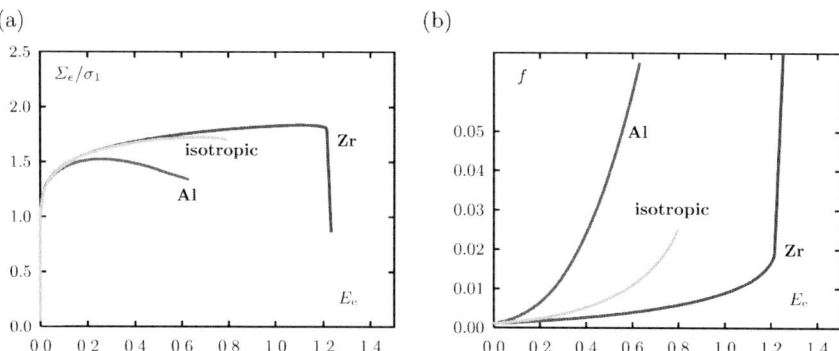

Fig. 4 Results of unit-cell calculations for three orthotropic matrices. (a) Effective stress, Σ_e, normalized by the matrix yield stress in tension along x_1, versus effective strain, E_e. (b) void volume fraction versus E_e.

4 Homogenization Framework

With the kind of results of Figure 4 in mind, the objective is to develop an analytical model that captures pre- and post-coalescence mechanics as well as plastic anisotropy effects. To this end, recourse to classical homogenization is made, just as in the original Gurson model [21]. Consider a porous representative volume element (RVE) as shown in Figure 5, where Ω represents the volume of the RVE and ω the total volume of the voids. Denote the microscopic stress and deformation rate fields in the RVE by $\sigma(\mathbf{x})$ and $\mathbf{d}(\mathbf{x})$, respectively. Let the RVE be subjected to homogeneous boundary deformation rate $\mathbf{D}$, i.e.,

$$\forall \mathbf{x} \in \partial\Omega, \qquad \mathbf{v}(\mathbf{x}) = \mathbf{D} \cdot \mathbf{x} \qquad (2)$$

It follows that $\mathbf{D} = \langle \mathbf{d}(\mathbf{x}) \rangle_\Omega$, where $\langle \cdot \rangle_\Omega$ represents the volume average over the RVE. In the Hill–Mandel [22, 23] homogenization theory, the macroscopic stress is defined as $\boldsymbol{\Sigma} \equiv \langle \sigma(\mathbf{x}) \rangle_\Omega$. The Hill–Mandel lemma states that:

$$\boldsymbol{\Sigma} : \mathbf{D} \equiv \langle \sigma \rangle_\Omega : \langle \mathbf{d} \rangle_\Omega = \langle \sigma : \mathbf{d} \rangle_\Omega \qquad (3)$$

for any kinematically admissible $\mathbf{d}$ and statically admissible σ. In particular, $\mathbf{d}$ and σ need not be related by any specific constitutive relation. For a rigid-perfectly plastic behavior for the matrix, the microscopic plastic dissipation, for a given $\mathbf{d}$, is defined as

$$\pi(\mathbf{d}) \equiv \sup_{\sigma^* \in \mathcal{C}} \sigma^* : \mathbf{d} \qquad (4)$$

where the supremum is taken over all stresses that fall within the microscopic convex of rigidity, $\mathcal{C}$. Using (4) in conjunction with the Hill–Mandel lemma (3), one has $\boldsymbol{\Sigma} : \mathbf{D} = \langle \sigma : \mathbf{d} \rangle_\Omega \leq \langle \pi(\mathbf{d}) \rangle_\Omega$. This inequality holds for any kinematically admissible velocity field; hence

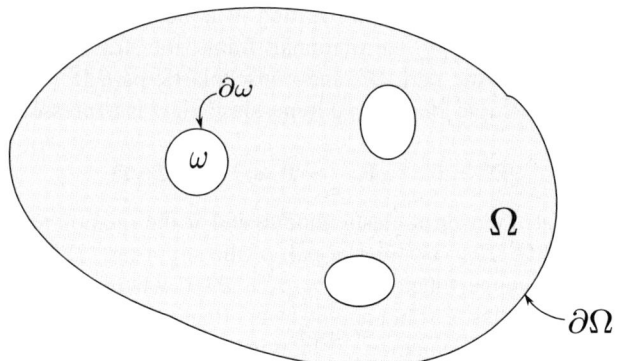

Fig. 5 Sketch of a porous representative volume element

$$\boldsymbol{\Sigma} : \mathbf{D} \leq \inf_{\mathbf{d} \in \mathcal{K}(\mathbf{D})} \langle \pi(\mathbf{d}) \rangle_{\Omega} \equiv \Pi(\mathbf{D}) \tag{5}$$

where $\mathcal{K}(\mathbf{D})$ denotes the set of microscopic deformation fields kinematically admissible with $\mathbf{D}$. $\Pi(\mathbf{D})$ is termed the macroscopic plastic dissipation associated with $\mathbf{D}$ [24]. Equation (5) represents a half-space in the macroscopic stress space. It can be shown that the macroscopic yield locus in stress space is the envelope of the hyperplanes given by (5) [21, 24]. Since $\Pi(\mathbf{D})$ is positively homogeneous of degree one the parametric equation of the macroscopic yield locus is given by, in view of the Euler relation,

$$\boldsymbol{\Sigma} = \frac{\partial \Pi}{\partial \mathbf{D}}(\mathbf{D}) \tag{6}$$

where the five ratios of the components of $\mathbf{D}$ act as the parameters. Elimination of the parameters between the six equations (6) yields the explicit equation of the yield locus.

In general, the set of appropriate velocity fields to be used in the above variational procedure, must change with the onset of void coalescence. This is consistent with the unit-cell calculations. Keralavarma and Benzerga [17] have followed the above procedure to derive an approximate yield locus for a porous material containing non-spherical voids embedded in an orthotropic Hill matrix under axisymmetric loading conditions. The effective behavior during void coalescence is modeled below following Benzerga [12, 13].

5 Computational Methodology

The micromechanics of Section 4 is used to obtain closed form expressions of a yield criterion, associated flow rule and evolution laws of structural variables, in the rate-independent limit. In what follows, the actual constitutive framework used is that of a progressively cavitating anisotropic viscoplastic solid [7, 15, 16] where the analytical form of the yield function (Section 4) is heuristically used as a plastic potential. In the objective (e.g. co-rotational) frame, the strain rate tensor is written as the sum of an elastic part, $\mathbf{D}^e$, and a viscoplastic part, $\mathbf{D}^p$. Assuming small elastic strains and isotropic elasticity, a hypo-elastic law is expressed using the rotated stress $\mathbf{P}$

$$\mathbf{D}^e = \mathbf{C}^{-1} : \dot{\mathbf{P}}, \qquad \mathbf{P} = J\, \boldsymbol{\Omega}^T \cdot \boldsymbol{\Sigma} \cdot \boldsymbol{\Omega} \tag{7}$$

where $\mathbf{C}$ is the rotated tensor of elastic moduli and $\boldsymbol{\Omega}$ the rotation tensor, e.g. identified with the spin $\mathbf{Q}$. The viscoplastic part of the strain rate, $\mathbf{D}^{(p)}$, is obtained by normality from the gauge function: $\phi = \sigma_\star - \bar{\sigma}(p)$ where $\bar{\sigma}$ is the matrix flow stress, p the effective plastic strain and $\sigma_\star$ is an effective matrix stress which is implicitly defined through an equation of the type $\mathcal{F}(\boldsymbol{\Sigma}, f, S, \mathbf{e}_z, \mathbf{H}, \sigma_\star) = 0$ with f the porosity, S the shape parameter (logarithm of the void aspect ratio W), $\mathbf{e}_z$ the void axis and $\mathbf{H}$ Hill's tensor. The potential $\mathcal{F}$ admits two different expressions, $\mathcal{F}^{(c-)}$ and $\mathcal{F}^{(c+)}$, prior to and after the onset of void-coalescence, respectively. The flow potential prior to coalescence is given by [6, 15]

$$\mathcal{F}^{(c-)} = \frac{3\,\mathbf{\Sigma}':\mathbf{H}:\mathbf{\Sigma}'}{2\,\sigma_\star^2} + 2q_w f \cosh\left(\frac{\kappa}{h}\frac{\mathbf{A}:\mathbf{\Sigma}}{\sigma_\star}\right) - 1 - q_w^2 f^2 = 0 \quad (8)$$

where $()'$ refers to the deviator, h is a factor calculated using Hill coefficients, expressed in the basis $[\mathbf{e}] = (\mathbf{e}_L, \mathbf{e}_T, \mathbf{e}_S)$ pointing onto the principal directions of orthotropy [8], as

$$h = \left[\frac{2}{5}\frac{h_L + h_T + h_S}{h_L h_T + h_T h_S + h_S h_L} + \frac{1}{5}\left(\frac{1}{h_{TS}} + \frac{1}{h_{LS}} + \frac{1}{h_{LT}}\right)\right]^{1/2}, \quad (9)$$

and $\mathbf{A}$ is the void anisotropy tensor expressed in the basis $[\mathbf{e}'] = (\mathbf{e}_x, \mathbf{e}_y, \mathbf{e}_z)$ associated with the void

$$\mathbf{A} = \alpha_2 (\mathbf{e}_x \otimes \mathbf{e}_x + \mathbf{e}_y \otimes \mathbf{e}_y) + (1 - 2\alpha_2)\,\mathbf{e}_z \otimes \mathbf{e}_z, \quad (10)$$

with α_2 in (10) and κ in (8) being scalar functions of both f and S. Eq. (8) for the potential corresponds to prolate voids. The general expression can be found in [17].

The flow potential after the onset of coalescence is given by [12, 13]:

$$\mathcal{F}^{(c+)}(\mathbf{\Sigma}, \chi, S, \mathbf{H}, \sigma_\star) = \frac{\sqrt{3/2\mathbf{\Sigma}':\mathbf{H}:\mathbf{\Sigma}'}}{\sigma_\star} + \frac{1}{2}\frac{|\mathbf{I}:\mathbf{\Sigma}|}{\sigma_\star} - \frac{3}{2}(1-\chi^2)\,\mathcal{C}_f(\chi, S) \leq 0 \quad (11)$$

where χ is the ligament size ratio and

$$\mathcal{C}_f(\chi, S) = 0.1\left(\frac{\chi^{-1}-1}{W^2 + 0.1\chi^{-1} + 0.02\chi^{-2}}\right)^2 + 1.3\sqrt{\chi^{-1}}; \quad W = e^S \quad (12)$$

At the onset of coalescence $\mathcal{F}^{(c-)} = \mathcal{F}^{(c+)} = 0$. As $\chi \to 1$ the material loses all stress carrying capacity.

To complete the constitution, strain-hardening and strain-rate effects are incorporated using standard viscoplastic laws, as appropriately cast for the type of material modeled. The evolution laws of the microstructural variables prior to coalescence are given by

$$\dot{f} = (1-f)\,\mathbf{I}:\mathbf{D}^{(p)}, \quad (13)$$

$$\dot{S} = \frac{3}{2}\left[1 + g(\mathcal{T})(1-\sqrt{f})^2\frac{\alpha_1 - \alpha_1^G}{1 - 3\alpha_1}\right]D_{zz}^{'(p)} + \left(\frac{1-3\alpha_1}{f} + 3\alpha_2 - 1\right)\mathbf{I}:\mathbf{D}^{(p)}, \quad (14)$$

where $g(\mathcal{T})$ is a function of the stress triaxiality ratio and is the result of fits to unit-cell calculations, Also, α_1 and α_1^G are analytical functions of f and S [16].

After the onset of coalescence the relevant variables are λ, χ and W; their rates are given by [13]

$$\dot{\lambda} = \frac{3}{2}\lambda\,D_{zz}^{'(p)}, \quad (15)$$

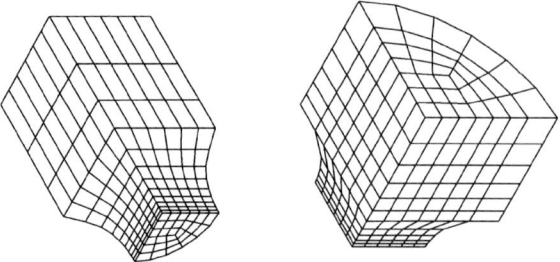

Fig. 6 3D meshes used for notched specimens.

which holds regardless of the spatial distribution of voids,

$$\dot{\chi} = \frac{3}{4} \frac{\lambda}{W} \left[\frac{3}{2\chi^2} - 1 \right] D_{zz}^{'(p)}, \tag{16}$$

which results from plastic incompressibility of the matrix material, and

$$\dot{W} = \frac{9}{4} \frac{\lambda}{\chi} \left[1 - \frac{1}{2\chi^2} \right] D_{zz}^{'(p)} \tag{17}$$

Assuming that voids rotate with the material, the evolution of void orientation is given by

$$\dot{\mathbf{e}}_z = \dot{\mathbf{\Omega}} \cdot \mathbf{\Omega}^T \cdot \mathbf{e}_z \tag{18}$$

where $\mathbf{\Omega}$ is the rotation used in (7) so that (18) follows from the objective frame description. If the co-rotational space frame is used then $\dot{\mathbf{\Omega}} \cdot \mathbf{\Omega}^T$ is simply the spin $\mathbf{W}$. In this case the rotation is determined by integration of $\dot{\mathbf{Q}} = \mathbf{W} \cdot \mathbf{Q}$ with the initial condition $\mathbf{Q}_{t_0} = \mathbf{I}$. Experimental evidence supports the general form (18) if the loading axes are initially aligned with the void axes [e'].

6 Predictive Modeling of Fracture in Notched Bars

The above constitutive equations were implemented in the finite element code ZEBULON [15, 16]. Incorporation of more recent developments [17] that follow upon the general scheme outlined in Section 4, is underway.

3D calculations were carried out using round tensile notched bars. Notch acuity is identified by the parameter ζ equal to 10 times the ratio of notch radius to the minimal section diameter Φ_0. For example, the meshes of Figure 6 correspond to $\zeta = 10$ and $\zeta = 2$. 20-node quadratic sub-integrated quadrilaterals were used. Because the analyses are restricted to loadings along principal axes, only one eighth of a bar is meshed. Lateral surfaces are free of normal tractions and symmetry conditions are enforced. A uniform displacement U is prescribed on the top surface.

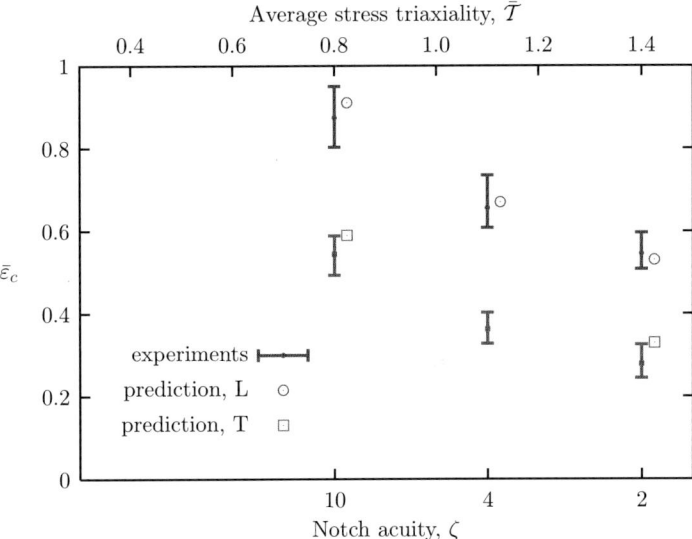

Fig. 7 Measured and predicted average strain to failure initiation, $\bar{\varepsilon}_c$, in various notched bars for two loading orientations.

Key material parameters include the initial values of void volume fraction, f_0, void aspect ratio, W_0, and relative void spacing λ_0. These were identified based on data related to second phase particles, noting that void nucleation is nearly complete at low plastic strains, especially at MnS inclusions [18]. In addition, plastic anisotropy was identified based on strain ratios such as those shown in Figure 2. Further details are provided in [16].

To minimize mesh-size effects, notch ductility, for given notch geometry and loading orientation, is identified with the value taken by the average strain to failure *initiation* in each bar. This strain is denoted by $\bar{\varepsilon}_c$. Comparison between measured and computed values is shown in Figure 7 for loading parallel to the rolling axis of the parent plate (L orientation) and transverse to it (T orientation).

The prediction of ductility is excellent, notably under L loading. The following is worth emphasizing, which highlights the progress made in terms of predictive capability with respect to state-of-the-art methodologies [3]. In the GTN approach, two adjustable parameters are employed: the critical void volume fraction (f_c) and the final void volume fraction. First, no such adjustable parameters are used here. Second, the orientation dependence of failure strains is well captured and, for a given loading orientation, the development of oval cross-sections due to induced anisotropy is also well predicted as discussed in [16]. It is important to note, however, that some key microstructural observations were needed to guide the use of appropriate initial values of the microstructural variables f_0, W_0 and λ_0.

In order to further qualify the predictive capability of the new approach, the computed values of void volume fraction at the onset of crack initiation in the bars, f_i, were tabulated. These values, were then compared with the maximum values meas-

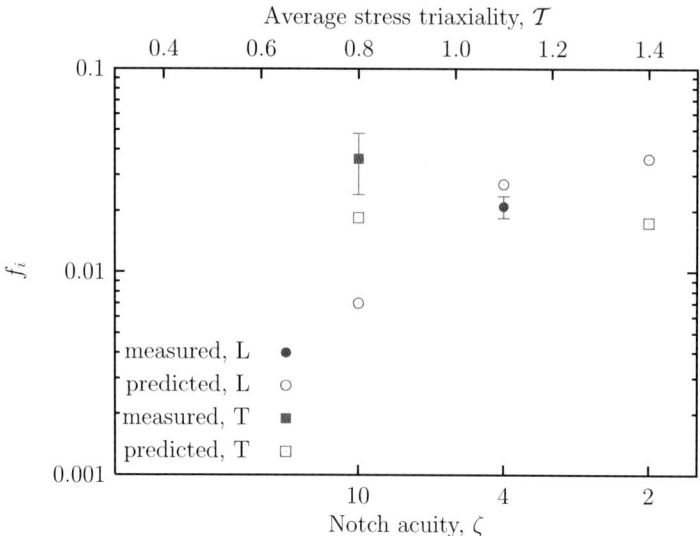

Fig. 8 Measured and predicted values of void volume fraction at crack initiation, f_i, in notched bars.

ured beneath fracture surfaces in actual experiments. Details of the experimental procedure followed may be found in [18]. The measurements therein rely on an objective definition of local porosity based on the construction of Dirichlet tessellations associated with a distribution of about 100 voids per specimen analyzed. Figure 8 summarizes the comparison between computed and measured values. The two are found to be very close to each other. Note that the values are about 1 to 3%, i.e., much smaller than the values of critical porosity f_c that are typically used in theoretical analyses employing the GTN approach. By the same account, these values are one order of magnitude larger than typical values of f_c needed to obtain a good match between measured and computed strains to failure when the GTN model is used; e.g., see [25]. The above observations reinforce the need for incorporating more microstructural information in ductile fracture predictions, just as proposed in the present paper.

7 Concluding Remarks

The computational ductile fracture methodology recently proposed by Benzerga and co-workers was critically reviewed. The new methodology is based on a set of micromechanical models and extends the widely accepted methodology established in previous works. Notable among the novel features of the theory is a proper account for initial and induced anisotropy, albeit in a restricted sense, and an enriched description of initial and evolving microstructure. The illustrations given here show the promise for improving structure-property relationships in the context of ductile fracture of structural materials. Focus was laid on prediction of notch ductility, where mesh size effects are small. Application to toughness predictions, i.e., in the presence of strong gradients in the macroscopic fields, will inevitably face the additional challenge of objectivity with mesh size in the softening regime. While this is an outstanding challenge, remediation of pathological mesh-size effects could benefit from micromechanics analyses in ways yet to be demonstrated.

References

1. V. Tvergaard. *Int. J. Frac.*, 18:237–252, 1982.
2. C. Chu and A. Needleman. *J. Eng. Mat. Tech.*, 102:249–256, 1980.
3. V. Tvergaard and A. Needleman. *Acta Metall.*, 32:157–169, 1984.
4. A. L. Gurson. *J. Eng. Mat. Tech.*, 99:2–15, 1977.
5. M. Gologanu, J.-B. Leblond, and J. Devaux. *J. Mech. Phys. Solids*, 41(11):1723–1754, 1993.
6. M. Gologanu, J.-B. Leblond, G. Perrin, and J. Devaux. Recent extensions of Gurson's model for porous ductile metals. In P. Suquet (Ed.), *Continuum Micromechanics*, CISM Lectures Series, pp. 61–130. Springer, New York, 1997.
7. A. A. Benzerga, J. Besson, and A. Pineau. *J. Eng. Mat. Tech.*, 121:221–229, 1999.
8. A. A. Benzerga and J. Besson. *Eur. J. Mech.*, 20(3):397–434, 2001.
9. J. Koplik and A. Needleman. *Int. J. Solids Structures*, 24(8):835–853, 1988.
10. T. Pardoen and J. W. Hutchinson. *J. Mech. Phys. Solids*, 48:2467–2512, 2000.
11. M. Gologanu. *Etude de quelques problèmes de rupture ductile des métaux*. PhD thesis, Université Paris 6, 1997.
12. A. A. Benzerga. *Rupture ductile des tôles anisotropes*. PhD thesis, Ecole Nationale Supérieure des Mines de Paris, 2000.
13. A. A. Benzerga. *J. Mech. Phys. Solids*, 50:1331–1362, 2002.
14. T. Pardoen and J. W. Hutchinson. *Acta Mater.*, 51:133–148, 2003.
15. A. A. Benzerga, J. Besson, R. Batisse, and A. Pineau. *Modelling Simul. Mater. Sci. Eng.*, 10:73–102, 2002.
16. A. A. Benzerga, J. Besson, and A. Pineau. *Acta Mater.*, 52:4639–4650, 2004.
17. S. M. Keralavarma and A. A. Benzerga. An approximate yield criterion for anisotropic porous media, *Comptes Rendus Mécanique*, 2008 (to appear).
18. A. A. Benzerga, J. Besson, and A. Pineau. *Acta Mater.*, 52:4623–4638, 2004.
19. J. Besson and R. Foerch. *Comput. Methods Appl. Mech. Engrg*, 142:165–187, 1997.
20. E. Riks. *Int. J. Solids Structures*, 15:529–551, 1979.
21. J.-B. Leblond. *Mécanique de la rupture fragile et ductile*. Hermes Science Publications, Lavoisier, 2003.
22. R. Hill. *J. Mech. Phys. Solids*, 15:79–95, 1967.
23. J. Mandel. Contribution théorique à l'étude de l'écrouissage et des lois d'écoulement plastique. In *Proceedings 11th International Congress on Applied Mechanics*, pp. 502–509. Springer, Berlin, 1964.
24. P. Suquet. *Plasticité et homogénéisation*. Thèse d'Etat, Université Pierre et Marie Curie – Paris VI, 1982.
25. K. Decamp, L. Bauvineau, J. Besson, and A. Pineau. *Int. J. Frac.*, 88:1–18, 1997.

Multiscale Methods for Fracturing Solids

Stefan Loehnert and Dana S. Mueller-Hoeppe

*Institute of Continuum Mechanics, Leibniz Universität Hannover,
Appelstr. 11, 30167 Hannover, Germany
E-mail: loehnert@ikm.uni-hannover.de*

Abstract. In this contribution a multiscale projection method to simulate localizing phenomena with influences on multiple scales such as macrocracks and their interaction with microcracks or heterogeneities in 2D and 3D is presented. The effect of crack shielding and crack amplification can be captured. The multiscale technique allows for the accurate simulation of fine scale fracture processes in large scale structures which is important for the prediction of the failure of the structure. The method uses the extended finite element method to simulate discontinuities on the macroscale as well as on the microscale.

Key words: multiscale, fracture, cracks, extended finite element method.

1 Introduction

Crack propagation in brittle materials in general strongly depends on the microstructure of the material. In some materials e.g. microcracks develop in the vicinity of the front of a propagating macrocrack leading to crack shielding and crack amplification. The macrocrack may coalesce with microcracks or it may propagate into a different direction due to the microcracks. Hence, microcracks in the vicinity of the crack front of a macrocrack have a significant influence on the macrocrack and have to be considered in the simulation of crack propagation in some materials.

Since microcracks usually are orders of magnitude smaller than a macrocrack, the interaction between micro- and macrocracks clearly is a multiscale phenomenon that usually cannot be captured in a single scale numerical analysis. Thus, for the calculation of a macroscopic fracturing structure an adaptive multiscale strategy is necessary.

There exists a huge variety of different multiscale strategies like the Variational Multiscale Method [3, 4], the Homogenized Dirichlet Projection Method [15] and FE^2 type methods [2]. However, multiscale methods based on homogenization processes usually are inappropriate for the calculation of localization phenomena such as cracks, shear bands and softening material behavior. For a homogenization based multiscale strategy a representative volume element (RVE) is required, and in case

B.D. Reddy (ed.), IUTAM Symposium on Theoretical, Modelling and Computational Aspects of Inelastic Media, 79–87.
© *Springer Science+Business Media B.V. 2008*

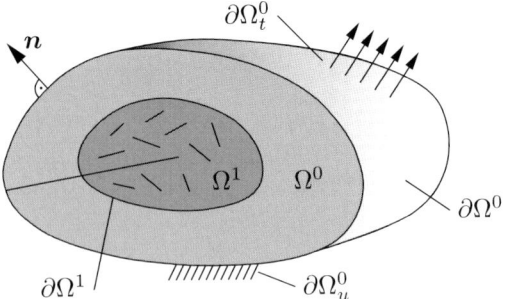

Fig. 1 Coarse and fine scale domain.

of localization phenomena the representativeness of the RVE may be lost. In general this coincides with the loss of material stability. In such cases it is necessary to apply a multiscale method that is capable of capturing the fine scale behavior accurately and map this microstructural response in an effective manner to the coarse scale computation.

In this contribution we present a three dimensional extension of a multiscale projection method originally developed for the two dimensional case by Loehnert and Belytschko [6]. The method is based on the projection of fine scale high resolution field quantities onto the coarse scale mesh. The fine scale solution is computed only in a small domain around the crack front of a macrocrack. The boundary conditions for the fine scale computations are provided by the coarse scale solution. In this paper we restrict ourselves to two scales. However, the presented multiscale method can easily be extended to more scales.

The simulation of cracks and fracture processes in two and three dimensions is a challenging task. There exist several approaches based on adaptive remeshing techniques, the strong discontinuity approach (SDA) [11], the partition of unity method (PUM) [7], the extended finite element method (XFEM) [8] or the generalized finite element method (GFEM) [13]. Since the XFEM is a rather flexible, robust and accurate method for the simulation of discontinuities, we employ the XFEM in our multiscale approach.

2 The Multiscale Projection Strategy

We consider a macroscopic body Ω^0 with an elastic macrocrack Γ_D^0. Around the crack front of the macrocrack we define a fine scale domain Ω^1 (see Figure 1). In the fine scale domain we resolve the microstructure consisting of microcracks accurately. The displacement field of the body is defined by

$$\boldsymbol{u} = \boldsymbol{u}^0 + \bar{\boldsymbol{u}}^1 \tag{1}$$

where u^0 is the coarse scale displacement field and $\bar{u}^1$ is the fluctuation of the displacements due to microstructural effects. On the boundary $\partial\Omega^1$ of the fine scale domain as well as in the domain $\Omega^0 \setminus \Omega^1$ the fluctuation of the fine scale displacement field $\bar{u}^1$ is assumed to vanish.

To obtain the weak form of the balance of momentum we introduce the test functions δu^0 on the coarse scale and δu^1 on the fine scale with $\delta u^0 = \mathbf{0}$ on the Dirichlet boundary $\partial\Omega_u^0$ of the coarse scale domain and $\delta u^1 = \mathbf{0}$ on the entire boundary $\partial\Omega^1$ of the fine scale domain.

In the present work, we use small deformation theory and isotropic linear elasticity for both, the coarse and fine scale computations. Then, the weak form of equilibrium on the coarse scale is

$$\int_{\Omega^0} \sigma(\varepsilon^1) : \delta\varepsilon^0 \, d\Omega - \int_{\Omega^0} f \cdot \delta u^0 \, d\Omega - \int_{\partial\Omega_t^0} t \cdot \delta u^0 \, d\Gamma = 0 \qquad (2)$$

where $\delta\varepsilon^0 = \mathrm{grad}^{\mathrm{sym}}\delta u^0$ and $\varepsilon^1 = \varepsilon(u^0 + \bar{u}^1)$ is the linearized strain tensor including the fluctuations due to microcracks.

In equation (2) the fine scale solution for strain field ε^1 is required. This fine scale strain field is calculated by solving the weak form

$$\int_{\Omega^1} \sigma(\varepsilon^1) : \delta\varepsilon^1 \, d\Omega - \int_{\Omega^1} f \cdot \delta u^1 \, d\Omega = 0 \qquad (3)$$

for the fine scale displacement field $u^1 = u^0 + \bar{u}^1$. Since it is assumed that the fine scale fluctuations $\bar{u}^1$ vanish on the boundary of the fine scale domain, the boundary conditions for the solution of equation (3) are given as pure Dirichlet boundary conditions by the coarse scale displacements on the entire boundary of the fine scale domain.

$$\bar{u}^1 = \mathbf{0} \quad \Leftrightarrow \quad u^1 = u^0 \quad \text{on} \quad \partial\Omega^1 \qquad (4)$$

In equation (3) we omit any contribution due to external tractions since we apply pure displacement boundary conditions only. However, this can only be done if the crack surfaces are traction free, i.e. there are no cohesive forces along the crack surfaces and cracks do not close. Cohesive tractions as well as crack closure can easily be incorporated in the multiscale projection method by changing equations (2) and (3), respectively. However, for simplicity we exclude these effects in the present work.

3 XFEM in Conjunction with Level Sets in 3D

An efficient way of describing the crack geometry implicitly in two and three dimensions is using level set functions [9, 12]. In three dimensions we use two level set functions, one describing the crack surface $\phi(X) = 0$ and one describing the crack front $\psi(X) = 0$. On the coarse as well as on the fine scale the two level set functions are given by nodal values and standard finite element interpolations

$$\phi^j = \sum_{I=1}^{n_n^j} N_I^j \phi_I^j \qquad \psi^j = \sum_{I=1}^{n_n^j} N_I^j \psi_I^j \qquad (5)$$

Here, j denotes the respective scale and N_I^j are the standard nodal shape functions. We choose the level set function to be a signed distance function such that $|\phi|$ is the smallest distance to the crack surface, sign(ϕ) corresponds to the respective side of the crack and $|\psi|$ is the smallest distance to the crack front. If $\psi > 0$ the point is located ahead of the crack front, if $\psi < 0$ it is located behind the crack front. Then, in a simple way it is possible to compute the distance r of a point to the crack front and its angle θ with the crack surface.

$$r = \sqrt{\phi^2 + \psi^2} \qquad \theta = \arctan\left(\frac{\phi}{\psi}\right) \qquad (6)$$

Following Moës et al. [8] when using the XFEM on all scales, the discretized displacement field can be written in the form

$$\boldsymbol{u}_h^j = \sum_{I=1}^{n_n^j} N_I^j \left(\boldsymbol{u}_I^j + \sum_{k=1}^{n_{\text{enr}}} f_k \boldsymbol{a}_{kI}^j\right) \qquad (7)$$

Nodes belonging to elements that are cut by the crack front are enriched by the four crack front enrichment functions

$$f_1(r,\theta) = \sqrt{r} \sin\left(\frac{\theta}{2}\right) \qquad (8)$$

$$f_2(r,\theta) = \sqrt{r} \cos\left(\frac{\theta}{2}\right) \qquad (9)$$

$$f_3(r,\theta) = \sqrt{r} \sin\left(\frac{\theta}{2}\right) \sin(\theta) \qquad (10)$$

$$f_4(r,\theta) = \sqrt{r} \cos\left(\frac{\theta}{2}\right) \sin(\theta) \qquad (11)$$

introduced for the 2D XFEM in [1] and for the 3D XFEM in [14]. These enrichment functions have the advantage that they form a basis for the asymptotic fields of mode 1, 2 and 3 cracks. They also allow for a crack front located inside an element. Note that only the enrichment function f_1 is discontinuous.

Nodes belonging to elements that are completely cut by the crack are enriched by the Heaviside step function

$$f_1(\phi) = H(\phi) = \begin{cases} +1, & \phi \geq 0 \\ -1, & \phi < 0 \end{cases} \qquad (12)$$

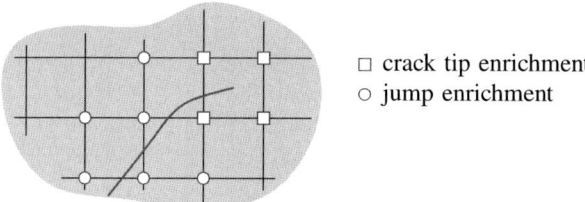

Fig. 2 Crack tip enrichments and jump enrichments

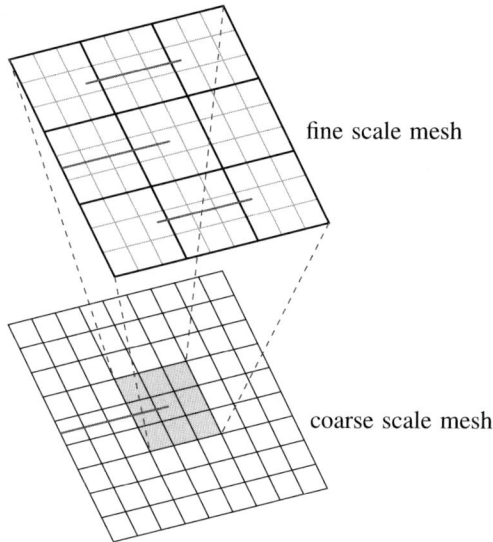

Fig. 3 Coarse and fine scale meshes

A 2D sketch of the type of enrichments used is shown in Figure 2.

The test functions are set up in a similar way as the displacements

$$\boldsymbol{\eta}_h^j = \sum_{I=1}^{n_n^j} N_I^j \left(\boldsymbol{\eta}_I^j + \sum_{k=1}^{n_{\text{enr}}} f_k \boldsymbol{b}_{kI}^j \right). \tag{13}$$

To simplify the notation, the displacements and test functions are rewritten

$$\boldsymbol{u}_h^j = \sum_{I=1}^{n_n^j} \hat{N}_I^j \hat{\boldsymbol{u}}_I^j \qquad \boldsymbol{\eta}_h^j = \sum_{I=1}^{n_n^j} \hat{N}_I^j \hat{\boldsymbol{\eta}}_I^j \tag{14}$$

where

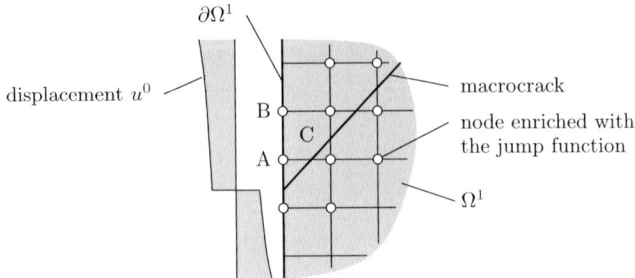

Fig. 4 Enrichments along the boundary of the fine scale domain.

$$\hat{\boldsymbol{N}}_I^j = \begin{pmatrix} N_I^j \\ N_I^j f_1 \\ \vdots \\ N_I^j f_{n_{enr}} \end{pmatrix}, \quad \hat{\boldsymbol{u}}_I^j = \begin{pmatrix} u_I^j \\ a_{1I}^j \\ \vdots \\ a_{n_{enr} I}^j \end{pmatrix}, \quad \hat{\boldsymbol{\eta}}_I^j = \begin{pmatrix} \eta_I^j \\ b_{1I}^j \\ \vdots \\ b_{n_{enr} I}^j \end{pmatrix} \quad (15)$$

Using the definition $\boldsymbol{B}_I^j = \text{grad}^{\text{sym}} \hat{\boldsymbol{N}}_I^j$, the weak form for the discretized system on the coarse scale becomes

$$\sum_{I=1}^{n_n^0} (\hat{\boldsymbol{\eta}}_I^0)^T \left(\int_{\Omega^0} \boldsymbol{B}_I^{0^T} : \sigma(\varepsilon^1) \, d\Omega - \int_{\Omega^0} \hat{\boldsymbol{N}}_I^0 \cdot \boldsymbol{f} \, d\Omega - \int_{\partial \Omega_t^0} \hat{\boldsymbol{N}}_I^0 \cdot \boldsymbol{t} \, d\Gamma \right) = 0 \quad (16)$$

and the discretized weak form for the fine scale problem is

$$\sum_{I=1}^{n_n^1} (\hat{\boldsymbol{\eta}}_I^1)^T \left(\int_{\Omega^1} \boldsymbol{B}_I^{1^T} : \sigma(\varepsilon^1) \, d\Omega - \int_{\Omega^1} \hat{\boldsymbol{N}}_I^1 \cdot \boldsymbol{f} \, d\Omega \right) = 0. \quad (17)$$

The fine scale domain mesh is chosen to be congruent with the coarse scale mesh in the domain where it is defined (Figure 3). On the entire boundary of the fine scale domain Dirichlet boundary conditions are prescribed according to equation (4). Even though the meshes on both scales are congruent, the nodal enrichments in general are different. Thus, it is not possible to simply interpolate the coarse scale displacements to determine the nodal degrees of freedom on the boundary of the fine scale mesh. It is necessary to project the coarse scale displacement field onto the fine scale mesh. We choose a least squares projection method

$$\int_{\Omega^1} \left(\sum_{I=1}^{n_n^1} \hat{\boldsymbol{N}}_I^1 \cdot \hat{\boldsymbol{u}}_I^1 - \sum_{J=1}^{n_n^0} \hat{\boldsymbol{N}}_J^0 \cdot \hat{\boldsymbol{u}}_J^0 \right) \cdot \left(\sum_{K=1}^{n_n^1} \hat{\boldsymbol{N}}_K^1 \cdot \hat{\boldsymbol{\eta}}_K^1 \right) d\Omega = 0. \quad (18)$$

Even though only the degrees of freedom of the nodes along the fine scale boundary are required, the integral in (18) has to be carried out over the entire fine scale domain to avoid non-uniquenesses of the solution. This is illustrated in Figure 4.

If a crack cuts the element C but not the boundary of the fine scale domain between the nodes A and B, then the jump enrichments and the standard degrees of freedom of the nodes A and B would be linear dependent if one carried out the integral in (18) only over the surface of the fine scale domain. Those degrees of freedom only lose their linear dependence if the integral is carried out over the entire element C.

4 Numerical Examples

The principal properties of the presented multiscale projection method in 2D were studied by Loehnert and Belytschko [6]. Herein, the influence of the coarse and fine scale mesh resolution as well as the chosen size of the fine scale domain on the accuracy of the numerical solution was investigated. A mode 1 crack problem of one macrocrack with two microcracks symmetrically placed around the tip of the macrocrack was computed using the multiscale strategy and compared to analytical solutions presented in [10].

It could be shown that there is hardly any influence of the coarse scale mesh resolution on the stress intensity factor of the macrocrack if the fine scale mesh resolution as well as the fine scale domain size is kept constant.

The main influence on the accuracy of the method comes from the fine scale mesh resolution. Since microcracks are modeled explicitly on the fine scale only, local effects dominate the fine scale solution. Thus, the fine scale mesh resolution has to be chosen such that the local effects can be resolved accurately.

The size of the fine scale domain has to be chosen such that the fine scale fluctuations have no significant effect on the field variables on the boundary of the fine scale domain. Additionally, since the coarse scale finite element mesh naturally is much stiffer than the fine scale mesh, it is necessary to have a minimum number of coarse scale elements covering the fine scale domain. Otherwise the coarse scale solution will impose artificially stiff boundary conditions on the fine scale problem.

The presented multiscale projection method can be used to compute arbitrary microcrack/macrocrack configurations in two and three dimensions. An elliptical macrocrack and two symmetrically placed microcracks in the vicinity of the crack front of the main crack are shown in Figure 5. A similar example has been treated analytically in [5]. In Figure 5 one can see the part of the cracks that are defined in the fine scale domain and the part that is defined only in the coarse scale domain. One can see the strong crack shielding effect of the two microcracks. In this case, the microcracks would propagate rather than the macrocrack.

In Figure 6 an elliptical microcrack is placed right in front of the crack front of a straight macrocrack. Also this example has been treated analytically in [5]. Here, a strong crack amplification effect occurs.

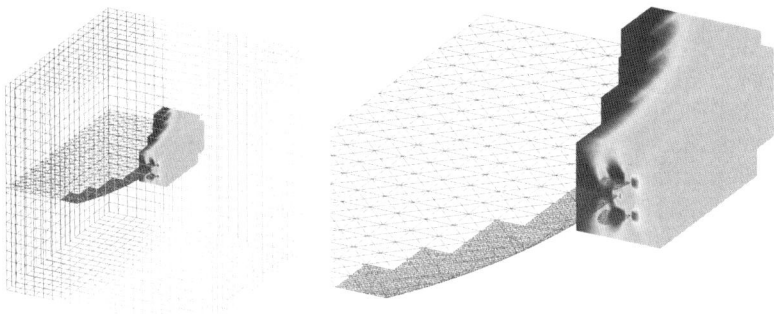

Fig. 5 Stress distribution in the coarse and fine scale domain of an elliptical main crack with two microcracks.

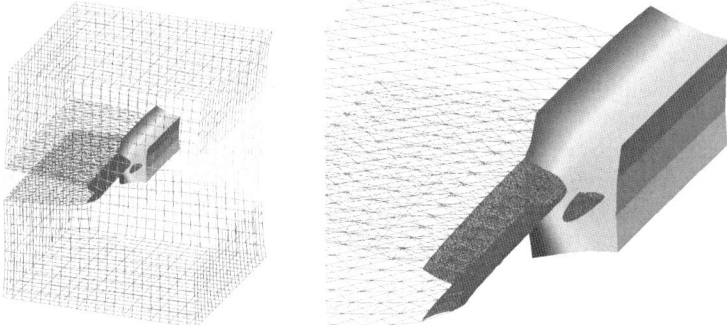

Fig. 6 Deformation of the coarse and fine scale domain of a straight main crack with one elliptical microcrack.

5 Conclusions

The presented multiscale projection method enables the accurate computation of the interaction of macrocracks with microcracks in two and three dimensions. This microcrack/macrocrack interaction can be of significant importance for the crack propagation behavior in some brittle materials. The method reduces the computational cost significantly, since the fine scale computations can be done separately from the coarse scale computations. Especially in three dimensions the numerical simulation of macroscopic fracture problems influenced by microstructural effects would not be efficiently possible in a single scale analysis.

We use the extended finite element method which facilitates the calculation of complex crack patterns in three dimensions. However, the presented multiscale method is also extendible to other applications such as shear bands, dislocations or heterogeneities. Hence, the presented multiscale approach within the context of the XFEM is an accurate and highly efficient possibility to compute the interaction of an arbitrary number of cracks on several scales.

References

1. T. Belytschko and T. Black. Elastic crack growth in finite elements with minimal remeshing. *International Journal for Numerical Methods in Engineering*, 45(5):601–620, 1999.
2. F. Feyel and J.L. Chaboche. FE^2 multiscale approach for modelling the elastoviscoplastic behaviour of long fibre sic/ti composite materials. *Computer Methods in Applied Mechanics and Engineering*, 183:309–330, 2000.
3. T.J.R. Hughes. Multiscale phenomena: Green's functions, the Dirichlet-to-Neumann formulation, subgrid scale models, bubbles and the origins of stabilized methods. *Computer Methods in Applied Mechanics and Engineering*, 127:387–401, 1995.
4. T.J.R. Hughes, G.R. Feijo, L. Mazzei, and J.B. Quincy. The variational multiscale method – A paradigm for computational mechanics. *Computer Methods in Applied Mechanics and Engineering*, 166:3–24, 1998.
5. J.P. Laures and M. Kachanov. Three-dimensional interactions of a crack front with arrays of penny-shaped microcracks. *International Journal of Fracture*, 48:255–279, 1991.
6. S. Loehnert and T. Belytschko. A multiscale projection method for macro/microcrack simulations. *International Journal for Numerical Methods in Engineering*, 71:1466–1482, 2007.
7. J.M. Melenk and I. Babuška. The partition of unity finite element method: Basic theory and applications. *Computer Methods in Applied Mechanics and Engineering*, 139:289–314, 1996.
8. N. Moës, J. Dolbow, and T. Belytschko. A finite element method for crack growth without remeshing. *International Journal for Numerical Methods in Engineering*, 46(1):131–150, 1999.
9. N. Moës, A. Gravouil, and T. Belytschko. Non-planar 3d crack growth by the extended finite element and level sets – Part I: Mechanical model. *International Journal for Numerical Methods in Engineering*, 53:2549–2568, 2002.
10. L.R.F. Rose. Microcrack interaction with a main crack. *International Journal of Fracture*, 31:233–242, 1986.
11. J. C. Simo, J. Oliver, and F. Armero. An analysis of strong discontinuities induced by strain-softening in rate-independent inelastic solids. *Computational Mechanics*, 12:277–296, 1993.
12. M. Stolarska, D.L. Chopp, N. Moës, and T. Belytschko. Modelling crack growth by level sets in the extended finite element method. *International Journal for Numerical Methods in Engineering*, 51:943–960, 2001.
13. T. Strouboulis, K. Copps, and I. Babuška. The generalized finite element method. *Comp. Meth. App. Mech. Eng.*, 190:4081–4193, 2001.
14. N. Sukumar, N. Moës, B. Moran, and T. Belytschko. Extended finite element method for three-dimensional crack modelling. *International Journal for Numerical Methods in Engineering*, 48:1549–1570, 2000.
15. T.I. Zohdi, J.T. Oden, and G.J. Rodin. Hierarchical modeling of heterogeneous bodies. *Computer Methods in Applied Mechanics and Engineering*, 138:273–298, 1996.

A Regularized Brittle Damage Model Solved by a Level Set Technique

N. Moës, N. Chevaugeon and F. Dufour

*GeM Institute, Ecole Centrale de Nantes/Université de Nantes/CNRS,
1 Rue de la Noë, 44321 Nantes, France
E-mail: {nicolas.moes, nicolas.chevaugeon, frederic.dufour}@ec-nantes.fr*

Abstract. We consider an elastic brittle damage model in which a degradation front is propagating based on a criteria depending both on the energy along the front and the curvature of the front. This model was introduced about 20 years ago but to our knowledge was not yet exploited numerically. The contribution of this paper is to solve this model using a level set technique coupled to the eXtended Finite Element Method (X-FEM).

Key words: brutal damage, front curvature, level set, eXtended Finite Element Method (X-FEM).

1 Introduction

About 20 years ago a paper was published describing a brittle damage model in which a degradation front propagates in a media [1]. The damage is called brittle because a virgin zone (in which the damage is $d = 0$) is seperated by a front from a totally damaged zone in which $d = 1$.

The model incorporates a Griffith type criteria. The front is growing provided the energy released is equal to a critical energy plus a term involving the front curvature. The presence of the curvature is essential for the model to avoid spurious localization.

To our knowledge this model has not been tackled numerically. We propose here an algorithm to solve it using a level set technique coupled to the eXtended Finite Element Method.

The fact that some areas in the domain are completely damaged and no longer provide any stiffnesses is taken care of by the eXtended Finite Element Method. Basically a zone with no stiffness behaves as a hole and is treated using the work presented in [2].

Finally, note that in the case of fatigue loading, a level set algorithm was already proposed in [3] to propagate a brittle damage front. In this paper the front speed is only curvature driven.

*B.D. Reddy (ed.), IUTAM Symposium on Theoretical, Modelling and Computational Aspects
of Inelastic Media, 89–96.
© Springer Science+Business Media B.V. 2008*

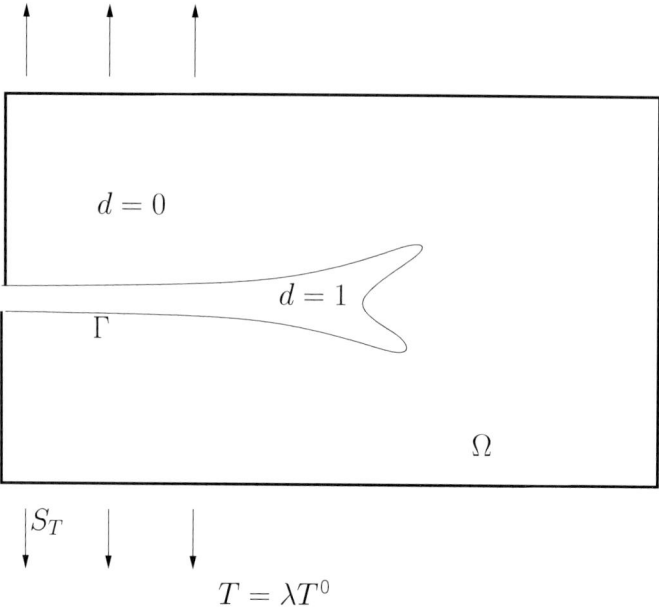

Fig. 1 Model problem showing a fully damaged zone progressing in an undamaged zone.

2 Description of the Mechanical Model

The model considered is the one introduced in [1]. We consider an elastic domain Ω submitted to imposed loads $T_d(x, \lambda)$ and displacements $u_d(x, \lambda)$ on the part of the boundary S_T and S_u, respectively. The parameter λ is the loading parameter and we assume that the imposed loads and displacements depend linearly on λ, i.e. $T_d(x, \lambda) = \lambda T^0(x)$ and $u_d(x, \lambda) = \lambda u^0(x)$. Note that a single loading parameter is considered. The extension of the level set update detailed below for multiple loading parameters needs more thinking.

The space of admissible displacement field is denoted as U and the space of admissible displacement field to zero is denoted as U_0.

The complementary part of the boundary delimiting the completely damaged and virgin material is denoted Γ. The potential energy in the system is given by

$$E(\Gamma, u, \lambda) = \int_\Omega e \, d\Omega - \int_{S_T} \lambda T^0 \cdot u \, dS$$

$$\text{with } e = \frac{1}{2} \epsilon(u) : E : \epsilon(u) \text{ and } u = \lambda u^0 \text{ on } S_u \quad (1)$$

where E is Hooke's tensor. The displacement field $u \in U$ is obtained through the stationarity of the functional

$$\int_\Omega \epsilon(u) : E : \epsilon(u^*) d\Omega = \int_{S_T} \lambda T^0 \cdot u^* dS \quad \forall u^* \in U_0 \tag{2}$$

The solution of which depends on the current degradation front location and load factor:

$$u = u(\Gamma, \lambda) \tag{3}$$

The free energy of the system at equilibrium is denoted W

$$W(\Gamma, \lambda) = E(\Gamma, u(\Gamma, \lambda), \lambda) \tag{4}$$

Assuming a normal velocity q_n modifying the location of the front, the free energy will be altered. Assuming the front is regular, the directional derivative of the free energy with respect to the velocity q_n is

$$DW[q_n] = -\int_\Gamma e q_n \, dS \tag{5}$$

The dual quantity to q_n on the front is the energy release rate e. The brittle propagation law is given by

- If $e < Yc + \gamma_c/\rho$, the propagation is impossible.
- If $e = Yc + \gamma_c/\rho$, the propagation is possible.

The above may be rewritten as

$$q_n \geq 0 \quad f = e - \left(Y_c + \frac{\gamma_c}{\rho}\right) \leq 0 \quad f q_n = 0 \tag{6}$$

These relations are very similar to the ones appearing in plasticity (taking q_n as the plastic strain rate and e as for instance the von Mises stress). The degradation front will move forward if the elastic energy on the front is superior to a critical energy Y_c plus a critical surface energy times the front curvature. The second term is essential to avoid spurious localization. The motivation to use the model is that it does not require a specific transition from damage to fracture.

Note that the presentation above is based on the paper by Nguyen et al. [1]. The only difference is that we have considered the curvature term to be part of the propagation law whereas in [1] it is part of the free energy through a surface energy term.

In order to give a geometrical interpretation to the dissipation process, consider a recently created damaged zone in grey in Figure 2. The free energy lost is given by

$$D_{12} = Y_c A + \gamma_c (L_2 - L_1) \tag{7}$$

To give more insight in the dissipation process, we consider in Figure 3 a velocity q_n on the front. This velocity will yield a change in length of the front, $\dot{L}$, a change in area of the damaged zone, $\dot{A}$, and a dissipation D:

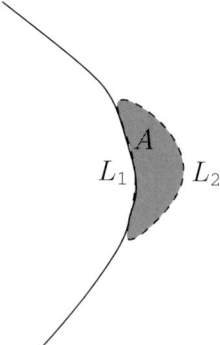

Fig. 2 The grey zone is the new damaged zone. The energy lost in the domain is given by $Y_c A + \gamma_c(L_2 - L_1)$ where A is the newly created area and $L_2 - L_1$ is the change in front length.

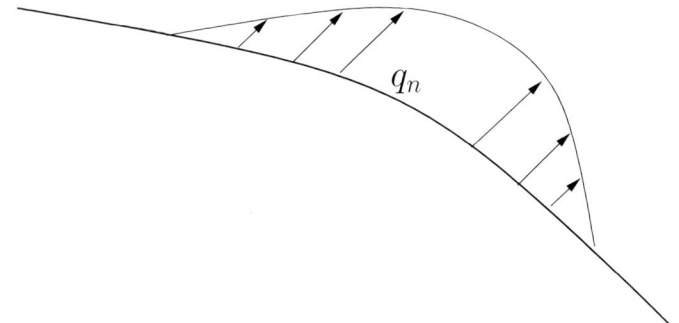

Fig. 3 A virtual velocity q_n on the front.

$$\dot{A} = \int_\Gamma q_n \, dS \qquad (8)$$

$$\dot{L} = \int_\Gamma \frac{q_n}{\rho} \, dS \qquad (9)$$

$$D = \int_\Gamma Y_c q_n + \gamma_c \frac{q_n}{\rho} \, dS \qquad (10)$$

3 Description of the Algorithm

In order to solve the mechanical model, we store the current location of the damaged front by the iso-zero of a level denoted ls. We solve only the elasticity equations in the positive part of the level set. The mesh used does not necessarily have to conform

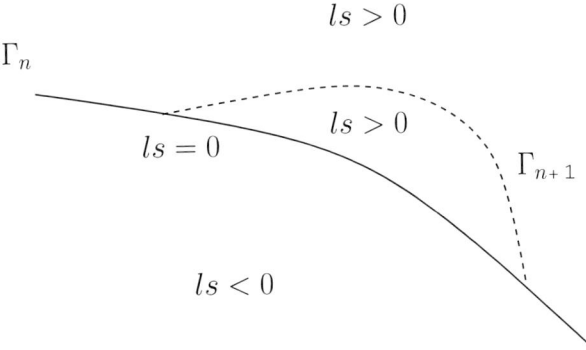

Fig. 4 The converged front (solid line) at time t_n. The damaged zone is located by the negative part of the level set. The front advance at time t_{n+1} is given by the dashed line. The dashed line must lay in the positive zone of the level set because the degradation front may not go backward in between one time step and the next one.

to the location of the damaged front because we use the extended finite element method to solve for the elasticity problem (see [2]). We consider in Figure 4 the location of the front Γ_n at time t_n, we look for the new location of the front Γ_{n+1} at time t_{n+1} and the corresponding loading factor λ_{n+1}. The new location of the front must fulfill the following relations

$$ls(\Gamma_{n+1}) \geq 0 \quad f_{n+1} = e_{n+1} - (Y_c + \frac{\gamma_c}{\rho_{n+1}}) \leq 0 \quad ls(\Gamma_{n+1})f_{n+1} = 0 \quad (11)$$

$$e_{n+1} = \lambda_{n+1}^2 \bar{e}_{n+1} \quad (12)$$

in which $\bar{e}_{n+1}$ is the front energy for a unit loading parameter.

Note that we do not consider the loading factor as given since the dissipation evolves in the domain, this factor may not evolve monotonically to maintain a stable solution. We monitor the problem using dissipation. We use a dissipation control algorithm generalizing the so-called crack-length control introduced in [4]. The amount of dissipated energy in the system is indeed a monotonically increasing function.

In order to solve the above equations (11–12), we iterate from an initial guess Γ_0, illustrated in Figure 5. The initial guess is obtained through a predictor step which will be described in a forthcoming paper. The correction step iterates on the front location until equations (11–12) are met. During the iterations, we make sure that the energy in the system is held fixed. In other words, it is the initial guess by its area and length changes which defines the energy lost over the time step $[t_n, t_{n+1}]$. At iteration i, the front velocity is computed on each point of the front using the following equations

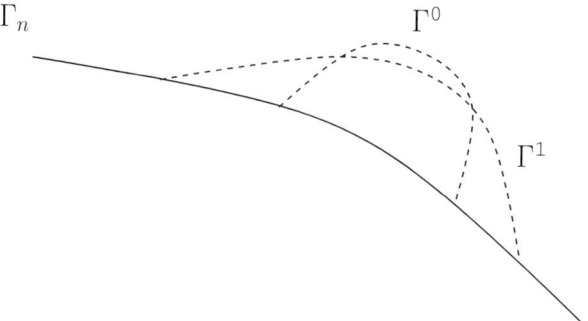

Fig. 5 The converged front (solid line) at time t_n. The initial guess Γ^0 to find the front at t_{n+1} as well as an illustration of the first iterate Γ^1.

$$\text{If } ls(\Gamma^i) > 0 \quad q^i = e^i - (Y_c + \frac{\gamma_c}{\rho^i}) \tag{13}$$

$$\text{If } ls(\Gamma^i) = 0 \quad q^i = <e^i - (Y_c + \frac{\gamma_c}{\rho^i})>_+ \tag{14}$$

$$e^i = (\lambda^i)^2 \bar{e}^i \tag{15}$$

$$\int_{\Gamma^i} q^i (Y_c + \frac{\gamma_c}{\rho^i}) dS = 0 \tag{16}$$

The iterations stop when the velocity is zero everywehere on the front. In practice, we compute the L2 norm and prescribe some tolerance. Equation (14) ensures that the current front will never be inside the previous converged front at time t_n. Equation (16) ensures that the dissipation is zero through the iterations. Note that the system above provides also the corresponding loading factor.

The velocity on the front q^i is used to update the level set location using standard level set procedures.

4 Basic Numerical Experiments

In Figure 6 we consider a plate with a notch under tension (tractions are imposed). We consider only the growth over one time step. The initial guess is circular and we "optimize" its shape through iterations using the equations of the preceding section. The converged shape is shown in Figure 7. In this example, the ratio $l_c = \gamma_c/Y_c = 1$.

The second example, depicted in Figure 8 aims at modeling the branching of a transverse crack in a composite. The notch is located in the soft area. The initial guess is again circular. The final shape is given for $l_c = \gamma_c/Y_c = 1$ and for $l_c = \gamma_c/Y_c = 0.1$. In both cases, the damage zone tends to depart from a straight propagation. We also note that the smaller l_c yields a more abrupt departure.

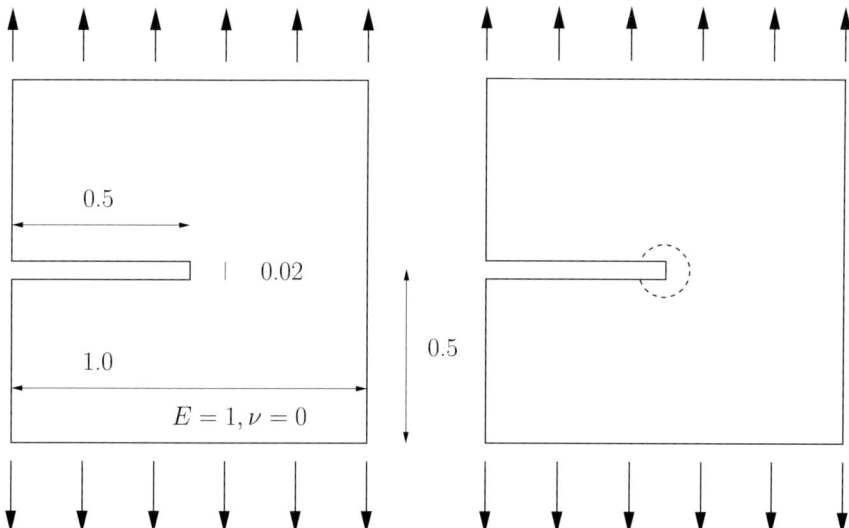

Fig. 6 A notched plate under tension (left) and the initial guess (dashed line on the right) for the degradation front.

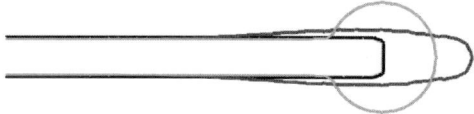

Fig. 7 The initial guess for the damage front location (circle) and the converged front location.

5 Conclusions

This paper develops some preliminary work on the resolution of a brittle damage model using a level set formulation. A Griffith type model is considered and solved using an iterative strategy. It is coupled to the extended finite element method allowing one to manage the mesh quite independently of the degradation front location. The interesting feature of the model for the future is to incorporate complex cracking patterns (since a single level set is used to locate the damaged zone). Further studies are also needed to analyze the influence of the parameter l_c ratio between surface and volumetric cracking energies.

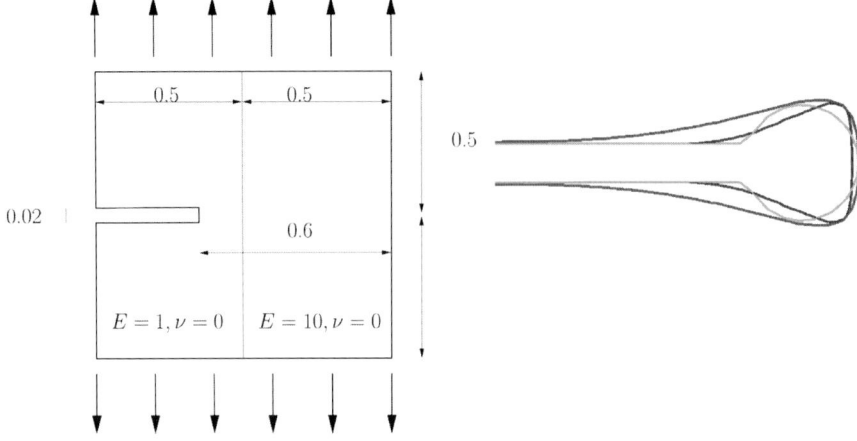

Fig. 8 A notched plate under tension composed of two materials (left). The initial guess is circular. The outer contour is the converged solution for $l_c = 1$ and the inner contour is the converged solution for $l_c = 0.1$ (right).

References

1. Q.S. Nguyen, R.M. Pradeilles, and C. Stolz. Sur une loi de propagation régularisante en rupture et endommagement fragile. *Compte-Rendus Acad. Sci. Paris, Série II*, 309:1515–1520, 1989.
2. C. Daux, N. Moës, J. Dolbow, N. Sukumar, and T. Belytschko. Arbitrary branched and intersecting cracks with the eXtended Finite Element Method. *International Journal for Numerical Methods in Engineering*, 48:1741–1760, 2000.
3. D. Salac and W. Lu. A level set approach to model directed nanocrack patterns. *Computational Materials Science*, 39(4):849–856, 2007.
4. P. Bocca, A. Carpinteri, and S. Valente. Mixed mode fracture of concrete. *Int. J. Solids Structures*, 27:1139–1153, 1991.

Gradient and Other
Non-Local Theories

A Counterpoint to Cermelli and Gurtin's Criteria for Choosing the 'Correct' Geometric Dislocation Tensor in Finite Plasticity

Amit Acharya

Civil and Environmental Engineering, Carnegie Mellon University, Pittsburgh, PA 15213-3890, USA
E-mail: acharyaamita@cmu.edu

Abstract. The criteria in [4] (Cermelli and Gurtin, 2001, *J. Mech. Phys. Solids*, **49**, 1539–1568) for choosing a geometric dislocation tensor in finite plasticity are reconsidered. It is shown that physically reasonable alternate criteria could just as well be put forward to select other measures; overall, the emphasis should be on the connections between various physically meaningful measures as is customary in continuum mechanics and geometry, rather than on criteria to select one or another specific measure. A more important question is how the geometric dislocation tensor should enter a continuum theory and it is shown that the inclusion of the dislocation density tensor in the specific free energy function in addition to the elastic distortion tensor is not consistent with the free energy content of a body as predicated by classical dislocation theory. Even in the case when the specific free energy function is meant to represent some spatial average of the actual microscopic free energy content of the body, a dependence on the average dislocation density tensor cannot be adequate.

Key words: dislocations, gradient plasticity, finite deformation.

1 Is a Unique Geometric Dislocation Tensor Necessary?

With regard to the question of characterizing measures of dislocation/Burgers vector density (**G**) appropriate for a continuum theory, Cermelli and Gurtin (C&G) [4] indicate that "the problem is not the absence of such a field, but rather the plethora of fields that have appeared in the literature." They then commit themselves to the task of showing that "there is but a single measure of geometrically necessary dislocations consistent with physically motivated requirements." Thus, the entire justification of this choice seems to rest on these "physically motivated requirements."

We proceed now to explore these requirements and provide alternative points of view to each of these requirements:

(C&G) (i) – **G** should measure the local Burgers vector in the microstructural configuration, per unit area in that configuration;

- While in itself this is a reasonable physical requirement for the use of some measure, it appears unreasonable to use this to rule out other measures of dislocation density. For instance, one could as well stipulate that the intended tensor should measure local, undeformed Burgers vector, per unit area of the current configuration (this also being the only physically available configuration for measurements). Under this definition, the C&G measure would not qualify but the two-point tensor field [2, 9]

$$-\text{curl}\, \mathbf{F}^{e-1}; \quad (\text{curl}\, \mathbf{F}^{e-1})_{ij} := \varepsilon_{jmn} F^{e-1}_{in,m} \qquad (1)$$

(with the 'curl' operation being with respect to the current configuration and a transpose of the C&G definition and the minus sign relates to the definition of the 'true' Burgers vector in the FS convention – see [9]) would, but such stipulation is, of course, vacuous in spirit for the student of continuum mechanics. To draw an analogy, this is not unlike the situation with the Cauchy and the First Piola–Kirchhoff stress tensors, the latter with respect to some (arbitrary) reference configuration. The former measures forces acting on surface elements of the current configuration, measured per unit area of the same area elements; the latter measures the same forces, measured per unit area of the image area element in the reference configuration. Both have clear physical meanings and preference of one or the other based solely on these definitions is arbitrary.

(C&G) (ii) – G should, at any point, be expressible in terms of the field $\mathbf{F}^p$ in a neighborhood of the point, since, by fiat, $\mathbf{F}^p$ characterizes the defect structure near the point in question;

- It may be argued that there is no physically distinguished reference configuration in the context of plasticity; in fact, as is well known, classical, finite deformation plasticity theory, whether of the single crystal variety or the 'microstructure-less' type (e.g. J-2 plasticity theory) can all be formulated without reference to the tensor $\mathbf{F}^p$ under the additive strain rate decomposition (it can be shown that elementary averaging procedures applied to a finite deformation field theory of dislocation mechanics consistent with classical dislocation theory [2] naturally lead to the additive decomposition of the velocity gradient). Thus, this second requirement of C&G may be argued to be of dubious physical origin.

(C&G) (iii) – G should be invariant under superposed compatible elastic deformations and also under compatible local changes in reference configuration, since these – being compatible – should not result in an intrinsic change in the distribution of GNDs near any point.

- Like (i), the invariance under superposed compatible elastic deformations is a physically reasonable requirement, but not one that can be used to rule out other measures of dislocation *density*. Being a measure of physical quantities per unit area, it stands to reason that if this area changes under compatible, elastic deformation, the per unit area, density measure should change. Consider, for example, three dislocation lines threading an area patch (at the minimum scale of resolution) in the current configuration. Roughly speaking, if under further deformation this area patch is stretched but it still contains only those three dislocation lines that do not move with respect to the material (i.e. elastic deformation), then the Burgers vector content *per unit area* of the patch should change; what should remain invariant is the local (unstretched) Burgers vector content of the patch, obtained by integrating the density measure over the area patch, as the patch deforms. The measure $\boldsymbol{\alpha} := -\mathrm{curl}\, \mathbf{F}^{e-1}$ satisfies this requirement and thus would be a good enough measure of dislocation density under the physical definition that it be the Burgers vector density, measured per unit area of the current configuration, just as the tensor $\mathbf{G} := \det(\mathbf{F}^e)(\mathrm{curl}\, \mathbf{F}^{e-1})\mathbf{F}^{e-T}$ is, under the definition of C&G.

In fact, the change in the $\boldsymbol{\alpha}$ field in arbitrary material area patches under superposed elastic motions of the current configuration for which no dislocations are nucleated in, or transported into, the patches can be made precise. One only need consider

$$\frac{d}{dt} \int_{A(t)} \boldsymbol{\alpha}\, \mathbf{n}\, da = \mathbf{0} \qquad (2)$$

for any deforming, but material, surface patch $A(t)$ and the answer is obtained by setting the appropriate convected rate [2, 5] to zero;

$$(\mathrm{div}\, \mathbf{v})\boldsymbol{\alpha} + \dot{\boldsymbol{\alpha}} - \boldsymbol{\alpha}\mathbf{L}^T = \mathbf{0} \quad \Rightarrow \quad \dot{\boldsymbol{\alpha}} = \boldsymbol{\alpha}\mathbf{L}^T - (\mathrm{div}\, \mathbf{v})\boldsymbol{\alpha}. \qquad (3)$$

Here, a superposed dot represents a material time derivative, $\mathbf{v}$ represents the material velocity, $\mathbf{L}$ the velocity gradient, and the result may be referred to as a transport theorem for areas. Clearly, if one wished to elevate the measure $\boldsymbol{\alpha}$ to be a unique measure, one could easily adjust the definition of the required measure to include the result (3) stipulating changes in $\boldsymbol{\alpha}$ in superposed elastic motions (it should be clear that a static condition corresponding to (3) is also easily derived involving the deformation between two elastically related configurations).

As for frame-indifference, the tensor $\boldsymbol{\alpha}$ being a kinematical quantity (as opposed to a constitutive response function), its definition itself suggests how it should transform under a superposed rigid body motion and thus it trivially satisfies the transformations expected of it under superposed rigid motions.

In summary, it is our opinion that while physically reasonable criteria can always be specified to single out one measure of dislocation density or another, in essence, such distinctions are superficial. *In particular, the measure* $\boldsymbol{\alpha} := -\mathrm{curl}\, \mathbf{F}^{e-1}$ *is an equally physically-valid measure of Burgers vector content as the* $\mathbf{G}$ *tensor of C&G*, as is the measure $\boldsymbol{\alpha}^* := \mathbf{F}^e \boldsymbol{\alpha}$ – a local, elastically stretched Burgers vector per unit area measure in the current configuration – that arises naturally in thermodynamic considerations leading to the definition of the finite deformation Peach–Koehler force[1] [2]. Moreover, what is of fundamental importance is the non-vanishing 'curl'

---

[1] This energetic force (per unit volume) depends explicitly on the Cauchy stress in contrast to the object called by the same name in [6] that depends on the defect stress in that theory.

of the inverse elastic tensor (or $\mathbf{F}^p$ on the reference configuration, if one chooses to work with this measure) characterizing the presence of defects in the neighborhood of a point [3]. Knowing the elastic and plastic distortions, the relationships between all the measures discussed by C&G are clearly established, and this is all that is important at a kinematical level.

To draw an historical analogy, denying this fact would be, in essence, similar to attributing fundamental importance to the discussion and controversy that plasticity theory has seen in the past regarding the question of the choice of the 'physically correct' objective stress rate for posing rate-type constitutive equations for the stress.[2]

2 Dependence of Stored Energy on the Dislocation Tensor; Is this Necessary?

Field equations and a class of constitutive equations that allow the prediction of stress fields of a prescribed dislocation density field (including individual dislocations) on a given configuration are given by [1, 2, 9]:

$$\operatorname{curl} \mathbf{F}^{e-1} = -\alpha$$

$$\operatorname{div} \mathbf{T} = \mathbf{0}; \quad \mathbf{T} = \hat{\mathbf{T}}(\mathbf{F}^e). \tag{4}$$

In this theory, *that is consistent with classical dislocation theory*, the tensor α plays a fundamental role whereas under the C&G physical requirements it would be an invalid measure.

It is also true that this theory (4), along with its counterpart for small deformations introduces a stored energy function that depends only on the elastic distortion and no additional dependence on the dislocation density tensor is introduced. In contrast, Gurtin's theory [6] (and many other following works in the literature) depends critically on the dependence of the stored energy function on the dislocation tensor to produce non-classical effects.

Restricting the discussion to small deformations to make the essential point, it is a well-known standard result (see, e.g., [7, sec. 4]) that the strain energy density of a dislocation distribution (including individual discrete dislocations) depends only on the elastic strain that results from the dislocation density distribution and not on *both* the elastic strain and the dislocation density. Furthermore, this strain energy density function corresponds to the linear elastic stress and strain fields of the dislocation distribution involved, as the latter fields are understood in classical dislocation theory (e.g., [8]). In what follows in this section, this result is illustrated

---

[2] The important matter in this case is the determination of the physically correct rate response from experiment or micromechanical theory which would naturally satisfy frame-indifference; once accomplished, it is a trivial matter to adjust this statement appropriately to pose it in terms of one objective rate or another so that the rate response implied by the adjusted statement is identical to the micromechanical/experimentally determined physically correct rate response.

in the context of standard procedures for solving boundary value problems in continuum inelasticity theory.

Consider the following question: we are interested in determining the state of internal stress and the strain energy in a linear elastic body of given geometry, for a prescribed dislocation density field $\boldsymbol{\alpha}$. For definiteness, the prescribed dislocation density field can be thought of as representing a screw dislocation along a straight cylinder representing its core. While the main conclusion of this section applies at all instants of time in the deformation of a body, it suffices to demonstrate the idea at any one instant and we choose the initial instant for definiteness. Thus, the problem may be thought of as determining the initial condition on the plastic distortion field in a conventional elastoplasticity calculation, where the plastic distortion field, $\mathbf{U}^p$, has to satisfy

$$\operatorname{curl} \mathbf{U}^p = -\boldsymbol{\alpha}. \tag{5}$$

Thus, we need to solve the equations

$$\operatorname{div}\left[\mathbf{C}\mathbf{U}^e\right] = \mathbf{0}$$
$$\mathbf{U}^e + \mathbf{U}^p = \operatorname{grad} \mathbf{u}$$
$$\operatorname{curl} \mathbf{U}^e = -\operatorname{curl} \mathbf{U}^p = \boldsymbol{\alpha}, \tag{6}$$

and since we are talking about initial conditions (for an elastoplasticity calculation), the displacement $\mathbf{u} \equiv \mathbf{0}$ at the initial time so that $\mathbf{U}^e = -\mathbf{U}^p$ at this time. Here, $\mathbf{C}$ is the possibly anisotropic linear elastic moduli with major and minor symmetries and $\mathbf{U}^e$ is the elastic distortion. In classical elastoplasticity, (6) is appended with an evolution equation for $\mathbf{U}^p$ or its symmetric part. For this problem (6) at the initial time when the displacement is known, we consider statically consistent traction boundary conditions (possibly vanishing) to be specified. The paper of [9] shows that solving these equations amounts to solving the problem of internal stress in classical dislocation theory corresponding to the prescribed dislocation density field. The existence of a non-trivial initial dislocation density distribution in the body is an eminently physical statement; the associated possibility of a non-trivial, *initial* plastic distortion in any theory where the fundamental relation (5) is active reflects the physical fact that the instantaneous dislocation density distribution encodes information of some portion of the past history of dislocation motion/nucleation in the body.

To solve the problem in the format of continuum inelasticity, we first note that it can be shown that there is at most one solution to the problem of calculating the initial distribution of stress, $\mathbf{T} = \mathbf{C}\boldsymbol{\varepsilon}^e$, where $\boldsymbol{\varepsilon}^e$ is the symmetric part of $\mathbf{U}^e$. Thus, as long as we can solve (6) and the associated boundary conditions by any procedure, the resulting solution would be the correct one. To find this solution, represent the plastic distortion as a sum of a gradient of a vector field and a tensor field whose 'curl' does not vanish as

$$\mathbf{U}^p := -\boldsymbol{\chi} + \operatorname{grad} \mathbf{z} \tag{7}$$

so that

$$-\operatorname{curl} \mathbf{U}^p = \boldsymbol{\alpha} \Rightarrow \operatorname{curl} \boldsymbol{\chi} = \boldsymbol{\alpha}. \tag{8}$$

In order to solve for χ in a well-posed manner, we append the equations

$$\text{div } \chi = \mathbf{0}$$

$$\chi \mathbf{n} = \mathbf{0} \quad \text{on boundary} \tag{9}$$

to (8) to obtain a Poisson's equation

$$\text{div grad } \chi = -\text{curl } \alpha \tag{10}$$

for the components of the tensor χ with Dirichlet boundary conditions; this problem may be solved by standard methods of potential theory.

With a solution for χ in hand, one solves the equilibrium equation

$$\text{div}[-\mathbf{C}\text{ grad }\mathbf{z}] = \text{div}[\mathbf{C}\chi] \tag{11}$$

for the vector field $\mathbf{z}$ with Neumann boundary conditions inferred from the prescribed traction boundary condition and the boundary values of the field χ. This is a standard problem in linear elasticity theory. The solution for grad $\mathbf{z}$ is unique, and this is all that matters for the present purpose.

The fields grad $\mathbf{z}$ and χ in conjunction with (7) and $\mathbf{u} \equiv \mathbf{0}$ now deliver the solution to (6) and thus the unique elastic strain and stress fields corresponding to the prescribed dislocation density field α, including arbitrary discrete dislocations in finite, anisotropic, linear elastic bodies.

The corresponding elastic strain energy density distribution in the body, consistent with the classical elastic theory of dislocations, is given by

$$\bar{\psi}(\varepsilon^e) := \frac{1}{2}(\mathbf{C}\,\varepsilon^e) : \varepsilon^e; \quad \varepsilon^e := (-\text{grad }z + \chi)_{\text{sym}} = \mathbf{U}^e_{\text{sym}} \tag{12}$$

and, consequently, an assumption of the form

$$\psi = \bar{\psi}\left(\varepsilon^e\right) + \hat{\psi}\left(\alpha\right) \tag{13}$$

would be superfluous and physically inaccurate in this context where core effects are not taken into account.

Of course, it may be argued that a continuum theory of the type proposed in [6] is meant to be a model of dislocation plasticity at scales of resolution much coarser than where every dislocation is well-resolved, so the appropriate form of a stored energy function for such a situation becomes an issue. But even here, if the coarse free energy function is meant to represent the spatially averaged free energy content in a representative volume element for the coarse-scale model, then an added dependence on the average dislocation density tensor cannot be adequate. For it is well understood that for two different spatial distributions of microscopic dislocation density α_1 and α_2 within an averaging volume with identical averages $\bar{\alpha}_1 = \bar{\alpha}_2$, the averaged free-energy content is, in general, different, i.e. $\overline{\psi_1} \neq \overline{\psi_2}$, where ψ represents the microscopic (specific) free energy. But a model that proposes to account for the added strain energy of dislocations at the coarse-scale by a unique functional dependence on $\bar{\alpha}$, say $R(\bar{\alpha})$, alone is bound to fail since the evaluation of such a function for the two cases would have to be identical:

$$R(\bar{\alpha}_1) = R(\bar{\alpha}_2) \quad \text{but} \quad \overline{\psi_1} \neq \overline{\psi_2}. \tag{14}$$

Acknowledgment

I thank Amine Benzerga for his comments on the paper.

References

1. Acharya, A. (2001) A model of crystal plasticity based on the theory of continuously distributed dislocations. *Journal of the Mechanics and Physics of Solids*, **49**, 761–785.
2. Acharya, A. (2004) Constitutive analysis of finite deformation field dislocation mechanics. *Journal of the Mechanics and Physics of Solids*, **52**, 301–316.
3. Acharya, A. and Bassani, J.L. (2000) Lattice incompatibility and a gradient theory of crystal plasticity. *Journal of the Mechanics and Physics of Solids*, **48**, 1565–1595.
4. Cermelli, P. and Gurtin, M.E. (2001) On the characterization of geometrically necessary dislocations in finite plasticity, *Journal of Mechanics and Physics of Solids*, **49**, 1539–1568.
5. Fox, N. (1966) A continuum theory of dislocations for single crystals. *Journal of the Institute of Mathematics and Its Applications*, **2**, 285–298.
6. Gurtin, M.E. (2002) A gradient theory of single-crystal viscoplasticity that accounts for geometrically necessary dislocations. *Journal of the Mechanics and Physics of Solids*, **50**, 5–32.
7. Kröner, E. (1981) Continuum theory of defects, in *Physics of Defects*, R. Balian et al. (Eds.), North-Holland, Amsterdam, pp. 217–315.
8. Nabarro, F.R.N. (1987) *Theory of Crystal Dislocations*. Dover Publications Inc.
9. Willis, J.R. (1967) Second-order effects of dislocations in anisotropic crystals. *International Journal of Engineering Science*, **5**, 171–190.

On Stability for Elastoplasticity of Integral-Type

Francesco Marotti de Sciarra

*Department of Structural Engineering, University of Naples Federico II,
via Claudio 21, 80125 Napoli, Italy*
E-mail: marotti@unina.it

Abstract. The paper deals with a thermodynamically consistent formulation for nonlocal elastoplastic models. On the basis of recent observations, the proposed formulation introduces two internal variable fields related to plastic softening. The stability of the nonlocal model is proved. Then it is shown how the considered nonlocal formulation can be specialized to two recently proposed nonlocal models so that the stability analysis can be straightforwardly extended to the existing models.

Key words: nonlocal plasticity, thermodynamics, softening, internal variables.

1 Introduction

Many materials exhibit a softening behaviour which consists in a decrease of stress at increasing strain that is a loss of positive definiteness of the tangent stiffness operator. Such a behaviour is coupled with the strain localization which determines the growing of narrow regions where plastic strains tend to concentrate whereas the remainder part of the body unloads elastically. In classical plasticity theories, the deformation can localize in a zone which is infinitely small so that a displacement discontinuity can develop. In fact, classical plasticity does not contain information about the size of the localization zone which tends to become infinitely thin in the continuum approach or takes on the size of the smallest finite element in a finite element approach. One possibility to overcome these shortcomings with local plasticity consists in introducing an internal length scale parameter into the continuum model of plasticity. Nonlocal effects can then be modelled by defining suitably weighted averages of kinematic fields linked to inelastic processes. The purpose of the present paper consists in presenting a thermodynamically consistent formulation of a nonlocal plasticity model in the framework of the internal variable theories. The nonlinear stability analysis of the nonlocal problem is then carried out following the concept of nonexpansivity proposed in local plasticity.

2 Nonlocal Elastoplasticity

A nonlocal elastoplastic structural problem defined on a regular bounded domain Ω of an Euclidean space is considered. Strain ε and stresses σ belong to dual spaces $\mathcal{D}$ and $\mathcal{S}$ respectively. Hardening and softening phenomena are modelled by an internal variable approach [2]. The kinematic (strain-like) internal variables are denoted by $\kappa \in \mathcal{Y}$, $\alpha_1 \in \mathcal{Y}_1$, $\alpha_2 \in \mathcal{Y}_2$ and the dual (stress-like) static internal variables are $\chi \in \mathcal{Y}'$, $\chi_1 \in \mathcal{Y}'_1$, $\chi_2 \in \mathcal{Y}'_2$ respectively. The inner product $\langle \cdot, \cdot \rangle$ in the dual spaces has the mechanical meaning of the internal virtual work. The nonlocal field $\bar{\alpha}_2 \in \mathcal{Z}$ can be obtained as a spatial weighted average of the local variable $\alpha_2 \in \mathcal{Y}_2$ by the following parametric representation:

$$\bar{\alpha}_2(x) = (\mathbf{R}\alpha_2)(x) \tag{1}$$

where $\mathbf{R} : \mathcal{Y}_2 \to \mathcal{Z}$ denotes a linear regularization operator [6, 10, 3]. The kinematic internal variable $\bar{\alpha}_2$ turns out to be nonlocal since its value at the point x of the body Ω depends on the entire field α_2. Attention is focused on integral nonlocal average of the form:

$$\bar{\alpha}_2(x) = (\mathbf{R}\alpha_2)(x) = \int_{\Omega} \beta_x(y) \alpha_2(y) \, dy$$

where $\beta_x(y)$ is a spatial weighting function depending on a material parameter called the internal length scale. For the subsequent analysis no explicit expression of β has to be assumed.

The first principle of thermodynamics for isothermal processes and for a nonlocal behaviour is $\int_{\Omega} \dot{e} \, dx = \langle \sigma, \dot{\varepsilon} \rangle$ where $\mathbf{e}$ is the internal energy density depending on strain ε, entropy s and internal variables $(\alpha_1, \alpha_2, \kappa)$ related to inelastic phenomena. It can be written in the pointwise form $\dot{e} = \sigma \cdot \dot{\varepsilon} + P$ where the nonlocality residual function P takes into account the energy exchanges between neighbour particles and fulfils the insulation condition $\int_{\Omega} P \, dx = 0$. The explicit dependence on the point is dropped for simplicity if no confusion can arise. The Helmholtz free energy is defined by means of the Legendre transform $\phi = \mathbf{e} - sT$ where T is the absolute temperature. The free energy functional Φ of the body Ω is defined according to the relation:

$$\Phi(e, \alpha_1, \alpha_2, \kappa) = \int_{\Omega} \phi(e(x), \alpha_1(x), \alpha_2(x), \kappa(x)) \, dx.$$

The nonlocal free energy is a saddle (convex-concave) differentiable functional Φ which is additively decomposed in the form:

$$\Phi(e, \alpha_1, \alpha_2, \kappa) = \Phi_{el}(e) + \Phi_{in}(\alpha_1, \mathbf{R}\alpha_2, \kappa) \tag{2}$$

where the elastic energy $\Phi_{el}(e)$ is convex in the elastic strain e and the nonlocal part of the free energy $\Phi_{in}(\alpha_1, \mathbf{R}\alpha_2, \kappa)$ is convex in (α_1, α_2) and concave in κ.

The second principle of thermodynamics for isothermal processes, in the present nonlocal context is enforced in its classical pointwise form, i.e. $\dot{s}T \geq 0$ everywhere

On Stability for Elastoplasticity of Integral-Type

in Ω where $\dot{s}$ is the internal entropy production rate per unit volume. Accordingly, performing the time derivative of the Helmholtz free energy in connection with the second principle and assuming that the temperature is constant, the dissipation D at a given point of the body is:

$$D = \dot{s}T = \sigma \cdot \dot{\varepsilon} - \dot{\phi} + P \geq 0 \tag{3}$$

which represents the Clausius–Duhem inequality for isothermal processes differing from its classical format by the presence of the nonlocality residual function P to guarantee the non-negativeness of the dissipation and to account for material nonlocality. Since $\varepsilon = e + p$, where p is the plastic strain, expanding the inequality (3) and recalling the expression (2) of the free energy, it results:

$$D = \sigma \cdot (\dot{e} + \dot{p}) - d_e\phi \cdot \dot{e} - d_{\alpha_1}\phi \cdot \dot{\alpha}_1 - d_{\alpha_2}\phi \cdot \dot{\alpha}_2 - d_\kappa\phi \cdot \dot{\kappa} + P \geq 0. \tag{4}$$

Using standard arguments (see e.g. [4]) the following state laws hold in terms of local and nonlocal fields:

$$\sigma = d_e\Phi(e, \alpha_1, \alpha_2, \kappa), \qquad \chi_1 = d_{\alpha_1}\Phi(e, \alpha_1, \alpha_2, \kappa),$$
$$\bar{\chi}_2 = d_{\alpha_2}\Phi(e, \alpha_1, \alpha_2, \kappa), \qquad -X = d_\kappa\Phi(e, \alpha_1, \alpha_2, \kappa). \tag{5}$$

By substituting the relations (5) into the expression (4), the dissipation becomes:

$$D = \sigma \cdot \dot{p} - \chi_1 \cdot \dot{\alpha}_1 - \bar{\chi}_2 \cdot \dot{\alpha}_2 + X \cdot \dot{\kappa} + P \geq 0. \tag{6}$$

At every point where an irreversible mechanism develops, the dissipation can be assumed in the following bilinear form:

$$D = \sigma \cdot \dot{p} - \chi_1 \cdot \dot{\alpha}_1 - \chi_2 \cdot \dot{\eta}_2 + X \cdot \dot{\kappa} \geq 0 \tag{7}$$

where χ_2 is a local variable thermodynamically conjugated to the variable η_2 whose expression has to be identified. Note that the nonlocality residual function has disappeared from the pointwise expression (7) of the dissipation. By comparing (6) and (7), the nonlocality residual function is given by:

$$P = \bar{\chi}_2 \cdot \dot{\alpha}_2 - \chi_2 \cdot \dot{\eta}_2 \tag{8}$$

and, employing the insulation condition, it results:

$$\langle \bar{\chi}_2, \dot{\alpha}_2 \rangle - \langle \chi_2, \dot{\eta}_2 \rangle = 0. \tag{9}$$

Substituting the equality $\langle \bar{\chi}_2, \dot{\alpha}_2 \rangle = \langle \chi_2, \bar{\dot{\alpha}}_2 \rangle$ in (9), it results $\langle \chi_2, \bar{\dot{\alpha}}_2 \rangle - \langle \chi_2, \dot{\eta}_2 \rangle = 0$ for any possible static variable χ_2. Hence the identification $\dot{\eta}_2 = \bar{\dot{\alpha}}_2$ hold true. Accordingly the nonlocality residual function (8) and the dissipation (7) can be given the following explicit expressions at a given point of the body Ω:

$$P = \bar{\chi}_2 \cdot \dot{\alpha}_2 - \chi_2 \cdot \dot{\bar{\alpha}}_2, \qquad D = \sigma \cdot \dot{p} - \chi_1 \cdot \dot{\alpha}_1 - \chi_2 \cdot \dot{\bar{\alpha}}_2 + X \cdot \dot{\kappa} \geq 0. \qquad (10)$$

If the nonlocal variables are constant in space, the regularization operator becomes the identity operator and the nonlocal variables turn out to be coincident to their local counterparts, i.e. $\dot{\bar{\alpha}}_2 = \dot{\alpha}_2$ and $\bar{\chi}_2 = \chi_2$. As a consequence the relation $(10)_1$ shows that the nonlocality residual function P vanishes, the inequality $(10)_2$ provides the expression of the dissipation in terms of local variables and the relations (5) yield the constitutive laws in terms of local fields. Hence the above analysis shows that the constant field requirement on the operator **R** and the insulation condition guarantee that the nonlocal plastic model behaves as a local one under uniform fields.

3 Formulation of the Constitutive Model

The constitutive model is formulated in terms of the following generalized vectors collecting together local and nonlocal variables:

$$\boldsymbol{\varepsilon} = [\varepsilon \quad 0 \quad 0]^T \qquad \boldsymbol{e} = [\varepsilon \quad \alpha_1 \quad \alpha_2]^T$$

$$\boldsymbol{p} = [p \quad -\alpha_1 \quad -\alpha_2]^T \qquad \boldsymbol{\sigma} = [\sigma \quad \chi_1 \quad \bar{\chi}_2]^T$$

where $\boldsymbol{\varepsilon}, \boldsymbol{e}, \boldsymbol{p}$ denote the generalized vectors of total strain, elastic strain and plastic strain and $\boldsymbol{\sigma}$ is the generalized stress. The scalar product between generalized vectors is $\prec \boldsymbol{\sigma}, \boldsymbol{e} \succ = \langle \sigma, \varepsilon \rangle + \langle \chi_1, \alpha_1 \rangle + \langle \bar{\chi}_2, \alpha_2 \rangle$. The term generalized will be omitted in the sequel if no confusion can arise. The constitutive relations (5) can be rewritten in terms of generalized variables as follows:

$$(\boldsymbol{\sigma}, -X) = d\Phi(\boldsymbol{e}, \kappa) \Leftrightarrow \begin{cases} \boldsymbol{\sigma} = d_e \Phi(\boldsymbol{e}, \kappa) \\ -X = d_\kappa \Phi(\boldsymbol{e}, \kappa). \end{cases} \qquad (11)$$

The elastic domain C is defined in the space of stresses and of static internal variables $(\boldsymbol{\sigma}, X)$ as the level set of a convex yield mode $G : \widehat{\mathcal{S}} \to \Re \cup \{+\infty\}$, where $\widehat{\mathcal{S}} = \mathcal{S} \times \mathcal{Y}'_1 \times \mathcal{Y}'_2 \times \mathcal{Y}'$ in the form:

$$C = \{(\boldsymbol{\sigma}, X) \in \widehat{\mathcal{S}} \; : \; G(\boldsymbol{\sigma}, X) \leq 0\} \qquad (12)$$

provided that the minimum of G is negative. From a mechanical standpoint the nonlocal yield mode G is usually written in terms of the convex yield function g in the form:

$$G(\boldsymbol{\sigma}, X) = G(\sigma, \chi_1, \bar{\chi}_2, X) = g(\sigma, \chi_1) - \bar{\chi}_2 - X - \sigma_o \leq 0 \qquad (13)$$

where σ_o represents the initial yield limit. The static internal variable χ_1 governs the kinematic hardening and the local and nonlocal static internal variables X and $\bar{\chi}_2$ control the size of the elastic domain. The generalized flow rule can be formulated in terms of the normal cone N_C to the elastic domain as follows:

$$(\dot{p}, \dot{\kappa}) \in N_C(\sigma, X) = \partial \sqcup_C (\sigma, X) \Leftrightarrow$$
$$\Leftrightarrow (\dot{p}, -\dot{\alpha}_1, -\dot{\alpha}_2, \dot{\kappa}) \in N_C(\sigma, \chi_1, \bar{\chi}_2, X) \tag{14}$$

being $\sqcup_C(\sigma, X)$ the indicator of the elastic domain which turns out to be zero if $(\sigma, X) \in C$ and $+\infty$ otherwise. The flow rule (14) can be reformulated in the following three equivalent forms:

$$(\dot{p}, \dot{\kappa}) \in N_C(\sigma, X) \Leftrightarrow (\sigma, X) \in \partial D(\dot{p}, \dot{\kappa}) \Leftrightarrow$$
$$\Leftrightarrow \sqcup_C (\sigma, X) + D(\dot{p}, \dot{\kappa}) = \prec \sigma, e \succ + \langle X, \dot{\kappa} \rangle \tag{15}$$

where the functional D has the mechanical meaning of dissipation associated with a given plastic rate $(\dot{p}, \dot{\kappa})$ and turns out to be the support functional of the elastic domain C. Accordingly the dissipative force is given by:

$$q = \bar{\chi}_2 + X = d_{\alpha_2} \Phi_{in}(\alpha_1, \alpha_2, \kappa) - d_\kappa \Phi_{in}(\alpha_1, \alpha_2, \kappa).$$

In the case of a linear elastic behaviour with linear hardening and softening, the free energy can be expressed as:

$$\Phi(e, \alpha_1, \alpha_2, \kappa) = \tfrac{1}{2} \langle \mathbf{E}e, e \rangle + \tfrac{1}{2} \langle \mathbf{H}_1 \alpha_1, \alpha_1 \rangle + \tag{16}$$
$$+ \tfrac{1}{2} \langle h_2 \mathbf{R}\alpha_2, \mathbf{R}\alpha_2 \rangle + \tfrac{1}{2} \langle h\kappa, \kappa \rangle$$

where $\mathbf{E}$ is the elastic stiffness, the kinematic hardening operator $\mathbf{H}_1$ is positive definite, the material modulus h_2 is positive and the softening modulus h is negative. Then the nonlocal constitutive relations (11) turn out to be:

$$\begin{cases} \sigma = \mathbf{E}e \\ \chi_1 = \mathbf{H}_1 \alpha_1 \\ \bar{\chi}_2 = \mathbf{R}' h_2 \mathbf{R} \alpha_2 \\ X = -h\kappa \end{cases} \tag{17}$$

where $\mathbf{R}'$ is the dual operator of $\mathbf{R}$. The expression $(17)_1$ represents the elastic relation, the expression $(17)_2$ provides the static internal variable which governs the kinematic hardening. The expression $(17)_3$ provides the nonlocal static internal force $\bar{\chi}_2$, conjugate to α_2, which is the $\mathbf{R}'$-transformed of the dissipative force $\chi = h_2 \bar{\alpha}_2$. The expression $(17)_4$ yields the static internal force X conjugate to κ. Then the dissipative force is given by:

$$q = \bar{\chi}_2 + X = \mathbf{R}' h_2 \mathbf{R}\alpha_2 - h\kappa = \mathbf{R}' h_2 \mathbf{R}\alpha_2 + h\alpha_2$$

since the flow rule (14) in connection with the expression (13) provide the equality $\dot{\alpha}_2 = -\dot{\kappa}$ so that, being in the elastic range $\alpha_2 = \kappa = 0$, it results $\alpha_2 = -\kappa$. The size of the elastic domain is then controlled by the kinematic internal variables α_2 which can be assumed as the driving variable of the softening law. It is worth noting that different expressions can be assumed for the constitutive relation between $\bar{\chi}_2$ and α_2 by a suitable modification of the expression of the free energy. Existing models can be cast in the present theory as shown in the subsection 4.1.

4 Stability

The property of nonexpansivity as a suitable measure of nonlinear stability of the evolution equations was introduced in [5]. Subsequently it is shown in [7, 9] that the evolution equations in local hardening plasticity exhibits the property of nonexpansivity.

Given a function $f : t \to f(t) = f_t \in \mathcal{X}$ and denoting by $f(t)$ and $\underline{f}(t)$ the flows corresponding to distinct initial conditions $f(0)$ and $\underline{f}(0)$ respectively, the flow f_t is said to be nonexpansive with respect to the scalar product generated by a positive-definite symmetric operator M if the following inequality holds:

$$\| f(t) - \underline{f}(t) \|_M \leq \| f(0) - \underline{f}(0) \|_M \qquad \text{for all } t \geq 0. \qquad (18)$$

The nonexpansivity condition (18) ensures that two flows generated by two nearby sets of initial conditions will be, at any time t, at least as near to each other as they were at the initial time. It is worth noting that the condition (18) is a nonlinear stability condition and no linearization will be carried out in the sequel in order to assess its validity.

A sufficient condition for (18) to hold is given by the condition:

$$\frac{d}{dt} \| f(t) - \underline{f}(t) \|_M^2 \leq 0 \qquad \text{for all } t \geq 0$$

or, equivalently, by the inequality:

$$\langle f(t) - \underline{f}(t), \dot{f}(t) - \underline{\dot{f}}(t) \rangle_M \leq 0 \qquad \text{for all } t \geq 0. \qquad (19)$$

The evolutive analysis of the nonlocal constitutive problem is performed by a preliminary subdivision of the load history in finite increments of times. The relevant finite increment of the unknown variables corresponding to a given increment of the total strain ε can be evaluated when their values are assigned at the beginning of the step. In the sequel Δp and $\Delta \kappa$ denote the increments of the variables p and κ in the step. For a given strain history $\varepsilon(t)$, the nonexpansivity of the nonlocal finite-step elastoplastic model is analysed for linear elasticity, hardening and softening behaviour. The relations pertaining to the nonlocal finite-step elastoplastic constitutive model follow from (17) and (15)$_2$ in the form:

$$\begin{cases} \varepsilon = e + p \\ (\sigma, -X) = d\Phi(e, \kappa) \\ (\sigma, X) \in \partial D(\Delta p, \Delta \kappa) \end{cases} \Leftrightarrow \begin{cases} \sigma = E(\varepsilon - p) \\ \chi_1 = \mathbf{H}_1 \alpha_1 \\ \bar{\chi}_2 = \mathbf{R}' h_2 \mathbf{R} \alpha_2 \\ X = -h\kappa \\ (\sigma, \chi_1, \bar{\chi}_2, X) \in \\ \quad \in \partial D(\Delta p, -\Delta \alpha_1, -\Delta \alpha_2, \Delta \kappa) \end{cases} \quad (20)$$

in which all the state variables are functions of time.

By virtue of the relations between convex and saddle functional, the dissipation D can be expressed in terms of the saddle functional $D_X : \mathcal{D} \times \mathcal{Y}_1 \times \mathcal{Y}_2 \times \mathcal{Y}' \to \bar{\mathfrak{R}}$, convex with respect to Δp and concave with respect to X, in the form:

$$D(\Delta p, \Delta \kappa) = -\inf_X \{\langle X, -\Delta \kappa \rangle - D_X(\Delta p, X)\}$$

so that the finite-step flow rule $(20)_2$ can be equivalently rewritten as follows:

$$(\sigma, X) \in \partial D(\Delta p, \Delta \kappa) \iff (\sigma, -\Delta \kappa) \in \partial D_X(\Delta p, X). \quad (21)$$

The subdifferential relation $(21)_2$ is then equivalent to state:

$$\sigma \in \partial_1 D_X(\Delta p, X) \qquad -\Delta \kappa \in \partial_2 D_X(\Delta p, X) \quad (22)$$

where $\partial_1 D_X$ denotes the subdifferential of D_X with respect to the variable p and $\partial_2 D_X$ denotes the superdifferential of D_X with respect to the variable X. As a consequence, the multivalued maps $\partial_1 D_X$ and $-\partial_2 D_X$ are cyclically monotone and hence monotone (Rockafellar 1970).

The monotonicity of $\partial_1 D_X$ and the relation $(22)_1$ yield:

$$\prec \sigma - \underline{\sigma}, \Delta p - \Delta \underline{p} \succ \geq 0 \quad (23)$$

where the pairs (σ, p) and $(\underline{\sigma}, \underline{p})$ denote the state variables arising from two non-local finite-step elastoplastic problems characterized by two distinct initial conditions. Substitution of the stress and of the static internal variables $(20)_{1-3}$ in (23) gives, after some algebra, the inequality:

$$\langle \mathbf{E}(p - \underline{p}), \Delta p - \Delta \underline{p} \rangle + \langle \mathbf{H}_1(\alpha_1 - \underline{\alpha}_1), \Delta \alpha_1 - \Delta \underline{\alpha}_1 \rangle + \\ + \langle \mathbf{R}' h_2 \mathbf{R}(\alpha_2 - \underline{\alpha}_2), \Delta \alpha_2 - \Delta \underline{\alpha}_2 \rangle \leq 0. \quad (24)$$

If the operators $\mathbf{H}_1$ and $\mathbf{R}' h_2 \mathbf{R}$ are definite positive, the inequality (24) can be rewritten in the form:

$$\prec p - \underline{p}, \Delta p - \Delta \underline{p} \succ_{\mathbb{M}} \leq 0$$

where $\mathbb{M} = \text{diag} [\mathbf{E}, \mathbf{H}_1, \mathbf{R}' h_2 \mathbf{R}]$. Hence the evolution of the plastic strain p and of the kinematic internal variables (α_1, α_2) is nonexpansive with respect to the scalar product induced by $\mathbb{M}$.

The nonexpansivity in terms of generalized stresses $\boldsymbol{\sigma} = (\sigma, \chi_1, \bar{\chi}_2)$ follows from the inequality (24) which can be rewritten in terms of stresses and static internal variables by means of the constitutive relations (20)$_{1-3}$ to get:

$$\langle (\sigma - \underline{\sigma}), \mathbf{E}^{-1}(\Delta\sigma - \Delta\underline{\sigma}) \rangle + \langle \chi_1 - \underline{\chi}_1, \mathbf{H}_1^{-1}(\Delta\chi_1 - \Delta\underline{\chi}_1) \rangle + \\ + \langle \bar{\chi}_2 - \underline{\bar{\chi}}_2, (\mathbf{R}'h_2\mathbf{R})^{-1}(\Delta\bar{\chi}_2 - \Delta\underline{\bar{\chi}}_2) \rangle \leq 0 \qquad (25)$$

or equivalently:

$$\prec \boldsymbol{\sigma} - \underline{\boldsymbol{\sigma}}, \Delta\boldsymbol{\sigma} - \Delta\underline{\boldsymbol{\sigma}} \succ_{\mathbb{M}^{-1}} \leq 0$$

which ensures the nonexpansivity of the generalized stresses $\boldsymbol{\sigma} = (\sigma, \chi_1, \bar{\chi}_2)$ with respect to the scalar product induced by $\mathbb{M}^{-1}$.

Let us now analyse the question of whether the evolution of the kinematic internal variable κ is nonexpansive. By virtue of the relations between convex and saddle functionals, the dissipation D can be expressed in terms of a saddle functional $D_\chi : \mathcal{D} \times \mathcal{Y}_1 \times \mathcal{Y}_2' \times \mathcal{Y} \to \mathfrak{R}$, convex with respect to $(\Delta p, -\Delta\alpha_1, \Delta\kappa)$ and concave with respect to $\bar{\chi}_2$, in the form:

$$D(\Delta p, -\Delta\alpha_1, -\Delta\alpha_2, \Delta\kappa) = -\inf_{\bar{\chi}_2}\left\{ \langle \bar{\chi}_2, \Delta\alpha_2 \rangle - D_\chi(\Delta p, -\Delta\alpha_1, \bar{\chi}_2, \Delta\kappa) \right\}$$

so that the finite-step flow rule (20)$_4$ can be equivalently rewritten as follows:

$$(\sigma, \chi_1, \Delta\alpha_2, X) \in \partial D_\chi(\Delta p, -\Delta\alpha_1, \bar{\chi}_2, \Delta\kappa). \qquad (26)$$

Hence the subdifferential relation (26)$_2$ is equivalent to state:

$$(\sigma, \chi_1, X) \in \partial_1 D_\chi(\Delta p, -\Delta\alpha_1, \bar{\chi}_2, \Delta\kappa) \qquad (27)$$

$$\Delta\alpha_2 \in \partial_2 D_\chi(\Delta p, -\Delta\alpha_1, \bar{\chi}_2, \Delta\kappa)$$

where $\partial_1 D_\chi$ denotes the subdifferential of D_χ with respect to the variables $(\Delta p, -\Delta\alpha_1, \Delta\kappa)$ and $\partial_2 D_\chi$ denotes the superdifferential of D_χ with respect to $\bar{\chi}_2$. The monotonicity of the multivalued map $-\partial_2 D_\chi$ implies:

$$\langle \bar{\chi}_2 - \underline{\bar{\chi}}_2, \Delta\alpha_2 - \Delta\underline{\alpha}_2 \rangle \leq 0. \qquad (28)$$

Such an inequality can, also, be assessed starting from (23)$_1$ by considering two nonlocal finite-step elastoplastic problems such that $\sigma = \underline{\sigma}$ and $\chi_1 = \underline{\chi}_1$.

Recalling the relation (20)$_3$ and the equality $\alpha_2 = -\kappa$, the inequality (28) yields:

$$\langle \mathbf{R}'h_2\mathbf{R}(\kappa - \underline{\kappa}), \Delta\kappa - \Delta\underline{\kappa} \rangle \leq 0. \qquad (29)$$

Then the kinematic internal variable κ is nonexpansive with respect to the scalar product induced by $\mathbf{R}'h_2\mathbf{R}$.

The nonexpansivity in terms of the static internal variable X follows from the inequality (29) and the relation (20)$_4$:

$$\langle X - \underline{X}, (\mathbf{R}'h\mathbf{R})^{-1}(\Delta X - \Delta \underline{X})\rangle \leq 0$$

which ensures the nonexpansivity of the static internal variable X with respect to the scalar product induced by $(\mathbf{R}'h\mathbf{R})^{-1}$.

4.1 Existing Models

Let the free energy (2) be written in the form:

$$\Phi(e, \alpha_1, \alpha_2, \kappa) = \tfrac{1}{2}\langle Ee, e\rangle + \tfrac{1}{2}\langle \mathbf{H}_1\alpha_1, \alpha_1\rangle + \qquad (30)$$

$$+ \tfrac{1}{2}\langle h_2(\mathbf{R}\alpha_2 - \alpha_2), (\mathbf{R}\alpha_2 - \alpha_2)\rangle + \tfrac{1}{2}\langle h\kappa, \kappa\rangle.$$

- The related nonlocal constitutive relations (11) turn out to be:

$$\begin{cases} \sigma = Ee \\ \chi_1 = \mathbf{H}_1\alpha_1 \\ \bar{\chi}_2 = (\mathbf{R}' - I)\,h_2(\mathbf{R}\alpha_2 - \alpha_2) = \\ \qquad = \mathbf{R}'h_2\mathbf{R}\alpha_2 - \mathbf{R}'h_2\alpha_2 - h_2\mathbf{R}\alpha_2 + h_2\alpha_2 \\ X = -h\kappa \end{cases} \qquad (31)$$

where I is the identity operator. If the material modulus h_2 is constant in space, the relation $(31)_3$ can be rewritten in the form $\bar{\chi}_2 = h_2 \mathbf{R}_*\alpha_2$. Hence the nonlocal variable $\bar{\chi}_2$ is obtained by means of the modified self-adjoint regularization operator $\mathbf{R}_* = \mathbf{R}'\mathbf{R} - \mathbf{R}' - \mathbf{R} + I$ and the corresponding constitutive relation adopted in Svedberg and Runesson (1998) is recovered. The dissipative force is then given by:

$$q = \bar{\chi}_2 + X = \mathbf{R}'h_2\mathbf{R}\alpha_2 - \mathbf{R}'h_2\alpha_2 - h_2\mathbf{R}\alpha_2 + h_2\alpha_2 - h\kappa =$$
$$= h_2\mathbf{R}_*\alpha_2 + h\alpha_2$$

and the size of the elastic domain depends on the kinematic internal variables α_2.
- If no kinematic hardening is considered in the model so that the internal variable α_1 is dropped in the expression (30) of the free energy and defining the modified regularization operator in the form $\mathbf{R}_* = \mathbf{R} - I$, the constitutive relations (11) become:

$$\begin{cases} \sigma = Ee \\ \bar{\chi}_2 = (\mathbf{R}' - I)\,h_2(\mathbf{R}\alpha_2 - \alpha_2) = \mathbf{R}'_* h_2\xi = \mathbf{R}'_*\chi_2 \\ X = -h\kappa \end{cases}$$

where $\chi_2 = h_2\xi$ and $\xi = \mathbf{R}\alpha_2 - \alpha_2 = \mathbf{R}_*\alpha_2$. Accordingly the elastic domain C is given by $g(\sigma) - \bar{\chi}_2 - X - \sigma_o \leq 0$ and the nonlocal model presented in [1] is then recovered.

By interchanging the regularization operator **R** with $\mathbf{R}_*$, the stability of the existing models of Svedberg and Runesson [11] and of Borino and Failla [1] follows.

5 Conclusion

The analysis of a constitutive nonlocal elastoplastic model is provided in the framework of the internal variable theories. The nonlocal model turns out to be rather versatile due to its thermodynamic basis and can be used to model different material behaviours such as nonlocal elasticity, nonlocal elastoplasticity and, in general, any material behaviour which can be formulated in terms of internal variable theories. The stability analysis is provided and it is shown that a standard procedure adopted for local plasticity can be successfully adopted for the proposed model. This work is currently in progress and the findings will be reported in the near future.

Acknowledgement

The financial support of the project MACE between Italian Government and Regione Campania is gratefully acknowledged.

References

1. Borino G, Failla B (2000) Thermodynamic consistent plasticity models with local and nonlocal internal variables. In: *European Congress an Computational Methods in Applied Sciences and Engineering*, CD-Rom Proceedings.
2. Halphen B, Nguyen QS (1975) Sur les matériaux standards généralisés. *J. Mech. Theor.* **14**:39–63.
3. Jirásek M, Rolshoven S (2003) Comparison of integral-type nonlocal plasticity models for strain-softening materials. *Int. J. Engng. Sci.* **41**:1553–1602.
4. Lemaitre J, Chaboche JL (1994) *Mechanics of Solids Materials*. Cambridge University Press, Cambridge, UK
5. Nguyen QS (1977) On elastic-plastic initial boundary value problem and its numerical integration. *Int. J. Numer. Methods Engrg.* **11**:817–832.
6. Pijaudier-Cabot G, Bažant ZP (1987) Nonlocal damage theory. *J. Eng. Mech. (ASCE)* **113**, 127–144.
7. Reddy BD, Martin JB (1991) Algorithms for the solution of internal variable problems in plasticity. *Comp. Meth. Appl. Mech. Eng.* **93**:253–273.
8. Rockafellar RT (1970) *Convex Analysis*. Princeton, Princeton University Press.
9. Simo JC, Govindjee S (1991) Non-linear B-stability and symmetry preserving return mapping algorithms for plasticity and viscoplasticity. *Int. J. Numer. Methods Engrg.* **31**:151–176.
10. Strömber L, Ristinmaa M (1996) FE-formulation of a nonlocal plasticity theory. *Comp. Meth. Appl. Mech. Eng.* **136**:127–144.
11. Svedberg T, Runesson K (1998) Thermodynamically consistent nonlocal and gradient formulations of plasticity. In: *EUROMECH Colloquium 378*, Mulhouse, France.

On the Mathematical Formulations of a Model of Strain Gradient Plasticity

F. Ebobisse[1], A.T. McBride[2] and B.D. Reddy[1,2]

[1]*Department of Mathematics and Applied Mathematics, University of Cape Town, 7701 Rondebosch, South Africa*
E-mail: francois.ebobissebille@uct.ac.za
[2]*Centre for Research in Computational and Applied Mechanics, University of Cape Town, 7701 Rondebosch, South Africa*
E-mail: amcbride@ebe.uct.ac.za, daya.reddy@uct.ac.za

Abstract. In this paper we study two mathematical formulations of a model of infinitesimal strain gradient plasticity for plastically irrotational materials proposed by Gurtin and Anand. The first formulation which exploits the context of convex analysis leads to a variational inequality whose well-posedness is studied. Due to the inability of this formulation to accommodate softening behaviour, a promising second formulation is proposed following the energetic approach for rate-independent processes developed by Mielke.

Key words: strain gradient plasticity, variational inequality, softening behaviour, energetic formulation.

1 Introduction

The inability of classical or conventional plasticity to model material behaviour at nano/meso scale levels has led to the development of new theories of plasticity in which size dependence is captured through the inclusion of either higher order derivatives of internal variables or higher order stresses in the yield criterion. For instance, in the pioneering works of Aifantis [1], Mühlhaus and Aifantis [13] the classical yield condition is augmented by a term involving the Laplacian of the effective plastic strain. Many other approaches have been presented in the literature. We can mention among all the papers [5, 8, 9].

While most of the new theories are based on the classical assumption that the direction of plastic flow is governed by the deviatoric part of the Cauchy stress, Gudmunson [8] and then Gurtin and Anand [9] proposed a different approach by introducing a microstress T^p which is power conjugate to the plastic strain rate $\dot{E}^p$ and a third order micropolar stress $\mathbb{K}^p$ which is power conjugate to the gradient of the plastic strain rate $\nabla \dot{E}^p$. These two tensors are related to the deviatoric part of the Cauchy tensor T through a microscopical force balance (2) which supplements the classical equilibrium (1).

The purpose of this paper is to study the variational formulations of the model of strain gradient plasticity for isotropic, plastically irrotational material, under small deformation, by Gurtin and Anand [9]. In [15], using the convex analytical setting of classical plasticity developed in [10], the flow law is written in its primal form and hence, combining both macroscopical and microscopical force balances together with the flow law, the model is formulated as a variational inequality whose well-posedness is studied in the case of hardening behaviour. We present in the first part of this paper a brief account of the formulation in [15].

As discussed in [15, section 4], it is quite impossible to accommodate softening behaviour in the formulation as variational inequality. On the other hand, softening behaviour is difficult to handle even for classical plasticity when formulated as variational inequality. A different formulation which accounts for plasticity with softening behaviour has been proposed recently by Dal Maso et al. in [3] following the energetic approach for rate-independent processes developed by Mielke in [11, 12].

In a recent paper by Giaccomini and Lussardi [6], an existence result for the Gurtin–Anand's model within the Mielke's energetic-approach is obtained and it has also been proved that the model converges in a suitable sense to the formulation of classical perfect plasticity in [2] whenever the energetic and dissipative length scales L in (7) and ℓ in (12) respectively go to zero.

Inspired by the results in the papers [3, 6], we present in the second part of this paper (see Section 3) an energetic formulation for the quasi-static evolution of the Gurtin–Anand's model of strain gradient plasticity with softening behaviour. We refer the reader to the forthcoming paper [4] for the details of our analysis.

2 The Governing Equations for the Problem

Let $\Omega \subset \mathbb{R}^3$ be a bounded connected open set which Lipschitz boundary $\partial\Omega = \Gamma_D \cup \Gamma_N$ which is occupied by an elastoplastic body in its undeformed configuration. The body is assumed to undergo infinitesimal deformations.

The model proposed by Gurtin and Anand [9] for small-deformation gradient plasticity is characterized by the inclusion of a second order tensor T^p, the microstress and a (third order tensor) polar microstress $\mathbb{K}^p$ which are conjugate respectively to the plastic strain rate $\dot{E}^p$ and gradient of plastic strain rate $\nabla \dot{E}^p$. The set of equations and boundary conditions below are obtained in [9] as consequences of the balance of internal and external power expenditures. In addition, the theory assumes zero plastic rotation. We confine attention to the case of rate-independent plastic behaviour, though the theory in [9] is valid for viscoplastic materials.

2.1 Balance Equations and Boundary Conditions

The conventional macroscopic force balance leads to the equation of equilibrium

$$\operatorname{div} T + f = 0 \tag{1}$$

in which T is the symmetric Cauchy stress and f is the body force. Due to the microstresses, the equilibrium equation above is augmented by the microforce balance

$$T^D = T^p - \operatorname{div} \mathbb{K}^p . \tag{2}$$

Here and henceforth $T^D_{ij} = T_{ij} - \frac{1}{3}\delta_{ij}\operatorname{tr} T$ denotes the deviatoric part of the second order tensor T. The divergence of the third order polar microstress is the second order tensor with components

$$(\operatorname{div} \mathbb{K})_{ij} = K_{ijk,k} , \tag{3}$$

and in which a subscript following a comma denotes partial differentiation with respect to that spatial component.

The macroscopic boundary conditions are given by

$$u = 0 \quad \text{on } \Gamma_D, \qquad t := T\nu = \bar{t} \quad \text{on } \Gamma_N . \tag{4}$$

That is, we assume homogeneous Dirichlet boundary conditions on a part Γ_D of the boundary $\partial\Omega$, and a prescribed surface traction on the complement $\Gamma_N := \partial\Omega \setminus \Gamma_D$. The outward unit normal to $\partial\Omega$ is denoted by ν. It is assumed that $\Gamma_D \neq \emptyset$. It is necessary also to prescribe microscopic boundary conditions on $\partial\Omega$; these take the form

$$E^p = 0 \quad \text{on } \Gamma_H, \qquad \mathbb{K}^p \nu = 0 \quad \text{on } \Gamma_F . \tag{5}$$

The subscripts 'H' and 'F' denote 'hard' and 'free' parts of the boundary respectively (see [9, section 8]), and Γ_H and Γ_F are complementary subsets of $\partial\Omega$.[1] The microscopic boundary conditions in (5) implies $\int_{\partial\Omega} \mathbb{K}^p \nu : \dot{E}^p dV = 0$ which is called in [9] *null microscopic power expenditure* on the boundary $\partial\Omega$.

2.2 Constitutive Theory

The constitutive equations are obtained from a free energy imbalance together with a flow law that characterizes plastic behaviour. The total strain E is additively decomposed into elastic and plastic components E^e and E^p, so that

$$E = E^e + E^p \tag{6}$$

with the plastic strain E^p satisfying $\operatorname{tr} E^p = 0$.

The strain-displacement relation is given by $E = Eu = (\nabla u + \nabla u^T)/2$.

The Burgers tensor G is a tensor of second order, equal to the curl of the plastic strain: that is, $G = \operatorname{curl} E^p$ or $G_{ij} = \varepsilon_{ijk} E^p_{ik,j}$, where ε_{ijk} is the permutation symbol.

---

[1] See section 8 of G&A for the situation in which $\mathbb{K}^p_{\text{dis}} = 0$.

In this paper (see also [15]) we modify the model presented in [9] in order to account for linear isotropic hardening. Therefore following the approach in [10] for classical plasticity, we consider here a free energy of the form

$$\psi(E^e, E^p, \gamma) := \underbrace{\frac{1}{2} E^e : \mathcal{C} E^e}_{\text{elastic energy}} + \underbrace{\frac{1}{2} \mu L^2 |\text{Curl } E^p|^2}_{\text{defect energy}} + \underbrace{\frac{1}{2} k |\gamma|^2}_{\text{hardening energy}} \qquad (7)$$

where $\mathcal{C}$ is the elasticity tensor which for homogeneous isotropic media is defined by $\mathcal{C} E = \lambda(\text{tr} E) I + \mu(E + E^T)$ for any second order tensor E, λ and μ called the Lamé moduli, L is an energetic length scale and k is a positive constant.

The local free-energy imbalance states that

$$\dot{\psi} - T : \dot{E}^e - T^p : \dot{E}^p - \mathbb{K}^p : \nabla \dot{E}^p \leq 0 . \qquad (8)$$

Expansion of the first term and substitution of (7) gives the elastic relation

$$T = \mathcal{C} E^e . \qquad (9)$$

Now, if we set

$$P_{jqp} = \varepsilon_{ipq} \frac{\partial \hat{\psi}}{\partial G_{ij}}, \quad \mathbb{K}^p_{\text{dis}} = \mathbb{K}^p - \mathbb{P} \quad \text{and} \quad g = -\frac{\partial \hat{\psi}}{\partial \gamma} = k\gamma \qquad (10)$$

then the dissipation inequality becomes

$$T^p : \dot{E}^p + \mathbb{K}^p_{\text{dis}} : \nabla \dot{E}^p + g\dot{\gamma} \geq 0 . \qquad (11)$$

2.3 The Flow Law

Set $\Sigma^p = (T^p, \mathbb{K}^p_{\text{dis}}, g)$. We have the yield criterion

$$\phi(\Sigma^p) = \sqrt{|T^p|^2 + \ell^{-2} |\mathbb{K}^p_{\text{dis}}|^2} + g - S_Y \leq 0 \qquad (12)$$

where ℓ is a dissipative length scale.

So the set of admissible generalized stresses is

$$\mathcal{K} := \left\{ \Sigma^p : \sqrt{|T^p|^2 + \ell^{-2} |\mathbb{K}^p_{\text{dis}}|^2} + g - S_Y \leq 0 \right\} . \qquad (13)$$

If $\Gamma^p = (E^p, \nabla E^p, \gamma)$ then the maximum dissipation principle gives the normality law

$$\dot{\Gamma}^p \in N_\mathcal{K}(\Sigma^p) \qquad (14)$$

where $N_\mathcal{K}(\Sigma^p)$ denotes the normal cone to $\mathcal{K}$ at Σ^p, which is the set of generalized strains $\dot{\Gamma}^p$ that satisfy $(\overline{\Sigma} - \Sigma^p) : \dot{\Gamma}^p \leq 0$ for all $\overline{\Sigma} \in \mathcal{K}$.

Since $\partial \mathcal{K}$ is smooth, this gives

$$\dot{\Gamma}^p = \lambda \frac{\partial \phi}{\partial \Sigma^p}.$$

That is,

$$\dot{E}^p = \lambda \frac{T^p}{S_Y - g}, \quad \nabla \dot{E}^p = \lambda \ell^{-2} \frac{\mathbb{K}^p_{\text{dis}}}{S_Y - g} \quad \text{and} \quad \dot{\gamma} = \lambda.$$

Next we introduce

$$d^p(\dot{E}^p) := \sqrt{|\dot{E}^p|^2 + \ell^2 |\nabla \dot{E}^p|^2} = \dot{\gamma} = \lambda.$$

Using convex analysis we find that

$$\dot{\Gamma}^p \in N_\mathcal{K}(\Sigma^p) \quad \Leftrightarrow \quad \Sigma^p \in \partial \mathcal{D}(\dot{\Gamma}^p) \tag{15}$$

where $\mathcal{D}$ is the dissipation function which in this case is defined by

$$\mathcal{D}(M) := \begin{cases} S_Y d^p(Q) & \text{if } d^p(Q) \leq \xi, \\ \infty & \text{otherwise} \end{cases} \quad \forall M = (Q, \nabla Q, \xi) \tag{16}$$

and $\partial \mathcal{D}(\dot{\Gamma}^p)$ denotes the subdifferential of $\mathcal{D}$ evaluated at $\dot{\Gamma}^p$. That is,

$$\Sigma^p \in \partial \mathcal{D}(\dot{\Gamma}^p) \quad \Leftrightarrow \quad \mathcal{D}(M) \geq \mathcal{D}(\dot{\Gamma}^p) + \Sigma^p : (M - \dot{\Gamma}^p) \quad \text{for any } M. \tag{17}$$

2.4 The Formulation as Variational Inequality

Assume for simplicity that $\Gamma_N = \emptyset$ and that $\bar{u} = 0$. Taking the inner product of the equilibrium equation (1) with $v - \dot{u}$ where v is an arbitrary function v which satisfies the Dirichlet boundary condition $(4)_1$, integrating over Ω, and then integrating by parts, also substituting (9) for T, we obtain

$$\int_\Omega \mathcal{C}(Eu - E^p) : (Ev - E\dot{u}) \, dx = \int_\Omega f \cdot (v - \dot{u}) \, dx. \tag{18}$$

Next, expanding (17) and integrating over Ω we have

$$\int_\Omega \mathcal{D}(Q, \nabla Q, \mu)\, dx \geq \int_\Omega \mathcal{D}(\dot{E}^p, \nabla \dot{E}^p, \dot{\gamma})\, dx + \int_\Omega T^p : (Q - \dot{E}^p)\, dx$$
$$+ \int_\Omega \mathbb{K}^p_{\text{dis}} : \nabla(Q - \dot{E}^p)\, dx + \int_\Omega g(\mu - \dot{\gamma})\, dx$$
$$\geq \int_\Omega \mathcal{D}(\dot{E}^p, \nabla \dot{E}^p, \dot{\gamma})\, dx + \int_\Omega T : (Q - \dot{E}^p)\, dx$$
$$- \int_\Omega \mathbb{P} : \operatorname{curl}(Q - \dot{E}^p)\, dx + \int_\Omega g(\mu - \dot{\gamma})\, dx \quad (19)$$

where the last inequality follows from the microscopic force balance (2), the definition of $\mathbb{K}^p_{\text{dis}}$ and $\mathbb{P}$ in (10) respectively and the microscopic boundary conditions (5) satisfied by Q and E^p.

Now following the formulation developed in [10] for classical plasticity, we introduce the spaces of functions

$$\mathcal{V} = [H^1_0(\Omega)]^3, \quad \mathcal{Q} = \left\{ \begin{array}{l} Q = (Q_{ij}) \in [H^1(\Omega)]^{3\times 3} : Q_{ji} = Q_{ij} \\ \operatorname{tr} Q = 0 \text{ a.e. in } \Omega, \ Q = 0 \text{ on } \Gamma_H \end{array} \right\}, \quad \mathcal{M} = L^2(\Omega).$$

These are Hilbert spaces when endowed respectively with the norms

$$\|v\|_\mathcal{V} = \|\nabla v\|_0, \quad \|Q\|_\mathcal{Q} = \left(\|Q\|_0^2 + \|\nabla Q\|_0^2\right)^{1/2} \text{ and } \|\xi\|_\mathcal{M} = \|\xi\|_0. \quad (20)$$

For convenience the L^2-norm, whether for scalar-, vector- or tensor-valued functions, is denoted by $\|\cdot\|_0$. Now, set

$$\mathcal{Z} = \mathcal{V} \times \mathcal{Q} \times \mathcal{M} \text{ and } \mathcal{W} = \left\{ (v, Q, \xi) \in \mathcal{Z} : \sqrt{|Q|^2 + \ell^2|\nabla Q|^2} \leq \xi \text{ a.e. in } \Omega \right\}.$$

Then $\mathcal{Z}$ is a Hilbert space with norm $\|z\|_\mathcal{Z} = (\|v\|_\mathcal{V}^2 + \|Q\|_\mathcal{Q}^2 + \|\xi\|_\mathcal{M}^2)^{1/2}$ and $\mathcal{W}$ is a non-empty closed convex cone in $\mathcal{Z}$.

Now adding up (18) and (19), we find that $w = (u, E^p, \gamma)$ solves the variational inequality

$$a(w(t), z - \dot{w}(t)) + j(z) - j(\dot{w}(t)) - \ell(z - \dot{w}(t)) \geq 0 \quad \forall z = (v, Q, \xi) \in \mathcal{W},$$

where

$$a(w, z) = \int_\Omega \mathcal{C}(Eu - E^p) : (Ev - Q)\, dx + k\int_\Omega \gamma\xi\, dx + \alpha L^2 \int_\Omega \operatorname{curl} E^p : \operatorname{curl} Q\, dx \quad (21)$$

$$j(z) := \begin{cases} \int_\Omega \mathcal{D}(Q, \nabla Q, \xi)\, dx & \text{if } z \in \mathcal{W}, \\ +\infty & \text{otherwise,} \end{cases} \quad \text{and} \quad \ell(z) := \int_\Omega f \cdot v\, dx. \quad (22)$$

The well-posedness of the Gurtin–Anand model formulated as variational inequality is obtained in [15] applying [10, theorem 7.3].

3 Energetic Formulation for Softening Behaviour

The analysis in [15, section 4] has shown the impossibility for the Gurtin–Anand's model to accommodate softening behaviour (corresponding in this case to $k < 0$) in the formulation as variational inequality. In fact, we loose the coercivity of the bilinear form a on $\mathcal{Z}$ or $\mathcal{W}$ when $k < 0$.

We present in this section an energetic formulation in the spirit of [3, 6, 11, 12], which accommodates softening behaviour.

3.1 Prescribed Boundary Displacement

To simplify the problem, we assume that there are no volume forces nor traction prescribed. So, the evolution is only governed by a prescribed boundary displacement on $\partial\Omega$ which is given by the trace of a function $w : [0, \infty) \times \Omega \to \mathbb{R}^3$ which is locally absolutely continuous in time with values in $H^1(\Omega, \mathbb{R}^3)$. For instance, $w \in H^1_{\text{loc}}(0, \infty; H^1(\Omega, \mathbb{R}^3))$.

3.2 Energies and Potentials

The energetic formulation of the quasi-static evolution of Gurtin–Anand's model with softening behaviour involves

$$Q_1(E^e) := \frac{1}{2}\int_\Omega E^e : \mathcal{C}E^e dx, \quad Q_2(E^p) = \frac{1}{2}\mu L^2 \int_\Omega |\text{curl}E^p|^2 dx, \quad (23)$$

$$\mathcal{V}(\xi) := \int_\Omega V(\xi(x))dx \quad (24)$$

where the potential $V : \mathbb{R} \to \mathbb{R}$ is the softening potential which is assumed here to be a function of class C^2 such that there exists $M > 0$ such that

$$-M \le V''(\xi) \le 0 \quad \forall \xi, \quad -\frac{1}{2}S_Y < V'(+\infty) \le V'(-\infty) < \frac{1}{2}S_Y, \quad (25)$$

where S_Y is the yield constant.

Remark 1. A typical potential V is given by $V(\xi) = -c(1 + |\xi|^2)^{1/2}$ with $0 < c < \frac{1}{2}S_Y$.

We find that the dissipation function corresponding to softening behaviour is given by

$$\mathcal{D}(M) = S_Y \max\left(\sqrt{|Q|^2 + \ell^2|\nabla Q|^2}, \xi\right) \quad \forall M = (Q, \nabla Q, \xi). \quad (26)$$

Notice that in this case the functional $\mathcal{H}(Q,\xi) := \int_\Omega \mathcal{D}(Q, \nabla Q, \xi)dx$ is defined in $W^{1,1}(\Omega, M^{3\times 3}) \times L^1(\Omega)$. However, $\mathcal{H}$ will be involved in an analysis (the incremental problem) which employs the direct method of calculus of variations. Therefore, the process of relaxation with respect to weak convergence (see [7]) in the space $BV(\Omega, M^{3\times 3}) \times L^1(\Omega)$ will give for any $(Q, \xi) \in BV(\Omega, M^{3\times 3}) \times L^1(\Omega)$

$$\mathcal{H}(Q, \xi) = S_Y \int_\Omega \max\left(\sqrt{|Q(x)|^2 + \ell^2 |\nabla Q(x)|^2}, \xi(x)\right) dx + S_Y l |D^s Q|(\Omega), \quad (27)$$

where ∇Q is the so-called approximate gradient of Q and $D^s Q$ is the singular part of the measure DQ with respect to the Lebesgue measure $\mathcal{L}^3$.

Remark 2. Notice that $\mathcal{H}(Q_1 + Q_2, \xi_1 + \xi_2) \leq \mathcal{H}(Q_1, \xi_1) + \mathcal{H}(Q_2, \xi_2)$ and there exists $m > 0$ such that for any $(Q_1, \xi_1), (Q_2, \xi_2) \in BV(\Omega, M^{3\times 3}) \times L^1(\Omega)$

$$\mathcal{H}(Q_1 - Q_2, \xi_1 - \xi_2) + \mathcal{V}(\xi_1) - \mathcal{V}(\xi_2) \geq m\left[\|Q_1 - Q_2\|_{BV} + \|\xi_1 - \xi_2\|_{L^1}\right]. \quad (28)$$

The dissipation of a function $t \to (Q(t), \xi(t))$ on an interval $[a, b] \subset [0, \infty)$ is defined by

$$\mathcal{D}_\mathcal{H}(Q, \xi; a, b) = \sup\left\{\sum_{j=1}^k \mathcal{H}(Q(t_j) - Q(t_{j-1}), \xi(t_j) - \xi(t_{j-1}))\right\}, \quad (29)$$

where the supremum is taken over all partitions $t_0 = a < t_1 < \cdots < t_{k-1} < t_k = b$ of $[a, b]$.

When $t \to (Q(t), \xi(t))$ is smooth then $\mathcal{D}_\mathcal{H}(Q, \xi; a, b) = \int_a^b \mathcal{H}(\dot{Q}(t), \dot{\xi}(t))dt$.

3.3 Admissible Configurations

Let $w \in H^1(\Omega, \mathbb{R}^3)$ be a boundary data prescribed on $\partial\Omega$. An admissible configuration relative to w is given by (u, E^e, E^p, ξ) such that

$$u \in W^{1,3/2}(\Omega, \mathbb{R}^3), \quad E^e \in L^2(\Omega, M^{3\times 3}_{\text{sym}}), \quad E^p \in BV(\Omega, M^{3\times 3}_{\text{sym}}), \quad \xi \in H^1(\Omega) \quad (30)$$

$$u = w \text{ on } \partial\Omega, \quad Eu = E^e + E^p, \quad \text{curl} E^p \in L^2(\Omega, M^{3\times 3}) \quad (31)$$

Remark 3. $E^e \in L^2(\Omega, M^{3\times 3}_{\text{sym}})$ and $E^p \in BV(\Omega, M^{3\times 3}_{\text{sym}}) \subset L^{3/2}(\Omega, M^{3\times 3}_{\text{sym}})$ imply that $\Rightarrow Eu \in L^{3/2}(\Omega, M^{3\times 3}_{\text{sym}})$. Thus, since $w \in H^1(\Omega, \mathbb{R}^3)$, it follows from Korn's inequality that $u \in W^{1,3/2}(\Omega, \mathbb{R}^3)$.

On the other hand, from the expression of the functional $\mathcal{H}$ and from the growth assumptions (25) on the derivatives of the function V, it is sufficient to find $\zeta \subset L^1(\Omega)$. However, such a choice might lead to solutions where ξ is a measure. To avoid this, we will consider here a perturbation of the model with the term $\beta |\nabla \xi|^2$ ($\beta > 0$) for which $\xi \in H^1(\Omega)$.

The family of the admissible configurations for the boundary data w is denoted by

$$\mathbb{A}(w) := \{(u, E^e, E^p, \xi) \text{ such that (30)–(31) are satisfied}\} \quad (32)$$

3.4 The Quasi-Static Evolution Using Global Stability

The quasi-static evolution for the Gurtin–Anand model using the global stability is defined as follows. Let $T > 0$. Find a map

$$t \to (u(t), E^e(t), E^p(t), \xi(t)) \in \mathbb{A}(w(t))$$

from $[0, T]$ to $W^{1,3/2}(\Omega, \mathbb{R}^3) \times L^2(\Omega, M_{\text{sym}}^{3\times 3}) \times BV(\Omega, M_{\text{sym}}^{3\times 3}) \times H^1(\Omega)$ such that

(a) *global stability*: for every $t \in [0, T]$ and for every $(v, E, P, \eta) \in \mathbb{A}(w(t))$

$$Q_1(E^e(t)) + Q_2(E^p(t)) + \mathcal{V}(\xi(t)) + \frac{1}{2}\beta\|\nabla\xi(t)\|_0^2$$
$$\leq Q_1(E) + Q_2(P) + \mathcal{V}(\eta) + \frac{1}{2}\beta\|\nabla\eta\|_0^2 - \mathcal{H}(P - E^p(t), \eta - \xi(t)); \quad (33)$$

(b) *energy inequality*: for any $t \in [0, T]$

$$\mathcal{E}(t) + \mathcal{D}_{\mathcal{H}}(E^p, \xi; 0, t) \leq \mathcal{E}(0) + \int_0^t \langle T(t) : E\dot{w}\rangle dt \quad (34)$$

where $\mathcal{E}(t) = Q_1(E^e(t)) + Q_2(\text{curl} E^p(t)) + \mathcal{V}(\xi(t)) + \beta\|\nabla\xi(t)\|_0^2$
$T(t) = \mathcal{C}E^e(t)$ and $\mathcal{D}_{\mathcal{H}}$ is defined in (29).

Following [11, 12] the strategy to construct the evolution described above is to use a variational method based on time discretization and on the solutions of suitable incremental problems. Precisely, let $n \geq 1$ be an integer and let $t_n^i = \frac{i}{n}T$ with $i = 0, 1, \ldots, n$.

Set $u_{n,i} = u(t_n^i)$, $E_{n,i}^e = E^e(t_n^i)$, $E_{n,i}^p = E^p(t_n^i)$, $\xi_{n,i} = \xi(t_n^i)$.

The incremental problem reads as follows: given $(u_{n,i}, E_{n,i}^e, E_{n,i}^p, \xi_{n,i}) \in \mathbb{A}(w(t_n^i))$, find $(u_{n,i+1}, E_{n,i+1}^e, E_{n,i+1}^p, \xi_{n,i+1}) \in \mathbb{A}(w(t_n^{i+1}))$ as the solution of the minimum problem

$$\min \begin{cases} Q_1(E) + Q_2(\text{curl} P) + \mathcal{V}(\eta) + \beta\|\nabla\eta\|_0^2 + \mathcal{H}(P - E_{n,i}^p, \eta - \xi_{n,i}), \\ (v, E, P, \eta) \in \mathbb{A}(w(t_n^{i+1})) \end{cases} \quad (35)$$

However, since the energy functional in (35) is not convex, the global minimality criterion is not appropriate in this case. In fact, the lack of convexity usually allows the functional to have multiple wells. Therefore, a quasi-static evolution driven by a global minimality criterion might prescribe jumps on the solution from one well to another. So, we propose here a quasi-static evolution governed by a local minimality criterion. That is, we replace the global stability by a stationary point criterion which reads as follows: for any $t \in [0, T]$, there exist $T^p(t)$ second order tensor, $\mathbb{K}^p_{\text{dis}}(t)$, and $\mathbb{K}^p(t)$ third order tensors such that, for $T(t) = \mathcal{C}E^e(t)$ and $g(t) = -V'(\xi(t)) + \beta\Delta\xi(t)$ we have

(a) $\operatorname{div} T(t) = 0$ in Ω and $\begin{cases} T^p(t) = T(t)_D + \operatorname{div} \mathbb{K}^p(t) \text{ in } \Omega, \\ \mathbb{K}^p \nu = 0 \text{ on } \partial\Omega; \end{cases}$

(b) $\sqrt{|T^p(x,t)|^2 + \ell^{-2}|\mathbb{K}^p_{\text{dis}}(x,t)|^2} \leq S_Y - g(x,t)$ for a.e. $x \in \Omega$;

(c) $(\dot{E}^p(t,x), \nabla\dot{E}^p(t,x), \dot{\xi}(t,x)) \in N_{\mathcal{K}}(T^p(t,x), \mathbb{K}^p_{\text{dis}}(t,x), g(t,x))$ where $\mathcal{K}$ is the set of generalized stresses defined in (13).

To construct the quasi-static evolution based of the stationary point criterion above and the energy inequality (34), we follow the vanishing viscosity approach in [3]. Precisely, we introduce an ε-regularized evolution.

3.5 The Regularized Evolution

Let $T > 0$ and $w \in H^1([0,T]; H^1(\Omega, \mathbb{R}^3))$. Let $(u_0, E_0^e, E_0^p, \xi_0) \in \mathbb{A}(w(0))$ and let $\varepsilon > 0$. A solution of the ε-regularized evolution in the time interval $[0, T]$ with boundary datum w and initial condition $(u_0, E_0^e, E_0^p, \xi_0)$ is a function $(u_\varepsilon, E_\varepsilon^e, E_\varepsilon^p, \xi_\varepsilon)$ with

$$u_\varepsilon \in H^1([0,T]; W^{1,3/2}(\Omega, \mathbb{R}^3)), \quad E_\varepsilon^e \in H^1([0,T]; L^2(\Omega, M_{\text{sym}}^{3\times 3}))$$
$$E_\varepsilon^p \in L^2([0,T]; BV(\Omega, M_{\text{sym}}^{3\times 3})), \quad \dot{E}_\varepsilon^p \in L^2(([0,T]; L^2(\Omega, M_{\text{sym}}^{3\times 3}))$$
$$\xi_\varepsilon \in H^1([0,T]; H^1(\Omega))$$

such that for every $t \in [0, T]$, there exists $T_\varepsilon^p(t)$ second order tensor, $\mathbb{K}_\varepsilon^p(t)$ and $\mathbb{K}_{\text{dis},\varepsilon}^p(t)$ third order tensors such that, setting $g_\varepsilon(t) = -V'(\xi_\varepsilon(t)) + \beta\Delta\xi_\varepsilon(t)$ and $T_\varepsilon(t) = \mathcal{C}E_\varepsilon^e(t)$, we have

(a)$_\varepsilon$ $(u_\varepsilon(0), E_\varepsilon^e(0), E_\varepsilon^p(0), \xi_\varepsilon(0)) = (u_0, E_0^e, E_0^p, \xi_0)$;

(b)$_\varepsilon$ $\forall t \in [0,T], (u_\varepsilon(t), E_\varepsilon^e(t), E_\varepsilon^p(t), \xi_\varepsilon(t)) \in \mathbb{A}(w(t))$;

(c)$_\varepsilon$ $\forall t \in [0,T], \operatorname{div} T_\varepsilon(t) = 0$ in Ω and $\begin{cases} T_\varepsilon^p(t) = T(t)_{\varepsilon,D} + \operatorname{div} \mathbb{K}_\varepsilon^p(t) \text{ in } \Omega, \\ \mathbb{K}_\varepsilon^p \nu = 0 \text{ on } \partial\Omega; \end{cases}$

(d)$_\varepsilon$ $\sqrt{|T_\varepsilon^p(x,t)|^2 + \ell^{-2}|\mathbb{K}_{\text{dis},\varepsilon}^p(x,t)|^2} \leq S_Y - g_\varepsilon(x,t)$ for a.e. $x \in \Omega$;

(e)$_\varepsilon$ $(\dot{E}_\varepsilon^p(t,x), \nabla\dot{E}_\varepsilon^p(t,x), \dot{\xi}_\varepsilon(t,x)) \in N_{\mathcal{K}}^\varepsilon(T_\varepsilon^p(t,x), \mathbb{K}_{\text{dis},\varepsilon}^p(t,x), g_\varepsilon(t,x))$ where $N_{\mathcal{K}}^\varepsilon(\Sigma) = \frac{1}{\varepsilon}(\Sigma - P_{\mathcal{K}}(\Sigma))$, $P_{\mathcal{K}}$ is the projection onto $\mathcal{K}$.

3.6 The Incremental Problem

The incremental problem corresponding to the ε-regularized evolution above is given by

$$\min \left\{ \begin{array}{l} \mathcal{Q}_1(E) + \mathcal{Q}_2(P) + \mathcal{V}(\eta) + \frac{1}{2}\beta\|\nabla\eta\|_0^2 + \mathcal{H}(P - E_{n,i}^p, \eta - \xi_{n,i}) \\ + \frac{n\varepsilon}{2T}\|P - E_{n,i}^p\|_0^2 + \frac{n\varepsilon}{2T}\|\eta - \xi_{n,i}\|_0^2, \quad (v, E, P, \eta) \in \mathbb{A}(w(t_n^{i+1})) \end{array} \right\} \quad (36)$$

We prove in [4] that (36) has a unique solution for n large enough.

Now in the next step we define piecewise constant interpolants

$$u_{n,\varepsilon}(t) := u_{n,i}, \ E_{n,\varepsilon}^e(t) := E_{n,i}^e, \ E_{n,\varepsilon}^p(t) := E_{n,i}^p, \ \xi_{n,\varepsilon}(t) := \xi_{n,i}, \ \forall t \in [t_n^i, t_n^{i+1}).$$

The ε-regularized evolution will then be obtained by taking the limit as $n \to \infty$ of these piecewise constant interpolants. The final step will be to study the asymptotic behaviour of the ε-regularized evolution as the viscosity parameter ε goes to 0.

Remark 4. Following the paper [6], a further analysis would be on the asymptotic behaviour of the formulation above when the energetic and dissipative length scales L and ℓ go to zero.

On the other hand, it would be interesting to consider isotropic hardening for a model of strain gradient plasticity with plastic spin, see e.g., [14].

References

1. Aifantis, E.C. (1984) On the microstructural origin of certain inelastic models, *ASME J. Eng. Mater. Technol.* 106:326–330.
2. Dal Maso, G., De Simone, A., Mora, M.G. (2006) Quasistatic evolution problems for linearly elastic – perfectly plastic material, *Arch. Ration. Mech. Anal.* 180:237–291.
3. Dal Maso, G., De Simone, A., Mora, M.G., Morini, M. (2005) A vanishing viscosity approach to quasistatic evolution in plasticity with softening, *Arch. Ration. Mech. Anal.*, to appear.
4. Ebobisse, F., McBride, A., Reddy, B.D. (2008) An energetic formulation of strain-gradient plasticity for isotropic, plastically irrotational materials with softening behaviour, in preparation.
5. Fleck, N., Hutchinson, J. (2001) A reformulation of strain gradient plasticity, *J. Mech. Phys. Solids* 49:2245–2271.
6. Giacomini, A., Lussardi, L. (2007) A quasistatic evolution for a model in strain gradient plasticity, Preprint.
7. Goffman, C., Serrin, J. (1964) Sublinear functions of measures and variational integrals, *Duke Math. J.* 31:159–178.
8. Gudmundson, P. (2004) A unified treatment of strain gradient plasticity, *J. Mech. Phys. Solids* 52:1379–1406.
9. Gurtin, M.E., Anand, L. (2005) A theory of strain-gradient plasticity for isotropic, plastically irrotational materials. Part I: Small deformations, *J. Mech. Phys. Solids*, 53:1624–1649.
10. Han, W., Reddy, B.D. (1999) *Plasticity: Mathematical Theory and Numerical Analysis*, Springer-Verlag, New York.
11. Mielke, A. (2003) Energetic formulation of multiplicative elasto-plasticity using dissipation distances, *Cont. Mech. Thermodynamics* 15:351–382.
12. Mielke, A. (2005) Evolution of rate-independent processes. In: *Evolutionary Equations*, Vol. II, Handbook of Differential Equations, C.M. Dafermos and E. Feireisl (Eds.), pp. 461–559. Elsevier/North-Holland, Amsterdam.
13. Mühlhaus, H-B., Aifantis, E. (1991) A variational principle for gradient plasticity, *Int. J. Solids Struct.*, 28(7):845–857.

14. Neff, P., Chełmiński, K., Alber, H.D. (2008) Notes on strain gradient plasticity. Finite strain covariant modelling and global existence in the infinitesimal rate-independent case. *Math. Mod. Meth. Appl. Sci. (M3AS)*, to appear.
15. Reddy, B.D., Ebobisse, F., McBride, A. (2008) Well-posedness of a model of strain gradient plasticity for plastically irrotational materials, *Int. J. Plasticity* 24:55–73.

Uniqueness of Strong Solutions in Infinitesimal Perfect Gradient-Plasticity with Plastic Spin

Patrizio Neff

Fachbereich Mathematik, Technische Universität Darmstadt,
Schlossgartenstrasse 7, 64289 Darmstadt, Germany
E-mail: neff@mathematik.tu-darmstadt.de

Abstract. A strain gradient plasticity model is motivated based on infinitesimal kinematics. The free energy is augmented by the curl of the non-symmetric plastic strain as a measure for the plastic incompatibility. Flow rules are derived and uniqueness of classical solutions is established.

Key words: gradient plasticity, plastic spin, dislocation density.

1 Introduction

Here we discuss a model of infinitesimal strain gradient plasticity including phenomenological Prager type linear kinematical hardening and nonlocal kinematical hardening due to dislocation interaction. Based on the additive decomposition of the displacement gradient into non-symmetric elastic and plastic distortions the formulation features a thermodynamically admissible model of infinitesimal plasticity involving only the Curl of the infinitesimal plastic distortion p. The model is invariant w.r.t. superposed rigid infinitesimal rotations of the reference, intermediate and spatial configuration but the model is not spin-free due to the nonlocal dislocation interaction and cannot be reduced to a dependence on the infinitesimal plastic strain tensor $\varepsilon_p = \text{sym } p$. Uniqueness of strong solutions of the infinitesimal model is obtained if two non-classical boundary conditions on the non-symmetric plastic distortion p are introduced: skew $\dot{p}.\tau = 0$ on the microscopically hard boundary $\Gamma_D \subset \partial\Omega$ and $[\text{Curl } p].\tau = 0$ on $\partial\Omega$, where τ are the tangential vectors at the boundary $\partial\Omega$.

There is an abundant literature on gradient plasticity formulations, in most cases letting the yield-stress depend also on some higher derivative of a scalar measure of accumulated plastic distortion [3]. Experimentally, the dependence of the yield stress on plastic gradients is well-documented [2] and may become important for very small samples.

B.D. Reddy (ed.), IUTAM Symposium on Theoretical, Modelling and Computational Aspects of Inelastic Media, 129–140.
© *Springer Science+Business Media B.V. 2008*

From a numerical point of view the incorporation of plastic gradients serves the purpose of removing the mesh-sensitivity, either in the softening case, or, more difficult to observe numerically, already in classical Prandtl–Reuss plasticity (shear bands and slip lines with ill-defined band width).

This IUTAM-meeting has shown that gradient plasticity is of high current interest [4–6], but rigorous mathematical studies are still rare. Reddy [14] treats a geometrically linear irrotational (no-spin) model of Gurtin [4], different from my proposal since only symmetric plastic strains appear.

My contribution is organized as follows: first, I introduce the model and show its thermodynamic admissibility. Then I prove that strong solutions of the obtained model with general monotone, non-associative flow-rule together with suitable boundary conditions on the non-symmetric infinitesimal plastic distortion p are unique. The existence question of a weak reformulation is treated in [11]. There, also a finite-strain parent model is given and related invariance questions are investigated in [10]. The relevant notation is found in the Appendix.

2 The Geometrically Linear Gradient Plasticity Model

The model is introduced informally by considering the well known multiplicative decomposition of the deformation gradient $F = F_e F_p$ into elastic and plastic parts and expanding to highest order. Thus we expand $F = 1\!\!1 + \nabla u$, $F_p = 1\!\!1 + p + \cdots$, $F_e = 1\!\!1 + e + \cdots$ and the multiplicative decomposition turns into

$$1\!\!1 + \nabla u = (1\!\!1 + e + \cdots)(1\!\!1 + p + \cdots) \rightsquigarrow \nabla u \approx e + p + \cdots,$$
$$F_e^T F_e - 1\!\!1 = 1\!\!1 + 2\,\text{sym}\,e + e^T e - 1\!\!1 \rightsquigarrow 2\,\text{sym}\,e = 2\,\text{sym}(\nabla u - p). \quad (1)$$

Hence one obtains to highest order the *additive decomposition* [7] of the displacement gradient $\nabla u = e + p$ into nonsymmetric elastic and plastic distortion. Here $\text{sym}\,e = \text{sym}(\nabla u - p)$ the *infinitesimal elastic lattice strain*, $\text{skew}\,e = \text{skew}(\nabla u - p)$ the *infinitesimal elastic lattice rotation* and $\kappa_e = \nabla\,\text{axl}(\text{skew}\,e)$ the *infinitesimal elastic lattice curvature* and p the *infinitesimal plastic distortion*. We assume the quadratic energy to be given by

$$W(\nabla u, p, \text{Curl}\,p) = W_e^{\text{lin}}(\nabla u - p) + W_{\text{ph}}^{\text{lin}}(p) + W_{\text{curl}}^{\text{lin}}(\text{Curl}\,p),$$
$$W_e^{\text{lin}}(\nabla u - p) = \mu \|\text{sym}(\nabla u - p)\|^2 + \frac{\lambda}{2}\text{tr}\left[\nabla u - p\right]^2, \quad (2)$$
$$W_{\text{ph}}^{\text{lin}}(p) = \mu\,H_0\|\text{dev sym}\,p\|^2, \quad W_{\text{curl}}^{\text{lin}}(\text{Curl}\,p) = \frac{\mu\,L_c^2}{2}\|\text{Curl}\,p\|^2.$$

The used free energy coincides with that in [9, p. 1783] apart for the local kinematical hardening contribution. Note that the *infinitesimal plastic distortion* $p : \Omega \subset \mathbb{R}^3 \mapsto \mathbb{M}^{3\times 3}$ need *not* be *symmetric*, but that only its symmetric part, the *in-*

finitesimal plastic strain[1] sym p, contributes to the local elastic energy expression. The *infinitesimal plastic rotation* skew p does not locally contribute to the elastic energy neither contributes to the local plastic self-hardening but appears in the non-local hardening. The resulting elastic energy is invariant under infinitesimal rigid rotations $\nabla u \mapsto \nabla u + \overline{A}$, $\overline{A} \in \mathfrak{so}(3)$ of the body. The invariance of the curvature contribution needs the homogeneity of the rotations.

Provided that the infinitesimal plastic distortion p is known, (2) defines a linear elasticity problem with pre-stress for the displacement u. It remains to provide an evolution law for p which is consistent with thermodynamics. To this end we use a nonlocal (integral) version of the second law of thermodynamics.

For any "nice" subdomain $\mathcal{V} \subseteq \Omega$ consider for fixed $t_0 \in \mathbb{R}$ the rate of change of energy storage due to inelastic processes

$$\frac{d}{dt} \int_{\mathcal{V}} W(\nabla u(x, t_0), p(x, t), \mathrm{Curl}\, p(x, t))\, dV =$$

$$\int_{\mathcal{V}} 2\mu \left\langle \mathrm{sym}(\nabla u - p), -\frac{d}{dt} p \right\rangle + \lambda \mathrm{tr}\left[\mathrm{sym}(\nabla u - p)\right] \mathrm{tr}\left[-\frac{d}{dt} p\right]$$

$$+ 2\mu\, H_0 \left\langle \mathrm{dev\, sym}\, p, \mathrm{dev\, sym}\, \frac{d}{dt} p \right\rangle + \mu\, L_c^2 \left\langle \mathrm{Curl}\, p, \mathrm{Curl}\, \frac{d}{dt} p \right\rangle dV$$

$$= \int_{\mathcal{V}} 2\mu \left\langle \mathrm{sym}(\nabla u - p), -\frac{d}{dt} p \right\rangle + \lambda \mathrm{tr}\left[\mathrm{sym}(\nabla u - p)\right] \left\langle \mathbb{1}, -\frac{d}{dt} p \right\rangle$$

$$- 2\mu\, H_0 \left\langle \mathrm{dev\, sym}\, p, -\frac{d}{dt} p \right\rangle + \mu\, L_c^2 \left\langle \mathrm{Curl}\, p, \mathrm{Curl}\, \frac{d}{dt} p \right\rangle dV$$

$$= \int_{\mathcal{V}} \left\langle 2\mu\, \mathrm{sym}(\nabla u - p) + \lambda \mathrm{tr}\left[\nabla u - p\right] \mathbb{1} - 2\mu\, H_0\, \mathrm{dev\, sym}\, p, -\frac{d}{dt} p \right\rangle$$

$$+ \mu\, L_c^2 \left\langle \mathrm{Curl}[\mathrm{Curl}\, p], \frac{d}{dt} p \right\rangle + \underbrace{\sum_{i=1}^{3} \mathrm{Div}\, \mu\, L_c^2 \left(\frac{d}{dt} p^i \times (\mathrm{curl}\, p)^i\right)}_{\text{"extra energy flux"}\; q(p_t, \mathrm{Curl}\, p)} dV. \quad (3)$$

Here and in the following, p^i denotes the i.th row of the matrix p.[2] Choosing constitutively as extra energy flux

$$q^i = \mu\, L_c^2 \left(\frac{d}{dt} p^i \times (\mathrm{curl}\, p)^i\right), \quad i = 1, 2, 3, \quad (4)$$

shows that the extended (nonlocal) form of the reduced dissipation inequality at constant temperature [8] may be evaluated as follows

$$0 \geq \int_{\mathcal{V}} \frac{d}{dt} W(\nabla u(x, t_0), p(x, t), \mathrm{Curl}\, p(x, t)) - \mathrm{Div}\, q(p_t, \mathrm{Curl}\, p)\, dV$$

---

[1] The notation $\varepsilon_p \in \mathrm{Sym}(3)$ is reserved to the purely local theory and the irrotational theory.
[2] The extra energy flux term is needed to account for the possible nonlocal exchange of energy across $\partial \mathcal{V}$.

$$= \int_V \left\langle 2\mu \operatorname{sym}(\nabla u - p) + \lambda \operatorname{tr}[\nabla u - p]\mathbb{1} - 2\mu H_0 \operatorname{dev} \operatorname{sym} p, -\frac{d}{dt}p \right\rangle$$
$$+ \mu L_c^2 \left\langle \operatorname{Curl}[\operatorname{Curl} p], \frac{d}{dt}p \right\rangle dV$$
$$= \int_V \left\langle \underbrace{2\mu \operatorname{sym}(\nabla u - p) + \lambda \operatorname{tr}[\nabla u - p]\mathbb{1} - 2\mu H_0 \operatorname{dev} \operatorname{sym} p - \mu L_c^2 \operatorname{Curl}[\operatorname{Curl} p]}_{=:\Sigma}, -\frac{d}{dt}p \right\rangle dV$$
$$= \int_V \left\langle \sigma - 2\mu H_0 \operatorname{dev} \operatorname{sym} p - \mu L_c^2 \operatorname{Curl}[\operatorname{Curl} p], -\frac{d}{dt}p \right\rangle dV, \tag{5}$$

where Σ is the *linearized Eshelby stress tensor* in disguise which is the driving force for the plastic evolution. Taking

$$\frac{d}{dt}p = \mathfrak{f}(\Sigma), \tag{6}$$

where the function $\mathfrak{f} : \mathbb{M}^{3\times 3} \mapsto \mathbb{M}^{3\times 3}$ with $\mathfrak{f}(0) = \{0\}$ satisfies the *monotonicity in zero* condition

$$\forall; \Sigma \in \mathbb{M}^{3\times 3}: \quad \langle \mathfrak{f}(\Sigma) - \mathfrak{f}(0), \Sigma - 0 \rangle = \langle \mathfrak{f}(\Sigma), \Sigma \rangle \geq 0, \tag{7}$$

ensures the correct sign in (5) (positive dissipation) and thus the plastic evolution law (6) is thermodynamically admissible. In the large scale limit $L_c = 0$ this is just the *class of pre-monotone type* defined by Alber [1]. The driving term Σ has the dimension of stress and $\operatorname{Div}(p_t \times \operatorname{Curl} p) = 0$ for purely elastic processes $p_t \equiv 0$.

In the case of associated plasticity the function $\mathfrak{f}$ may be obtained as subdifferential $\partial \chi$ of a convex function χ. To this end, let us define the elastic domain in stress-space $K := \{\Sigma \in \mathbb{M}^{3\times 3} \mid \|\operatorname{dev} \Sigma\| \leq \sigma_y \}$ with yield stress σ_y, corresponding indicator function

$$\chi(\Sigma) = \begin{cases} 0 & \|\operatorname{dev} \Sigma\| \leq \sigma_y \\ \infty & \text{else}, \end{cases} \tag{8}$$

and subdifferential in the sense of convex analysis

$$\partial \chi(\Sigma) = \begin{cases} 0 & \|\operatorname{dev} \Sigma\| < \sigma_y \\ \mathbb{R}_0^+ \frac{\operatorname{dev} \Sigma}{\|\operatorname{dev} \Sigma\|} & \|\operatorname{dev} \Sigma\| = \sigma_y \\ \emptyset & \|\operatorname{dev} \Sigma\| > \sigma_y. \end{cases} \tag{9}$$

Choosing dev Σ *instead of* dev sym Σ *in* (8) *allows for plastic spin.*

The remaining divergence term which has to be evaluated in order for an *a priori* global energy inequality

$$\int_\Omega \frac{d}{dt} W(\nabla u(x,t_0), p(x,t), \operatorname{Curl} p(x,t)) \, dV \leq 0 \tag{10}$$

to hold over the entire body is given by the *global insulation condition*

$$\int_\Omega \sum_{i=1}^{3} \text{Div}\left(\frac{d}{dt}p^i \times (\text{curl } p)^i\right) dV = \int_{\partial\Omega} \sum_{i=1}^{3} \left\langle \frac{d}{dt}p^i \times (\text{curl } p)^i, \mathbf{n} \right\rangle dS = 0. \tag{11}$$

The last condition is satisfied, e.g., if in each point of the boundary $\partial\Omega$ the *localized insulation condition* holds, i.e.,

$$0 = \left\langle \frac{d}{dt}p^i \times (\text{curl } p)^i, \mathbf{n} \right\rangle, \quad x \in \partial\Omega, \quad i = 1, 2, 3, \tag{12}$$

which may be satisfied by postulating[3]

$$p(x,t).\tau = p(x,0).\tau, \quad x \in \Gamma_D \quad \left(\Rightarrow \frac{d}{dt}p(x,t).\tau = 0\right),$$
$$\text{Curl } p(x,t).\tau = 0, \quad x \in \partial\Omega \setminus \Gamma_D. \tag{13}$$

3 Strong Infinitesimal Gradient Plasticity with Plastic Spin

The infinitesimal strain gradient plasticity model reads now: find

$$u \in H^1([0,T]; H_0^1(\Omega, \Gamma_D, \mathbb{R}^3)), \quad \text{sym } p \in H^1([0,T]; L^2(\Omega, \mathfrak{sl}(3)),$$
$$\text{Curl } p(t) \in L^2(\Omega, \mathbb{M}^{3\times 3}), \quad \text{dev Curl Curl } p(t) \in L^2(\Omega, \mathbb{M}^{3\times 3}), \tag{14}$$

such that

$$\text{Div } \sigma = -f, \quad \sigma = 2\mu \, \text{sym}(\nabla u - p) + \lambda \, \text{tr}[\nabla u - p] \, \mathbb{1},$$
$$\dot{p} \in \partial \chi(\Sigma^{\text{lin}}), \quad \Sigma^{\text{lin}} = \Sigma_e^{\text{lin}} + \Sigma_{\text{sh}}^{\text{lin}} + \Sigma_{\text{curl}}^{\text{lin}},$$
$$\Sigma_e^{\text{lin}} = 2\mu \, \text{sym}(\nabla u - p) + \lambda \, \text{tr}[\nabla u - p] \mathbb{1} = \sigma, \tag{15}$$
$$\Sigma_{\text{sh}}^{\text{lin}} = -2\mu \, H_0 \, \text{dev sym } p, \quad \Sigma_{\text{curl}}^{\text{lin}} = -\mu \, L_c^2 \, \text{Curl(Curl } p),$$
$$u(x,t) = u_d(x), \quad p(x,t).\tau = p(x,0).\tau, \quad x \in \Gamma_D,$$
$$0 = [\text{Curl } p(x,t)].\tau, \quad x \in \partial\Omega \setminus \Gamma_D, \quad p(x,0) = p^0(x).$$

In general, $\Sigma_{\text{curl}}^{\text{lin}}$ is *not symmetric even if p is symmetric*. Thus, the plastic inhomogeneity is responsible for the plastic spin contribution in this rotationally

---

[3] It is not immediately obvious how a boundary condition on p at Γ_D can be posed. In Gurtin [6, 2.17] it is shown that the *microscopically hard condition* $\dot{p}.\tau_{|\Gamma_D} = 0$ has a precise physical meaning: there is no flow of the Burgers vector across the boundary Γ_D.

invariant formulation. Since $\partial \mathcal{X}$ is monotone, the formulation is thermodynamically admissible. This remains true if we replace $\partial \mathcal{X}$ with a general flow function $\mathfrak{f} : \mathbb{M}^{3\times 3} \mapsto \mathbb{M}^{3\times 3}$ which is only pre-monotone. The mathematically suitable space for symmetric p is the classical space $H_{\text{curl}}(\Omega) := \{v \in L^2(\Omega), \ \text{Curl } v \in L^2(\Omega)\}$. The boundary conditions on the plastic distortion p serve only the purpose to fix ideas.

In the large scale limit $L_c \to 0$ we recover a classical elasto-plasticity model with local kinematic hardening of Prager-type. Observe that the term $\Sigma^{\text{lin}}_{\text{curl}} = -\mu L_c^2 \, \text{Curl}(\text{Curl } p)$ acts as *nonlocal kinematical backstress* and constitutes a crystallographically motivated alternative to merely phenomenologically motivated backstress tensors. The term $-2\mu H_0 \, \text{dev sym } p$ is a *symmetric local kinematical backstress*. The model is therefore able to represent linear kinematic hardening[4] and Bauschinger-like phenomena. Moreover, the driving stress Σ is non-symmetric due to the presence of the second order gradients, while the local contribution σ, due to elastic lattice strains, remains symmetric.

Additionally, the infinitesimal local contributions are fully rotationally invariant (isotropic and objective) with respect to the transformation $(\nabla u, p) \mapsto (\nabla u + A(x), p + A(x))$ and the nonlocal dislocation potential is still invariant with respect to the infinitesimal rigid transformation $(\nabla u, p) \mapsto (\nabla u + \overline{A}, p + \overline{A})$ where $\overline{A}, A(x) \in \mathfrak{so}(3)$.

4 Uniqueness of Strong Solutions

Assume that strong solutions to the model (15) exist. I will show that these solutions are already unique. The aim of this paragraph is, moreover, to study the influence of the different boundary conditions for the plastic distortion p on the possible uniqueness. In that way it is intended to identify the weakest boundary condition which suffices for uniqueness. Possible boundary conditions (which are sufficient for the global insulation condition) are

$$\text{pure micro-free:} \quad \text{Curl } p.\tau = 0, \quad x \in \partial\Omega,$$

$$\text{micro-hard/free:} \quad \begin{cases} \text{Curl } p.\tau = 0, & x \in \partial\Omega \setminus \Gamma_D \quad \text{micro-free} \\ \dot{p}.\tau = 0, & x \in \Gamma_D \quad \text{micro-hard} \end{cases}$$

$$\text{spin micro-hard/free:} \quad \begin{cases} \text{Curl } p.\tau = 0, & x \in \partial\Omega \quad \text{micro-free} \\ [\text{skew } \dot{p}].\tau = 0, & x \in \Gamma_D \quad \text{spin micro-hard} \end{cases}$$

---

[4] Purely phenomenological Prager linear kinematic hardening can also be written as the system

$$\dot{\varepsilon}_p \in \partial \mathcal{X}(\sigma - b), \quad \dot{b} = 2\mu H_0 \dot{\varepsilon}_p, \tag{16}$$

with b the symmetric backstresss tensor and $H_0 > 0$ the constant hardening modulus. Assuming $b(x, 0) = 2\mu H_0 \varepsilon_p(x, 0)$ and integration yields the format given in (15).

pure micro-hard: $\dot{p}.\tau = 0$, $\quad x \in \partial\Omega$,

global insulation condition: $\int_{\partial\Omega} \sum_{i=1}^{3} \langle \dot{p}^i \times (\operatorname{curl} p)^i, \mathbf{n} \rangle \, dS = 0$. (17)

We note that the *global insulation condition is not additively stable*, i.e., the difference of two solutions $p_1 - p_2$ which satisfy each individually the insulation condition need not satisfy the insulation condition. Thus the global insulation condition is not a good candidate for establishing uniqueness.[5]

Here we follow closely the uniqueness proof given in [1, p.32], using the a priori energy estimate and the monotonicity for the difference of two solutions. We allow in this part the generality of a monotone flow function $\mathfrak{f}$ instead of $\partial\chi$. Assume that two strong solutions (u_1, p_2) and (u_2, p_2) of (15) exist (satisfying the same boundary and initial conditions), notably

$$\sigma_1.\mathbf{n} = \sigma_2.\mathbf{n} = 0, \quad x \in \partial\Omega \setminus \Gamma_D,$$
$$u_1 = u_2 = u_d, \quad x \in \Gamma_D.$$
(18)

Insert the difference of the solutions into the total energy W, integrate over Ω and consider the time derivative

$$\frac{d}{dt} \int_\Omega W(\nabla(u_1 - u_2), p_1 - p_2, \operatorname{Curl}(p_1 - p_2)) \, dV$$
$$= \int_\Omega \langle DW_e^{\operatorname{lin}}(\nabla(u_1 - u_2), p_1 - p_2), \nabla \dot{u}_1 - \nabla \dot{u}_2 \rangle$$
$$- \langle DW_e^{\operatorname{lin}}(\nabla(u_1 - u_2), p_1 - p_2), \dot{p}_1 - \dot{p}_2 \rangle$$
$$+ \langle DW_{\operatorname{ph}}^{\operatorname{lin}}(p_1 - p_2), \dot{p}_1 - \dot{p}_2 \rangle$$
$$+ \langle DW_{\operatorname{curl}}^{\operatorname{lin}}(\operatorname{Curl}(p_1 - p_2)), \operatorname{Curl} \frac{d}{dt}(p_1 - p_2) \rangle dV$$
$$= \int_\Omega \langle \sigma(\nabla(u_1 - u_2), p_1 - p_2), \nabla \dot{u}_1 - \nabla \dot{u}_2 \rangle$$
$$- \langle \sigma(\nabla(u_1 - u_2), p_1 - p_2), \dot{p}_1 - \dot{p}_2 \rangle$$
$$+ \langle 2\mu H_0 \operatorname{dev} \operatorname{sym}(p_1 - p_2), \dot{p}_1 - \dot{p}_2 \rangle$$
$$+ \langle \mu L_c^2 \operatorname{Curl}(p_1 - p_2), \operatorname{Curl} \frac{d}{dt}(p_1 - p_2) \rangle dV$$
$$= - \int_\Omega \langle \underbrace{\operatorname{Div} \sigma(\nabla(u_1 - u_2), p_1 - p_2)}_{=0}, \dot{u}_1 - \dot{u}_2 \rangle dV$$

---
[5] In the spirit of Gurtin [5] the insulation condition is motivated by imposing boundary conditions that result in a "null expenditure of microscopic power".

$$+ \underbrace{\int_{\partial\Omega} \langle \sigma(\nabla(u_1 - u_2), p_1 - p_2).\mathbf{n}, (u_1 - u_2)_t \rangle \, dS}_{=0 \text{ with (18)}}$$

$$- \langle \sigma(\nabla(u_1 - u_2), p_1 - p_2), \dot{p}_1 - \dot{p}_2 \rangle$$

$$+ \int_{\Omega} \langle 2\mu \, H_0 \, \text{dev sym}(p_1 - p_2), \dot{p}_1 - \dot{p}_2 \rangle$$

$$+ \left\langle \mu \, L_c^2 \, \text{Curl}(p_1 - p_2), \text{Curl} \frac{d}{dt}(p_1 - p_2) \right\rangle dV$$

$$= -0 + 0 + \int_{\Omega} \langle 2\mu \, H_0 \, \text{dev sym}(p_1 - p_2), \dot{p}_1 - \dot{p}_2 \rangle$$

$$- \langle \sigma(\nabla(u_1 - u_2), p_1 - p_2), \dot{p}_1 - \dot{p}_2 \rangle$$

$$+ \left\langle \mu \, L_c^2 \, \text{Curl Curl}(p_1 - p_2), \frac{d}{dt}(p_1 - p_2) \right\rangle dV$$

$$= \int_{\Omega} \left\langle \Sigma_2^{\text{lin}} - \Sigma_1^{\text{lin}}, \dot{p}_1 - \dot{p}_2 \right\rangle dV = - \int_{\Omega} \left\langle \Sigma_2^{\text{lin}} - \Sigma_1^{\text{lin}}, \dot{p}_2 - \dot{p}_1 \right\rangle dV$$

$$= - \int_{\Omega} \left\langle \Sigma_2^{\text{lin}} - \Sigma_1^{\text{lin}}, \mathfrak{f}(\Sigma_2^{\text{lin}}) - \mathfrak{f}(\Sigma_1^{\text{lin}}) \right\rangle dV \leq 0, \qquad (19)$$

due to the monotonicity of $\mathfrak{f}$. Hence, after integrating the last inequality in time we obtain also for the difference of two solutions

$$\int_{\Omega} W(\nabla(u_1 - u_2)(t)), (p_1 - p_2)(t), \text{Curl}(p_1 - p_2)(t)) \, dV$$

$$\leq \int_{\Omega} W(\nabla(u_1 - u_2)(0)), (p_1 - p_2)(0), \text{Curl}(p_1 - p_2)(0)) \, dV = 0. \qquad (20)$$

Thus we have

$$\int_{\Omega} \| \text{sym}(\nabla(u_1 - u_2)(t) - (p_1 - p_2)(t) \|^2 \, dV = 0,$$

$$\int_{\Omega} \| \text{dev sym}(p_1 - p_2)(t) \|^2 \, dV = 0,$$

$$\int_{\Omega} \| \text{Curl}(p_1 - p_2)(t) \|^2 \, dV = 0. \qquad (21)$$

Since $p_1, p_2 \in \mathfrak{sl}(3)$ it follows that $\text{sym}(p_1 - p_2) = 0$ almost everywhere, i.e., $p_1 - p_2 \in \mathfrak{so}(3)$. Moreover, from the micro-hard boundary condition $\dot{p}_1.\tau = \dot{p}_2.\tau = 0$ we obtain $p_1(x,t).\tau = p_2(x,t).\tau = p(x,0).\tau$ which implies that $(p_1 - p_2).\tau = 0$ on Γ_D for two linear independent tangential directions τ. Since a skew-symmetric matrix $A \in \mathfrak{so}(3)$ has either rank two or rank zero (in which case it is zero) we conclude that $p_1 - p_2 = 0$ on the Dirichlet-boundary Γ_D due to the skew-symmetry of the difference. However Curl controls all first partial derivatives on skew-symmetric matrices [12], i.e. it holds locally

$$\forall\, A(x) \in \mathfrak{so}(3): \quad \|\operatorname{Curl} A(x)\|^2 \geq \frac{1}{2}\|\nabla A(x)\|^2, \tag{22}$$

therefore $p_1 - p_2 = 0$ by Poincaré's inequality. Thus from Korn's first inequality we obtain uniqueness also for he displacement u. □

As a result: apart for the pure micro-free condition and the global insulation condition all mentioned boundary conditions in (17) ensure uniqueness of classical solutions. In the case of the pure micro-free condition, the skew-symmetric part of the difference of two solutions remains indetermined up to a constant skew-symmetric matrix. The spin micro-hard condition has the advantage of not imposing a Dirichlet boundary condition on the symmetric plastic strain tensor sym p which is also not present in the classical theory. Thus it is a candidate for the desired weakest boundary condition.

5 Irrotational Strong Infinitesimal Gradient Plasticity

For completeness let us also give the infinitesimal strain gradient plasticity system without plastic spin which is included as a special case in (15). To this end we specify the elastic domain in stress-space $K := \{\Sigma \in \mathbb{M}^{3\times 3} \mid \|\operatorname{dev} \operatorname{sym} \Sigma\| \leq \sigma_y\}$ with yield stress σ_y, corresponding indicator function

$$\chi^{\text{sym}}(\Sigma) = \begin{cases} 0 & \|\operatorname{dev} \operatorname{sym} \Sigma\| \leq \sigma_y \\ \infty & \text{else}, \end{cases} \tag{23}$$

and subdifferential in the sense of convex analysis

$$\partial \chi^{\text{sym}}(\Sigma) = \begin{cases} 0 & \|\operatorname{dev} \operatorname{sym} \Sigma\| < \sigma_y \\ \mathbb{R}_0^+ \frac{\operatorname{dev} \operatorname{sym} \Sigma}{\|\operatorname{dev} \operatorname{sym} \Sigma\|} & \|\operatorname{dev} \operatorname{sym} \Sigma\| = \sigma_y \\ \emptyset & \|\operatorname{dev} \operatorname{sym} \Sigma\| > \sigma_y. \end{cases} \tag{24}$$

Thus, $\partial \chi^{\text{sym}}(\Sigma)$ is symmetric, in which case p will remain symmetric, whenever the initial condition for p is symmetric. Hence, we may rename $\varepsilon_p := \operatorname{sym} p$ in the following. As boundary condition on p we use the candidate for the weakest boundary condition which ensures uniquess, i.e. the spin micro hard condition. It turns out that the local condition on the skew-symmetric part is automatically satisfied. The model reads: find

$$u \in H^1([0,T]; H_0^1(\Omega, \Gamma_D, \mathbb{R}^3)), \quad \varepsilon_p \in H^1([0,T]; L^2(\Omega, \mathfrak{sl}(3)),$$

$$\operatorname{Curl} \varepsilon_p(t) \in L^2(\Omega, \mathbb{M}^{3\times 3}), \quad \operatorname{dev} \operatorname{sym} \operatorname{Curl} \operatorname{Curl} \varepsilon_p(t) \in L^2(\Omega, \mathbb{M}^{3\times 3}), \tag{25}$$

such that

$$\text{Div}\,\sigma = -f\,,\quad \sigma = 2\mu\,(\varepsilon - \varepsilon_p) + \lambda\,\text{tr}\,[\varepsilon]\,1\!\!1\,,$$

$$\dot{\varepsilon}_p \in \partial \chi^{\text{sym}}(\Sigma^{\text{lin}})\,,\quad \Sigma^{\text{lin}} = \Sigma_{\text{e}}^{\text{lin}} + \Sigma_{\text{sh}}^{\text{lin}} + \Sigma_{\text{curl}}^{\text{lin}}\,,$$

$$\Sigma_{\text{e}}^{\text{lin}} = 2\mu\,(\varepsilon - \varepsilon_p) + \lambda\,\text{tr}\,[\varepsilon]\,1\!\!1 = \sigma\,,$$

$$\Sigma_{\text{sh}}^{\text{lin}} = -2\mu\,H_0 \varepsilon_p\,,\quad \Sigma_{\text{curl}}^{\text{lin}} = -\mu\,L_c^2\,\text{Curl}(\text{Curl}\,\varepsilon_p)\,,$$

$$u(x,t) = u_{\text{d}}(x)\,,\quad 0 = [\text{Curl}\,\varepsilon_p(x,t)].\tau\,,\quad x \in \partial\Omega\,,$$

$$\varepsilon_p(x,0) = \varepsilon_p^0(x) \in \text{Sym}(3) \cap \mathfrak{sl}(3)\,. \tag{26}$$

Again, classical solutions, if they exist, are unique. For this result to hold the higher order boundary conditions on ε_p are not needed!

6 Discussion

The classical elasto-perfectly plastic Prandtl–Reuss model with kinematic hardening has been extended to include a weak nonlocal interaction of the plastic distortion by introducing the dislocation density in the Helmholtz free energy. The evolution equation for plasticity follows by an application of the secod law of thermodynamics in the formulation proposed by Maugin [8] together with sufficient conditions guaranteeing the insulation condition.

With Gurtin and Anand [5] on gradient plasticity I can say: "Our goal is a theory that allows for constitutive dependencies on (the dislocation density tensor) G, but that otherwise does not depart drastically from the classical theory." This has been achieved, since

- The *large scale limit* $L_c \to 0$ with *zero local hardening* $H_0 = 0$ does coincide with the classical *Prandtl–Reuss* model with deviatoric von Mises flow rule.
- The *large scale limit* $L_c \to 0$ does determine the plastic distortion to be *irrotational*, i.e., only $\varepsilon_p := \text{sym}\,p$ appears (zero plastic spin).
- A weak reformulation of the model for $L_c > 0$ is *well-posed*. Existence and uniqueness are obtained in suitable Hilbert-spaces [11]. Uniqueness of classical solutions is also guaranteed.
- The model for $L_c > 0$ does contain *maximally second order derivatives* in the evolution law.
- The model for $L_c > 0$ is *linearized materially and spatially covariant* and *thermodynamically consistent* (in the extended sense).
- The model for $L_c > 0$ is *isotropic* with respect to both, the *referential and intermediate configuration*.
- The model for $L_c > 0$ does contain *first order boundary conditions at the hard Dirichlet boundary* $\Gamma_D \subset \partial\Omega$ for the plastic distortion p only in terms of the plastic spin skew p there.

- The symmetric plastic strains $\varepsilon_p := \text{sym}\, p$ remain free of first order (essential Dirichlet) boundary conditions as in classical elasto-plasticity.
- The model for $L_c > 0$ does contain *second order boundary conditions* on p like Curl $p.\tau = 0$ at the total external boundary $\partial\Omega$, motivated from thermodynamics and insulation conditions.

The proposed gradient plasticity model approximates formally the classical model in the large scale limit $L_c = 0$ since then the plastic distortion p remains symmetric and no boundary conditions are set. Plastic spin is purely a feature of the nonlocality of the model. Summarizing, for the elasto-plastic infinitesimal strain gradient model with spin the following has been obtained: uniqueness of strong solutions with micro-free/hard boundary conditions.

Currently, the dislocation based plasticity model is being implemented, however, only for the irrotational case (26) without plastic spin [13]. There, boundary conditions on the symmetric plastic strain ε_p need not be imposed.

Appendix: Notation

Let $\Omega \subset \mathbb{R}^3$ be a bounded domain with Lipschitz boundary $\partial\Omega$ and let Γ be a smooth subset of $\partial\Omega$ with non-vanishing 2-dimensional Hausdorff measure. We denote by $\mathbb{M}^{3\times 3}$ the set of real 3×3 second order tensors, written with capital letters. The standard Euclidean scalar product on $\mathbb{M}^{3\times 3}$ is given by $\langle X, Y\rangle_{\mathbb{M}^{3\times 3}} = \text{tr}\left[XY^T\right]$, and thus the Frobenius tensor norm is $\|X\|^2 = \langle X, X\rangle_{\mathbb{M}^{3\times 3}}$ (we use these symbols indifferently for tensors and vectors). The identity tensor on $\mathbb{M}^{3\times 3}$ will be denoted by $\mathbb{1}$, so that $\text{tr}[X] = \langle X, \mathbb{1}\rangle$. We let Sym and PSym denote the symmetric and positive definite symmetric tensors respectively. We adopt the usual abbreviations of Lie-algebra theory, i.e. $\mathfrak{so}(3) := \{X \in \mathbb{M}^{3\times 3}\,|\, X^T = -X\}$ are skew symmetric second order tensors and $\mathfrak{sl}(3) := \{X \in \mathbb{M}^{3\times 3}\,|\, \text{tr}[X] = 0\}$ are traceless tensors. We set $\text{sym}(X) = \frac{1}{2}(X^T + X)$ and $\text{skew}(X) = \frac{1}{2}(X - X^T)$ such that $X = \text{sym}(X) + \text{skew}(X)$. For $X \in \mathbb{M}^{3\times 3}$ we set for the deviatoric part $\text{dev}\, X = X - \frac{1}{3}\,\text{tr}[X]\,\mathbb{1} \in \mathfrak{sl}(3)$. For a second order tensor X we let $X.e_i$ be the application of the tensor X to the column vector e_i and X^i denotes the i.th row of X. The curl of a three by three matrix is defined to be the vector curl applied on the i.th row, written in the i.th row, i.e.,

$$\text{curl}\begin{pmatrix} p^{11} & p^{12} & p^{13} \\ p^{21} & p^{22} & p^{23} \\ p^{31} & p^{32} & p^{33} \end{pmatrix} = \begin{pmatrix} \text{curl}[p^{11}, p^{12}, p^{13}] \\ \text{curl}[p^{21}, p^{22}, p^{23}] \\ \text{curl}[p^{31}, p^{32}, p^{33}] \end{pmatrix}.$$

References

1. H.D. Alber. *Materials with Memory. Initial-Boundary Value Problems for Constitutive Equations with Internal Variables*, Lecture Notes in Mathematics, Vol. 1682. Springer, Berlin, 1998.
2. N.A. Fleck, G.M. Müller, M.F. Ashby, and J.W. Hutchinson. Strain gradient plasticity: Theory and experiment. *Acta. Metall. Mater.*, 42(2):475–487, 1994.
3. P. Gudmundson. A unified treatment of strain gradient plasticity. *J. Mech. Phys. Solids*, 52:1379–1406, 2004.
4. M.E. Gurtin. On the plasticity of single crystals: Free energy, microforces, plastic-strain gradients. *J. Mech. Phys. Solids*, 48:989–1036, 2000.
5. M.E. Gurtin and L. Anand. A theory of strain-gradient plasticity for isotropic, plastically irrotational materials. Part I: Small deformations. *J. Mech. Phys. Solids*, 53:1624–1649, 2005.
6. M.E. Gurtin and A. Needleman. Boundary conditions in small-deformation, single crystal plasticity that account for the Burgers vector. *J. Mech. Phys. Solids*, 53:1–31, 2005.
7. E. Kröner and G. Rieder. Kontinuumstheorie der Versetzungen. *Z. Phys.*, 145:424–429, 1956.
8. G. A. Maugin. *The Thermomechanics of Nonlinear Irreversible Behaviors.* Nonlinear Science, Vol. 27. World Scientific, Singapore, 1999.
9. A. Menzel and P. Steinmann. On the continuum formulation of higher gradient plasticity for single and polycrystals. *J. Mech. Phys. Solids*, 48:1777–1796, Erratum 49, (2001), 1179–1180, 2000.
10. P. Neff. Remarks on invariant modelling in finite strain gradient plasticity. *Technische Mechanik*, 28(1):13–21, 2008.
11. P. Neff, K. Chełmiński, and H.D. Alber. Notes on strain gradient plasticity. Finite strain covariant modelling and global existence in the infinitesimal rate-independent case. *Math. Mod. Meth. Appl. Sci. (M3AS)*, 19(2), 2009, to appear.
12. P. Neff and I. Münch. Curl bounds Grad on SO(3). *ESAIM: Control, Optimisation and Calculus of Variations*, 14(1):148–159, 2008, published online, DOI: 10.1051/cocv:2007050.
13. P. Neff, A. Sydow, and C. Wieners. Numerical approximation of incremental infinitesimal gradient plasticity. Preprint IWRM 08/01, http://www.mathematik.uni-karlsruhe.de/iwrmm/seite/preprints/media, *Int. J. Num. Meth. Engrg.*, 2008, to appear, DOI: 10.1002/nme.2420.
14. B.D. Reddy, F. Ebobisse, and A.T. McBride. Well-posedness of a model of strain gradient plasticity for plastically irrotational materials. *Int. J. Plasticity*, 24:55–73, 2008.

Algorithms and
Computational Aspects

SQP Methods for Incremental Plasticity with Kinematic Hardening

Christian Wieners

*Universität Karlsruhe, Institut für Angewandte und Numerische Mathematik,
Englerstr. 2, 76128 Karlsruhe, Germany
E-mail: wieners@math.uni-karlsruhe.de*

Abstract. We introduce a SQP method for the solution of the dual problem in incremental infinitesimal plasticity. We show that every SQP step corresponds to a primal plasticity problem with linearized flow rule. Moreover, we show the equivalence to a uniformly convex minimization problem which can be solved by a semi-smooth Newton method. The performance of the method is demonstrated by a 3-d benchmark example.

Key words: computational plasticity, SQP methods.

1 Introduction

The radial return together with a consistent linearization is the standard solution procedure for incremental plasticity (see, e.g., [10]). Although this class of algorithm is very general (with variants for a broad variety of plasticity models) and in most cases also efficient and reliable, the convergence analysis is restricted to simple situations, see, e.g., [1, 2].

Algorithms in mathematical programming have a long tradition in computational plasticity (see, e.g., [7]), in particular for the solution of local optimization problem for the internal variables. In the last years algorithms of numerical optimization where transferred to the full variational problem of incremental plasticity: semi-smooth Newton methods [3]), interior-point algorithms [6], and SQP methods [12].

In particular SQP methods appear to have a structural advantage: the SQP iterates $\{\sigma^k\}$ are a minimizing sequence for the dual minimization problem in incremental plasticity, whereas the iterates $\{u^k\}$ of the radial return method minimize a suitable primal functional. Since the dual problem for the stress is uniformly convex (which is not the case for the primal displacement problem), we expect (and indeed observe in examples) at least asymptotically better convergence for the SQP method. Nevertheless, in the application to perfect plasticity the solution of the quadratic problem in the single SQP step remains difficult and no uniform bounds are available.

In this contribution we extent the SQP method for perfect plasticity (presented in [12]) to a plasticity model with hardening. Then, hardening adds some regularity to the quadratic problem in the single SQP step, so that we now can prove global convergence for the semi-smooth Newton method which is used to solve this quadratic minimization problem with linearized constraints. The new algorithm is derived by a linearization of the flow rule which then leads to a sequence of linear variational problems with linear inequality constraints. The purpose of this paper is to show that this subproblem is equivalent to a quadratic minimization problem and thus the method is indeed equivalent to the SQP method.

2 Discrete Plasticity with Kinematic Hardening

For simplicity of the presentation, we discuss here only the case for kinematic hardening, where the back stress is given by $\boldsymbol{\beta} = H_0 \boldsymbol{\varepsilon}_p$. We restrict ourselves to the discrete model (see, e.g., [4] for the analysis of the continuous model).

Data. Let $\Omega \subset \mathbf{R}^3$ be the reference configuration, and let $\Gamma_D \cup \Gamma_N = \partial \Omega$ be a decomposition of the boundary. Let $[0, T]$ be a fixed time interval. We prescribe a displacement vector $\boldsymbol{u}_D(t)$ for the essential boundary conditions on Γ_D and a load functional

$$\ell(t, \delta \boldsymbol{u}) = \int_\Omega \boldsymbol{b}(t) \cdot \delta \boldsymbol{u} \, dx + \int_{\Gamma_N} \boldsymbol{t}_N(t) \cdot \delta \boldsymbol{u} \, da$$

depending on body force densities $\boldsymbol{b}(t)$ and traction force densities $\boldsymbol{t}_N(t)$.

Let $\mathrm{Sym}(3) = \{\boldsymbol{\tau} \in \mathbf{R}^{3,3} : \boldsymbol{\tau}^T = \boldsymbol{\tau}\}$ be the set of symmetric matrices. The elastic material properties (in the infinitesimal model) are determined by the isotropic elasticity tensor $\mathbb{C}: \mathrm{Sym}(3) \longrightarrow \mathrm{Sym}(3)$ defined by $\mathbb{C} : \boldsymbol{\varepsilon} = 2\mu \boldsymbol{\varepsilon} + \lambda \, \mathrm{trace}(\boldsymbol{\varepsilon}) \boldsymbol{I}$, depending on the Lamé constants $\lambda, \mu > 0$. On $\mathrm{Sym}(3)$, the elasticity tensor is symmetric and positive definite.

Plasticity is described by the yield function $\phi(\boldsymbol{\alpha}) = |\mathrm{dev}(\boldsymbol{\alpha})| - K_0$ depending on the yield stress $K_0 > 0$, where $\mathrm{dev}(\boldsymbol{\alpha}) = \boldsymbol{\alpha} - \frac{1}{3} \mathrm{trace}(\boldsymbol{\alpha}) \boldsymbol{I}$ is the deviatoric part of the stress. Since the yield function is convex, its derivative $D\phi$ is monotone, and $D^2\phi$ is positive semi-definite with

$$D\phi(\boldsymbol{\alpha}) = \frac{\mathrm{dev}(\boldsymbol{\alpha})}{|\mathrm{dev}(\boldsymbol{\alpha})|}, \quad D^2\phi(\boldsymbol{\alpha}) : \boldsymbol{\eta} = \frac{\mathrm{dev}(\boldsymbol{\eta})}{|\mathrm{dev}(\boldsymbol{\alpha})|} - \frac{\mathrm{dev}(\boldsymbol{\alpha}) : \mathrm{dev}(\boldsymbol{\eta})}{|\mathrm{dev}(\boldsymbol{\alpha})|^2} \frac{\mathrm{dev}(\boldsymbol{\alpha})}{|\mathrm{dev}(\boldsymbol{\alpha})|}$$

for $\boldsymbol{\alpha} \neq \boldsymbol{0}$. Finally, let $H_0 > 0$ be the kinematic hardening modulus.

Discretization in space. Let $V \subset C^{0,1}(\overline{\Omega}, \mathbf{R}^3)$ be a finite element space spanned by nodal basis functions. Let

$$V(\boldsymbol{u}_D) = \{\boldsymbol{v} \in V : \boldsymbol{v}(\boldsymbol{x}) = \boldsymbol{u}_D(\boldsymbol{x}) \text{ for } \boldsymbol{x} \in D\},$$

where $D \subset \Gamma_D$ is the set of all nodal points on Γ_D.

Let $\Xi \subset \Omega$ be quadrature points. We define the discrete spaces $\Lambda = \{\lambda \colon \Xi \longrightarrow \mathbb{R}\}$ for the return parameter, and $S = \{\eta \colon \Xi \longrightarrow \mathrm{Sym}(3)\}$ for stresses and strains. This gives $\varepsilon(v) \in S$ by $\varepsilon(v)(\xi) = \mathrm{sym}(Dv(\xi))$ for $v \in V$. The plastic strain is approximated in $E = \{\eta \colon \Xi \longrightarrow \mathrm{Sym}(3), \mathrm{trace}(\eta) = 0\}$.

The incremental model. Let $0 = t_0 < t_1 < \cdots < t_N = T$ be a time series. The backward Euler scheme reads as follows. For $n = 1, 2, 3, \ldots$ the next increment depends on the material history described by ε_p^{n-1} (starting with $\varepsilon_p^0 = 0$), the new load $\ell_n(\delta u) = \ell(t_n, \delta u)$ and the new Dirichlet boundary values $u_D^n = u_D(t_n)$. We compute the displacement $u^n \in V(u_D^n)$ satisfying the essential boundary conditions, the plastic strain $\varepsilon_p^n \in E$, and the plastic multiplier $\lambda^n \in \Lambda$ satisfying the following system of equations:

the equilibrium equation for the stress $\sigma^n = \mathbb{C} : (\varepsilon(u^n) - \varepsilon_p^n)$

$$\int_\Omega \sigma^n : \varepsilon(\delta u)\, dx = \ell_n(\delta u), \qquad \delta u \in V(0), \tag{1a}$$

the discretized flow rule for the relative stress $\alpha^n = \sigma^n - H_0 \varepsilon_p^n$

$$\varepsilon_p^n = \varepsilon_p^{n-1} + \gamma^n D\phi(\alpha^n), \tag{1b}$$

and the complementarity conditions

$$\gamma^n \phi(\alpha^n) = 0, \quad \gamma^n \geq 0, \quad \phi(\alpha^n) \leq 0. \tag{1c}$$

In the discrete model, (1b) and (1c) are evaluated only in the finite set Ξ.

3 Algorithmic Plasticity

The projection method. The standard approach for solving system (1) is obtained by evaluating (1b) and (1c) independently for every material point. This defines a projection (the "radial return") of the trial stress onto the set of admissible stresses, which is then inserted in (1a).

The linearized projection method. We propose a method which solves $(u^n, \varepsilon_p^n, \gamma^n)$ simultaneously by an iteration $(u^{n,k}, \varepsilon_p^{n,k}, \gamma^{n,k})$ for $k = 1, 2, \ldots$.

We start with $(\varepsilon_p^{n,0}, \gamma^{n,0}) = (\varepsilon_p^{n-1}, \gamma^{n-1})$ and some $u^{n,0} \in V(u_D^n)$. In every step $k \geq 1$ we define the linearized flow rule

$$\phi_{n,k}(\alpha) = \phi(\alpha^{n,k-1}) + D\phi(\alpha^{n,k-1}) : (\alpha - \alpha^{n,k-1}), \tag{2}$$

depending on $\alpha^{n,k-1} = \sigma^{n,k-1} - H_0 \varepsilon_p^{n,k-1}$. Then, the following problem is solved: find $(u^{n,k}, \varepsilon_p^{n,k}, \gamma^{n,k}) \in V(u_D^n) \times E \times \Lambda$ such that for the stress tensor $\sigma^{n,k} = \mathbb{C} :$

$(\varepsilon(u^{n,k}) - \varepsilon_p^{n,k})$ the equilibrium equation

$$\int_\Omega \sigma^{n,k} : \varepsilon(\delta u)\, dx = \ell_n(\delta u), \qquad \delta u \in V(0) \tag{3a}$$

holds, and for the relative stress $\alpha^{n,k} = \sigma^{n,k} - H_0 \varepsilon_p^{n,k}$ the linearized flow rule

$$\varepsilon_p^{n,k} = \varepsilon_p^{n-1} + \gamma^{n,k} D\phi(\alpha^{n,k-1}) + \gamma^{n,k-1} D^2\phi(\alpha^{n,k-1}) : (\alpha^{n,k} - \alpha^{n,k-1}) \tag{3b}$$

and the complementarity conditions

$$\gamma^{n,k} \phi_{n,k}(\alpha^{n,k}) = 0, \quad \gamma^{n,k} \geq 0, \quad \phi_{n,k}(\alpha^{n,k}) \leq 0 \tag{3c}$$

are satisfied.

Note that – due to the complementarity conditions for the linearized flow rule – this system is still nonlinear. Since problem (3) has the same structure as (1), it can also be solved by a return mapping algorithm, where the projection onto the admissible set $\{\phi(\alpha) \leq 0\}$ is now replaced by the projection onto the affine linear half space $\{\phi_{n,k}(\alpha) \leq 0\}$.

Depending on $u^{n,k}$, we introduce the elastic trial stress

$$\theta^{n,k} = \theta_n(u^{n,k}), \qquad \text{where } \theta_n(u) = \mathbb{C} : (\varepsilon(u) - \varepsilon_p^{n-1}) - H_0 \varepsilon_p^{n-1}.$$

Evaluating (3b) and (3c) yields

$$\gamma^{n,k} = \frac{\max\{0, \phi_{n,k}(\theta^{n,k})\}}{2\mu + H_0}$$

for the return parameter and $\alpha^{n,k} = P_{n,k}(\theta^{n,k})$ for the relative stress, where

$$P_{n,k}(\theta) = \theta - \mathbb{A}_{n,k} : \theta - \max\{0, \phi(\theta)\} D\phi(\alpha^{n,k-1})$$

$$\text{with } \mathbb{A}_{n,k} = \frac{(2\mu + H_0)\gamma^{n,k-1} |\mathrm{dev}(\alpha^{n,k-1})|}{(2\mu + H_0)\gamma^{n,k-1} + |\mathrm{dev}(\alpha^{n,k-1})|} D^2\phi(\alpha^{n,k-1})$$

is the projection onto the linearized admissible set (this can be verified by elementary calculus). Rewriting the stress in the form

$$\sigma^{n,k} = c_0 \mathbb{C} : \varepsilon(u^{n,k}) + (1 - c_0)\alpha^{n,k} \qquad \text{with } c_0 = \frac{H_0}{2\mu + H_0}$$

results in the stress response $S_{n,k}(u) = c_0 \mathbb{C} : \varepsilon(u) + (1 - c_0) P_{n,k}(\theta_n(u))$. Inserting the stress into (3a) yields a nonlinear variational equation for the displacement:

Lemma 1. *Let $u^{n,k} \in V(u_D^n)$ be a solution of*

$$\int_\Omega S_{n,k}(\boldsymbol{u}^{n,k}) : \boldsymbol{\varepsilon}(\delta\boldsymbol{u})\,d\boldsymbol{x} = \ell_n(\delta\boldsymbol{u}), \qquad \delta\boldsymbol{u} \in \boldsymbol{V}(\boldsymbol{0}). \tag{4}$$

Then, $\boldsymbol{u}^{n,k}$ together with $\boldsymbol{\varepsilon}_p^{n,k} = \boldsymbol{\varepsilon}(\boldsymbol{u}^{n,k}) - \mathbb{C}^{-1} : S_{n,k}(\boldsymbol{u}^{n,k}) \in E$ and $\gamma^{n,k} = |\boldsymbol{\varepsilon}_p^{n,k} - \boldsymbol{\varepsilon}_p^{n-1}| \in \Lambda$ solves system (3).

The analysis in the next section shows that the nonlinear problem (4) indeed has a unique solution.

More details on this method and the explicit construction of $P_{n,k}$ are explained in [12] for perfect plasticity and in [9] for a nonlocal model.

The consistent tangent operator. The projection $P_{n,k}(\cdot)$ is only nonlinear with respect to the term $\max\{0, \phi_{n,k}(\boldsymbol{\theta})\}$ and therefore a Lipschitz function. Thus, a generalized multi-valued derivative $\partial P_{n,k}(\boldsymbol{\theta})$ exists (see [5] for the definition and for properties of generalized derivatives). Since $\mathrm{sgn}(\max\{0, s\}) \in \partial \max\{0, s\}$ with $\mathrm{sgn}(s) = 0$ for $s = 0$ and $\mathrm{sgn}(s) = 1$ for $s > 0$, a consistent linearization of the stress response $S_{n,k}(\boldsymbol{\theta})$ is therefore obtained by the specific choice

$$\mathbb{C}_{n,k}(\boldsymbol{\theta}) = \mathbb{C} - (1 - c_0)\mathbb{A}_{n,k}$$
$$- (1 - c_0)\,\mathrm{sgn}\left(\max\{0, \phi_{n,k}(\boldsymbol{\theta})\}\right) D\phi(\boldsymbol{\alpha}^{n,k-1}) \otimes D\phi(\boldsymbol{\alpha}^{n,k-1}).$$

This defines now a generalized Newton method for the solution of (4): starting with $\boldsymbol{u}^{n,k,0} = \boldsymbol{u}^{n,k}$, we compute for $m = 1, 2, \ldots$ the residual

$$r_{n,k,m}(\delta\boldsymbol{u}) = \ell_n(\delta\boldsymbol{u}) - \int_\Omega S_{n,k}(\boldsymbol{u}^{n,k,m-1}) : \boldsymbol{\varepsilon}(\delta\boldsymbol{u})\,d\boldsymbol{x}.$$

If the residual is not small enough, we compute the Newton correction $\Delta\boldsymbol{u}^{n,k,m} \in \boldsymbol{V}(\boldsymbol{0})$ solving the linear system

$$\int_\Omega \boldsymbol{\varepsilon}(\Delta\boldsymbol{u}^{n,k,m}) : \mathbb{C}_{n,k}\left(\boldsymbol{\theta}_n(\boldsymbol{u}^{n,k,m-1})\right) : \boldsymbol{\varepsilon}(\delta\boldsymbol{u})\,d\boldsymbol{x} = r_{n,k,m}(\delta\boldsymbol{u}), \quad \delta\boldsymbol{u} \in \boldsymbol{V}(\boldsymbol{0}).$$

Then, we determine a suitable damping factor $\rho_{n,k,m} \in (0, 1]$ and we set

$$\boldsymbol{u}^{n,k,m} = \boldsymbol{u}^{n,k,m-1} + \rho_{n,k,m}\Delta\boldsymbol{u}^{n,k,m}.$$

4 Convergence Analysis

In the first step, we show that the problem is well-defined. Therefore, we define the elliptic bilinear form

$$a\bigl((\boldsymbol{u}, \boldsymbol{\varepsilon}_p), (\boldsymbol{v}, \boldsymbol{\eta})\bigr) = \int_\Omega (\boldsymbol{\varepsilon}(\boldsymbol{u}) - \boldsymbol{\varepsilon}_p) : \mathbb{C} : (\boldsymbol{\varepsilon}(\boldsymbol{v}) - \boldsymbol{\eta})\,d\boldsymbol{x} + H_0 \int_\Omega \boldsymbol{\varepsilon}_p : \boldsymbol{\eta}\,d\boldsymbol{x}.$$

Lemma 2. *A unique minimizer* $(\Delta \boldsymbol{u}^n, \Delta \boldsymbol{\varepsilon}_p^n) \in V(\Delta \boldsymbol{u}_D^n) \times E$ *of the primal functional*

$$\mathcal{J}_n(\Delta \boldsymbol{u}, \Delta \boldsymbol{\varepsilon}_p) = \frac{1}{2} a\big((\Delta \boldsymbol{u}, \Delta \boldsymbol{\varepsilon}_p), (\Delta \boldsymbol{u}, \Delta \boldsymbol{\varepsilon}_p)\big) + a\big((\boldsymbol{u}^{n-1}, \boldsymbol{\varepsilon}_p^{n-1}), (\Delta \boldsymbol{u}, \Delta \boldsymbol{\varepsilon}_p)\big)$$
$$+ K_0 \int_\Omega |\Delta \boldsymbol{\varepsilon}_p| \, d\boldsymbol{x} - \ell_n(\Delta \boldsymbol{u})$$

exists, and the minimizer is characterized by the solution of the system (1), *where* $\boldsymbol{u}^n = \boldsymbol{u}^{n-1} + \Delta \boldsymbol{u}^n$, $\boldsymbol{\varepsilon}_p^n = \boldsymbol{\varepsilon}_p^{n-1} + \Delta \boldsymbol{\varepsilon}_p^n$, *and* $\gamma^n = |\Delta \boldsymbol{\varepsilon}_p^n|$.

Proof. The proof follows [9, Th. 8]. Since the primal functional is uniformly convex in $V(\Delta \boldsymbol{u}_D^n) \times E$, a unique minimizer exists, which is therefore characterized as the unique critical point of $\mathcal{J}_n(\cdot)$. Variation with respect to $\boldsymbol{u}$ yields the equilibrium equation (1a), and variation with respect to $\boldsymbol{\varepsilon}_p$ gives the dissipation inequality in the form

$$K_0 \int_\Omega |\boldsymbol{\eta}| \, d\boldsymbol{x} \geq K_0 \int_\Omega |\Delta \boldsymbol{\varepsilon}_p| \, d\boldsymbol{x} + \int_\Omega \boldsymbol{\alpha}^n : (\boldsymbol{\eta} - \Delta \boldsymbol{\varepsilon}_p) \, d\boldsymbol{x}, \qquad \boldsymbol{\eta} \in E \qquad (5)$$

with $\boldsymbol{\alpha}^n = \mathbb{C} : (\boldsymbol{\varepsilon}(\boldsymbol{u}^n) - \boldsymbol{\varepsilon}_p^n) - H_0 \boldsymbol{\varepsilon}_p^n$. Since (5) is equivalent to the flow rule (1b) and the complementarity condition (1c), the unique critical point of $\mathcal{J}_n(\cdot)$ is a solution of (1). □

In the next step we show that the algorithm presented in the previous section is equivalent to the SQP method applied to the dual minimization problem for the generalized stress $(\boldsymbol{\sigma}, \boldsymbol{\beta}) \in S \times B$. We use for the back stress the space $B = \{\boldsymbol{\eta} : \Xi \longrightarrow \mathrm{Sym}(d), \, \mathrm{trace}(\boldsymbol{\eta}) = 0\}$, and we use the inner products

$$(\boldsymbol{\sigma}, \boldsymbol{\tau})_S = \int_\Omega \boldsymbol{\sigma} : \mathbb{C}^{-1} : \boldsymbol{\tau} \, d\boldsymbol{x}, \qquad \boldsymbol{\sigma}, \boldsymbol{\tau} \in S,$$
$$(\boldsymbol{\beta}, \boldsymbol{\eta})_B = \int_\Omega \frac{1}{H_0} \boldsymbol{\beta} : \boldsymbol{\eta} \, d\boldsymbol{x}, \qquad \boldsymbol{\beta}, \boldsymbol{\eta} \in B$$

and the corresponding norms $\|\cdot\|_S$ and $\|\cdot\|_B$, respectively. The flow function for the generalized stress is given by $\psi(\boldsymbol{\sigma}, \boldsymbol{\beta}) = \phi(\boldsymbol{\sigma} - \boldsymbol{\beta})$.

For simplicity, we restrict ourselves to homogeneous boundary conditions $\boldsymbol{u}_D = \boldsymbol{0}$ on Γ_D (see [12, Th. 6.1] for the extension to non-homogeneous Dirichlet boundary conditions).

Lemma 3. *Let* $(\boldsymbol{u}^n, \boldsymbol{\varepsilon}_p^n, \gamma^n)$ *be the solution of the system* (1). *Then, the generalized stress* $(\boldsymbol{\sigma}^n, \boldsymbol{\beta}^n)$ *with* $\boldsymbol{\sigma}^n = \mathbb{C} : (\boldsymbol{\varepsilon}(\boldsymbol{u}^n) - \boldsymbol{\varepsilon}_p^n)$ *and* $\boldsymbol{\beta}^n = H_0 \boldsymbol{\varepsilon}_p^n$ *minimizes the dual functional*

$$\mathcal{D}_n(\boldsymbol{\sigma}, \boldsymbol{\beta}) = \frac{1}{2} \|(\boldsymbol{\sigma}, \boldsymbol{\beta})\|_{S \times B}^2 - \big((\boldsymbol{\sigma}, \boldsymbol{\beta}), (\boldsymbol{\sigma}^{n-1}, \boldsymbol{\beta}^{n-1})\big)_{S \times B}$$

subject to the equilibrium constraint

$$\int_\Omega \sigma : \varepsilon(\delta u)\, dx = \ell_n(\delta u), \qquad \delta u \in V(0) \tag{6}$$

and the convex constraint

$$\psi(\sigma, \beta) \le 0. \tag{7}$$

Proof. Since the dual functional is uniformly convex in $S \times B$ and the admissible set determined by (6) and (7) is closed, convex and not empty, a unique solution exists.

The associated Lagrange functional has the form

$$\mathcal{L}_n(\sigma, \beta, u, \gamma) = \mathcal{D}_n(\sigma, \beta) - \int_\Omega \sigma : \varepsilon(u)\, dx + \ell_n(u) + \int_\Omega \gamma \psi(\sigma, \beta)\, dx.$$

Thus, $(\sigma^n, \beta^n, u^n - u^{n-1}, \gamma^n)$ is a KKT point of the constrained minimization problem if and only if

$$D_{(\sigma,\beta)} \mathcal{L}_n(\sigma^n, \beta^n, u^n - u^{n-1}, \gamma^n) = 0, \tag{8}$$

the equilibrium constraint (6), and the complementarity conditions

$$\psi(\sigma^n, \beta^n) \le 0, \qquad \gamma^n \ge 0, \qquad \gamma^n \psi(\sigma^n, \beta^n) = 0 \tag{9}$$

are satisfied. Note that (8) is equivalent to

$$\begin{pmatrix} \mathbb{C}^{-1}(\sigma^n - \sigma^{n-1}) - \varepsilon(u^n - u^{n-1}) \\ \frac{1}{H_0}(\beta^n - \beta^{n-1}) \end{pmatrix} + \gamma^n D\psi(\sigma^n, \beta^n) = 0. \tag{10}$$

Inserting

$$D\psi(\sigma, \beta) = \left(\frac{\sigma - \beta}{|\sigma - \beta|}, -\frac{\sigma - \beta}{|\sigma - \beta|} \right)$$

we observe that (6), (8), (9) together with $\beta^n = H_0 \varepsilon_p^n$ is equivalent to (1a), (1b) and (1c). This shows for a solution $(u^n, \varepsilon_p^n, \gamma^n)$ of (1) that the corresponding values $(\sigma^n, \beta^n, \Delta u^n, \gamma^n)$ satisfy the KKT condition and thus (σ^n, β^n) is the dual solution. $\square$

The SQP method for the solution of the dual problem is now defined as follows: Starting with some $(\sigma^{n,0}, \beta^{n,0}, \gamma^{n,0})$, we define iterates $(\sigma^{n,k}, \beta^{n,k}, \gamma^{n,k})$ for $k = 1, 2, 3, \ldots$ by the successive solution of suitable quadratic minimization problems with linear constraints. For every SQP step k, we define, depending on the previous iterate $(\sigma^{n,k-1}, \beta^{n,k-1}, \gamma^{n,k-1})$, the quadratic functional

$$\mathcal{D}_{n,k}(\Delta\sigma, \Delta\beta) = D_{(\sigma,\beta)} \mathcal{D}_{n,k}(\sigma^{n,k-1}, \beta^{n,k-1})[(\Delta\sigma, \Delta\beta)] \tag{11}$$
$$+ \frac{1}{2} D^2_{(\sigma,\beta)} \mathcal{L}_n(\sigma^{n,k-1}, \beta^{n,k-1}, \cdot, \gamma^{n,k-1})[(\Delta\sigma, \Delta\beta), (\Delta\sigma, \Delta\beta)]$$

(where $\mathcal{D}_{n,k}$ does not depend on $\boldsymbol{u}^{n,k-1}$) and the linearized flow function

$$\psi_{n,k}(\Delta\boldsymbol{\sigma}, \Delta\boldsymbol{\beta}) = \psi(\boldsymbol{\sigma}^{n,k-1}, \boldsymbol{\beta}^{n,k-1}) + D\psi(\boldsymbol{\sigma}^{n,k-1}, \boldsymbol{\beta}^{n,k-1})[\Delta\boldsymbol{\sigma}, \Delta\boldsymbol{\beta}]$$
$$= \phi_{n,k}(\Delta\boldsymbol{\sigma} - \Delta\boldsymbol{\beta}).$$

The SQP iterate is defined as the unique minimizer $(\Delta\boldsymbol{\sigma}^{n,k}, \Delta\boldsymbol{\beta}^{n,k})$ of the quadratic functional (11) subject to the linear constraint

$$\int_\Omega \Delta\boldsymbol{\sigma} : \boldsymbol{\varepsilon}(\delta\boldsymbol{u})\,dx = \ell_n(\delta\boldsymbol{u}) - \int_\Omega \Delta\boldsymbol{\sigma}^{n,k-1} : \boldsymbol{\varepsilon}(\delta\boldsymbol{u})\,dx, \quad \delta\boldsymbol{u} \in V(\boldsymbol{0}) \quad (12)$$

and the linearized constraint

$$\psi_{n,k}(\Delta\boldsymbol{\sigma}, \Delta\boldsymbol{\beta}) \leq 0. \quad (13)$$

Then, set $(\boldsymbol{\sigma}^{n,k}, \boldsymbol{\beta}^{n,k}) = (\boldsymbol{\sigma}^{n,k-1}, \boldsymbol{\beta}^{n,k-1}) + (\Delta\boldsymbol{\sigma}^{n,k}, \Delta\boldsymbol{\beta}^{n,k})$, and let $(\boldsymbol{u}^{n,k}, \gamma^{n,k})$ be the corresponding Lagrange multipliers.

Since the dual problem is convex, the SQP subproblem is admissible, and since the quadratic functional (11) is uniformly convex, the minimizer is unique.

In analogy to the previous Lemma, one can show in the same way the following result (cf. [12, sect. 7.3]).

Lemma 4. *The KKT system for the minimization problem in the SQP step is equivalent to system (3). Moreover, the solution $(\boldsymbol{u}^{n,k}, \boldsymbol{\varepsilon}_p^{n,k}, \gamma^{n,k})$ of the system (3) and the minimizer of the SQP subproblem $(\Delta\boldsymbol{\sigma}^{n,k}, \Delta\boldsymbol{\beta}^{n,k})$ are related by $\mathbb{C} : (\boldsymbol{\varepsilon}(\boldsymbol{u}^{n,k}) - \boldsymbol{\varepsilon}_p^{n,k}) = \boldsymbol{\sigma}^{n,k-1} + \Delta\boldsymbol{\sigma}^{n,k}$ and $H_0 \boldsymbol{\varepsilon}_p^{n,k} = \boldsymbol{\beta}^{n,k-1} + \Delta\boldsymbol{\beta}^{n,k}$.*

Thus, the algorithm which is presented in Section 3 is equivalent to the SQP method for uniformly convex minimization problems, and thus the standard convergence analysis for SQP methods applies.

It remains to analyze the generalized Newton method which is used for the solution of the single SQP step. This analysis is based on the following Lemma, which can be directly obtained by simple (but lengthy) calculations (see [8] where is is shown in detail for Cosserat plasticity).

Lemma 5. *The nonlinear variational equation (4) is the first variation of the functional*

$$J_{n,k}(\boldsymbol{u}) = \frac{c_0}{2}\int_\Omega \boldsymbol{\varepsilon}(\boldsymbol{u}) : \mathbb{C} : \boldsymbol{\varepsilon}(\boldsymbol{u})\,dx + \frac{1-c_0}{2}\int_\Omega \boldsymbol{\theta}_n(\boldsymbol{u}) : \mathbb{C}^{-1} : \boldsymbol{\theta}_n(\boldsymbol{u})\,dx$$
$$-\frac{1-c_0}{4\mu}\int_\Omega \left(\boldsymbol{\theta}_n(\boldsymbol{u}) : \mathbb{A}_{n,k} : \boldsymbol{\theta}_n(\boldsymbol{u}) + \max\{0, \phi_{n,k}(\boldsymbol{\theta}_n(\boldsymbol{u}))\}^2\right)dx - \ell_n(\boldsymbol{u}).$$

Moreover, the functional $J_{n,k}(\cdot)$ is bounded and uniformly convex satisfying

$$\frac{c_0}{2}\|\boldsymbol{u}\|_V^2 \leq J_{n,k}(\boldsymbol{u}) + \ell_n(\boldsymbol{u}) \leq \frac{1}{2}\|\boldsymbol{u}\|_V^2.$$

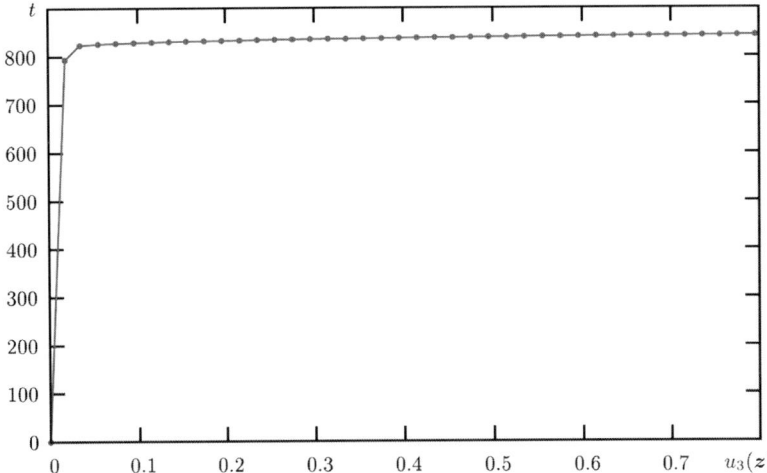

Fig. 1 Load-displacement curve for a sample computation with small hardening close to perfect plasticity. Here, the load depends linearly on t, and the displacement $u = (u_1, u_2, u_3)$ is shown at the corner point $z = (10, 1, 1)$.

Fig. 2 Final distribution of the plastic strain on the surface boundary.

As a consequence, the damped Newton iteration is monotonically decreasing and globally convergent. More precisely, one can show that fixed damping $\rho_{n,k,m} \equiv c_0$ results in the estimate

$$\frac{c_0}{2} \|\boldsymbol{u}^{n,k,m} - \boldsymbol{u}^{n,k}\|_V^2 \leq (1 - c_0^4)^k \left(J_{\text{red}}(\boldsymbol{u}^{n,k,0}) - J_{\text{red}}(\boldsymbol{u}^{n,k})\right).$$

Since for realistic material parameters $1 - c_0^4$ is very close to 1, this estimate is far too pessimistic. In particular, to this problem the analysis of semi-smooth Newton methods applies [5] which proves local super-linear convergence. In fact, numerically one always observes quadratic convergence for the final Newton steps [12].

5 A Numerical Experiment

We demonstrate the performance of the SQP method for the 3-d analog of the configuration defined in [12], and we use the reparametrization described in [9]. The computation is realized in the parallel finite element code M++ [11].

For the example we use the Lamé parameters $\mu = 80193.80$ [N/mm^2], $\lambda = 110743.82$ [N/mm^2], the yield stress $K_0 = 367.42$ [N/mm^2], and a very small hardening modulus $H_0 = 8.01938$ [N/mm^2]. The results are given for a hexahedral mesh with 63 376 unknowns and 40 loading steps.

A load-displacement curve is shown in Figure 1, the plastic strain distribution is illustrated in Figure 2. The SQP method turns out to be very robust near the limit load of the corresponding model of perfect plasticity (in average, three SQP iteration steps per load increment are required). Moreover, we always observe quadratic error reduction in the final SQP iteration step.

References

1. J. Alberty, C. Carstensen, and D. Zarrabi. Adaptive numerical analysis in primal elastoplasticity with hardening. *IMA Journal of Numerical Analysis*, 171:175–204, 1999.
2. R. Blaheta. Convergence of Netwon-type methods in incremental return mapping analysis of elasto-plastic problems. *Comput. Meth. Appl. Mech. Engrg.*, 147:167–185, 1997.
3. P. W. Christensen. A nonsmooth Newton method for elastoplastic problems. *Comput. Meth. Appl. Mech. Engrg.*, 191:1189–2119, 2002.
4. W. Han and B. D. Reddy. *Plasticity: Mathematical Theory and Numerical Analysis*. Springer-Verlag, Berlin, 1999.
5. R. Klatte and B. Kummer. *Nonsmooth Equations in Optimization*, Nonconvex Optimization and Its Applications, Vol. 60. Kluwer Academic Publishers, Dordrecht, 1993.
6. K. Krabbenhoft, A. V. Lyamin, S. W. Sloan, and P. Wriggers. An interior-point algorithm for elastoplasticity. *Int. J. Numer. Meth. Eng.*, 69:592–626, 2007.
7. G. Maier and D. Grierson. *Engineering plasticity by mathematical programming*. Pergamon Press, New York, 1979.

8. P. Neff, K. Chelminski, W. Müller, and C. Wieners. A numerical solution method for an inifinitesimal elasto-plastic cosserat model. *Mathematical Models and Methods in Applied Sciences (M3AS)*, 17:1211–1240, 2007.
9. P. Neff, A. Sydow, and C. Wieners. Numerical approximation of incremental infinitesimal gradient plasticity. *Int. J. Numer. Meth. Eng.*, 2008, to appear.
10. J. C. Simo and T. J. R. Hughes. *Computational Inelasticity*. Springer-Verlag, Berlin, 1998.
11. C. Wieners. Distributed point objects. A new concept for parallel finite elements. In R. Kornhuber, R. Hoppe, J. Périaux, O. Pironneau, O. Widlund, and J. Xu (Eds.), *Domain Decomposition Methods in Science and Engineering*, Lecture Notes in Computational Science and Engineering, Vol. 40, pp. 175–183. Springer-Verlag Berlin, 2004.
12. C. Wieners. Nonlinear solution methods for infinitesimal perfect plasticity. *Z. Angew. Math. Mech. (ZAMM)*, 87:643–660, 2007.

Simulation of Elastoplastic Forming Processes Using Overlapping Domain Decomposition and Inexact Newton Methods

Stephan Brunssen[1], Corinna Hager[1], Florian Schmid[2] and Barbara Wohlmuth[1]

[1]*IANS, University of Stuttgart, Pfaffenwaldring 57, 70569 Stuttgart, Germany*
E-mail: {brunssen, hager, wohlmuth}@ians.uni-stuttgart.de
[2]*FNB, Technical University of Darmstadt, Petersenstr. 30,*
64287, Darmstadt, Germany
E-mail: schmid@fnb.tu-darmstadt.de

Abstract. Incremental metal cold forming is a forming technique using simple shaped, flexible tools moving on a CNC-path. These processes are in general very difficult to control. The aim of this contribution is to provide a partitioned solution scheme that will help to accelerate the implicit, quasi-static Finite Element simulation of such processes.

Key words: elastoplasticity, coupled problems, forming process, two-scale, overlapping domain decomposition.

1 Introduction

Some of the major problems of a Finite Element (FE) simulation of incremental metal cold forming processes are:

- The forming zone is small but very mobile. The punch is contacting almost each point of the workpiece at some time during the process.
- The number of degrees of freedom has to be very large to resolve the final shape and all intermediate stages of the workpiece and many load steps are required.
- Two types of nonlinearities have to be taken into account. The material behavior is modeled by an elastoplastic constitutive law and the contact between the tool and the workpiece results in inequality constraints.

The main idea to tackle these challenges is to use a 'divide and conquer' approach: We employ an overlapping domain decomposition (ODDM), see Figure 1. The workpiece (Ω_H) is discretized with a relatively coarse mesh, whereas the forming zone ($\Omega_{h,n}$, depending on the position of the tool at time t_n) is meshed with a fine grid which is assumed to contain the zone of plastic deformation at any time. Of course, the two FE domains have to interchange information:

- Plastic data, mainly the tensor of plastic deformation: The basic idea is to use the additive split of the strain tensor ε into an elastic part ε^{el} and a plastic part ε^{pl},

to perform the plastic computation only on the fine grid and to store the resulting value of ε^{pl} globally.
- Contact data: The contact computation for the interaction of the tool and the workpiece is only performed on the fine grid, and the corresponding contact stresses are transferred to the coarse grid. A mortar operator is employed for this contact coupling.

Thus, the local nonlinear effects enter the coarse global grid computation only on the right hand side. The resulting system can very efficiently be solved by an inexact Newton method employing block Gauss–Seidel iterations for the coarse-fine transfer.

This strategy has the following advantages:

- All nonlinearities of the present problem can be separated from the coarse grid such that the coarse problem which contains most of the degrees of freedom is linear.
- The nonlinear computation on the small fine grid can efficiently be performed by a primal-dual active set strategy which applies to the contact computation as well as to the resolution of the plastic effects.
- Using the fine patch, the plastic zone and the contact problem can be adequately resolved without discretizing the workpiece with a global fine grid. Such procedure would lead to high computational cost, due to the large system of linear equations and the frequent reassembly of the tangential stiffness matrix. In addition, there is no need to perform an adaptive computation which would cause frequent and expensive remeshing due to the mobility of the forming zone.

The current formulation of the method is limited to small deformations to keep the notation simple and because incremental processes like deep rolling [14] can be accurately modeled in the regime of small deformations.

Review of the Literature

For an overview on adaptive discretization techniques, the reader is referred to [1] and the references therein. Advanced mesh superposition techniques have been around since the early nineties [10], often but not exclusively in the context of crack propagation. A very good overview on those methods can be found in the recent work [18]. In [8, 15], the coupling of two domains, one discretized with an h-version overlay mesh and one discretized with a p-version coarse mesh, is solved with a block Gauss–Seidel (GS) iteration. In [5], this idea is extended to a completely mobile patch and the incorporation of contact. Nonlinear complementary functions have been employed in [2, 13] for contact computations and in [6] for plasticity with hardening.

2 Constitutive Equations

In the rest of the paper, a body $\Omega \subset \mathbb{R}^3$ consisting of a material with quasistatic infinitesimal associative elastoplastic behavior and linear kinematic/isotropic hardening is considered, coming into contact with a fixed tool. Allowing for Tresca or Coulomb friction, the strong form of the governing equations are as follows:

$$-\operatorname{div} \boldsymbol{\sigma} = \mathbf{0}, \quad \text{on } \Omega, \qquad \boldsymbol{u} = \mathbf{0}, \quad \text{on } \Gamma_D,$$
$$\boldsymbol{\sigma} = \mathbf{C}^{\text{el}} \boldsymbol{\varepsilon}^{\text{el}} = \mathbf{C}^{\text{el}}(\boldsymbol{\varepsilon}(\boldsymbol{u}) - \boldsymbol{\varepsilon}^{\text{pl}}), \quad \text{on } \Omega, \qquad -\boldsymbol{\sigma}\boldsymbol{v} = \boldsymbol{\lambda}, \quad \text{on } \Gamma_C,$$
$$\boldsymbol{\varepsilon}(\boldsymbol{u}) = \tfrac{1}{2}(\nabla \boldsymbol{u} + \nabla \boldsymbol{u}^T), \quad \operatorname{tr} \boldsymbol{\varepsilon}^{\text{pl}} = \mathbf{0}, \quad \text{on } \Omega.$$

In these relations, $\mathbf{C}^{\text{el}}$ denotes the Hooke tensor, $\boldsymbol{\sigma}$ the Cauchy stress and $\boldsymbol{\varepsilon}(\boldsymbol{u})$ the linearized strain tensor. The boundary $\partial\Omega$ is partitioned into two nonoverlapping parts Γ_D and Γ_C; the body is clamped on Γ_D, and Γ_C denotes the potential contact boundary with the contact normal $\boldsymbol{v}$. The unknown contact stress $\boldsymbol{\lambda}$ and the plastic strain $\boldsymbol{\varepsilon}^{\text{pl}}$ are determined according to the following rules:

$$\boldsymbol{\lambda}_\tau := \boldsymbol{\lambda} - \lambda_v \boldsymbol{v}, \qquad \boldsymbol{\eta} := \operatorname{dev} \boldsymbol{\sigma} - \tfrac{2}{3} K \boldsymbol{\varepsilon}^{\text{pl}},$$
$$\dot{\boldsymbol{u}}_\tau = \mu^{\text{co}} \frac{\boldsymbol{\lambda}_\tau}{\|\boldsymbol{\lambda}_\tau\|}, \qquad \dot{\boldsymbol{\varepsilon}}^{\text{pl}} = \mu^{\text{pl}} \frac{\boldsymbol{\eta}}{\|\boldsymbol{\eta}\|},$$
$$\phi^{\text{co}}(\boldsymbol{\lambda}_\tau, \lambda_v) := \|\boldsymbol{\lambda}_\tau\| - (g_\tau + \mathfrak{F} \lambda_v), \qquad \phi^{\text{pl}}(\boldsymbol{\eta}, \alpha) := \|\boldsymbol{\eta}\| - \sqrt{\tfrac{2}{3}}(\sigma_0 + H\alpha),$$
$$\mu^{\text{co}} \geq 0, \qquad \mu^{\text{pl}} \geq 0,$$
$$\phi^{\text{co}}(\boldsymbol{\lambda}_\tau, \lambda_v) \leq 0, \qquad \phi^{\text{pl}}(\boldsymbol{\eta}, \alpha) \leq 0,$$
$$\mu^{\text{co}} \phi^{\text{co}}(\boldsymbol{\lambda}_\tau, \lambda_v) = 0, \qquad \mu^{\text{pl}} \phi^{\text{pl}}(\boldsymbol{\eta}, \alpha) = 0,$$
$$\lambda_v \geq 0,$$
$$g_v(\boldsymbol{u}) \leq 0, \qquad \dot{\alpha} = \sqrt{\tfrac{2}{3}} \mu^{\text{pl}},$$
$$\lambda_v g_v(\boldsymbol{u}) = 0.$$
(1)

On the left side of (1) the contact laws are listed, beginning with the split of $\boldsymbol{\lambda}$ into its normal part λ_v and its tangential part $\boldsymbol{\lambda}_\tau$. The friction law consists of the flow rule for the tangential displacement $\boldsymbol{u}_\tau$ and the complementary conditions with respect to the yield function ϕ^{co} and the consistency parameter μ^{co}. The case $\mathfrak{F} = 0$ corresponds to Tresca friction, whereas for $g_\tau = 0$ one gets the Coulomb friction law. The normal contact is described by the last three conditions enforcing the nonpenetration of Ω and the obstacle by means of the normal gap function $g_v(\boldsymbol{u})$ (see [13] for details). Comparing these relations with the plasticity laws on the right side of (1), many parallels can be seen. The plastic strain $\boldsymbol{\varepsilon}^{\text{pl}}$ and the relative deviatoric stress $\boldsymbol{\eta}$ fulfill a similar flow rule. Furthermore, the consistency parameter μ^{pl} and the yield function ϕ^{pl} satisfy the same complementary conditions. Only the evolution law for the parameter of isotropic hardening α is different from the complementary conditions of normal contact.

Each set of complementary conditions implies a corresponding admissible set for the boundary/relative stress:

$$E_\tau(\lambda_\nu) := \{\lambda_\tau : \phi^{co}(\lambda_\tau, \lambda_\nu) \leq 0\}, \qquad E^{pl}(\alpha) := \{\eta : \phi^{pl}(\eta, \alpha) \leq 0\},$$
$$E_\nu := \{\lambda_\nu : \lambda_\nu \geq 0\}.$$

Further, the orthogonal projectors onto these sets can be written as follows:

$$P_\tau(\lambda_\tau) := \min\left(1, \frac{\mathfrak{g}_\tau + \mathfrak{F}\lambda_\nu}{\|\lambda_\tau\|}\right)\lambda_\tau, \qquad P^{pl}(\eta) := \min\left(1, \sqrt{\frac{2}{3}}\frac{\sigma_0 + H\alpha}{\|\eta\|}\right)\eta,$$
$$P_\nu(\lambda_\nu) := \max(0, \lambda_\nu).$$

The time derivative in (1) is discretized by the backward Euler scheme and the increments from time t_n to t_{n+1} are denoted by $\Delta \cdot_{n+1} := (\cdot_{n+1} - \cdot_n)$. To simplify the notation, the time index $\cdot_{n+1}$ is omitted from now on. Further, we introduce trial values by choosing positive constants c^{pl}, c_τ, c_ν and setting

$$\lambda_\tau^{tr} := \lambda_\tau + c_\tau \Delta u_\tau, \qquad \eta^{tr} := \eta + c^{pl}\Delta\varepsilon^{pl},$$
$$\lambda_\nu^{tr} := \lambda_\nu + c_\nu g_\nu(u).$$

Using these definitions, the complementary conditions from (1) can be equivalently rewritten as (see [2,6]):

$$\lambda_\tau = P_\tau(\lambda_\tau^{tr}), \qquad \eta = P^{pl}(\eta^{tr}), \qquad (2)$$
$$\lambda_\nu = P_\nu(\lambda_\nu^{tr}).$$

Remark 1. From the definition of the relative deviatoric stress η, one can see that for the special choice $c^{pl} = 2\mu + \frac{2}{3}K$, the trial relative stress η^{tr} is independent of the actual plastic strain ε^{pl}. Furthermore, using the linear hardening laws in (1), the deviatoric stress can be expressed by

$$\text{dev}\,\sigma = \min\left(1, \frac{3\mu\theta_n + H + K}{3\mu + H + K}\right)\eta^{tr} + \frac{2}{3}K\varepsilon_n^{pl}, \qquad \theta_n := \sqrt{\frac{2}{3}}\frac{\sigma_0 + H\alpha_n}{\|\eta^{tr}\|},$$

(see [12] for details.) Hence, $\text{dev}\,\sigma$ at time t_{n+1} can be calculated using only the value ε_n^{pl} of the former time step. This leads to the well-known radial return algorithm with η^{tr} as the elastic predictor.

Using $\sigma = \text{dev}\,\sigma + \frac{1}{3}\text{tr}\,\sigma\,\text{Id}$, the constraints can be incorporated into the weak formulation of the equilibrium equation at time t_{n+1}: Find $u \in \mathbb{V}$ for an appropriate function space $\mathbb{V}$ such that for all $v \in \mathbb{V}$

$$\int_\Omega \left(\text{dev}\,\sigma + \frac{K}{3}\text{tr}\,\varepsilon(u)\right):\varepsilon(v)\,dx + \int_{\Gamma_C}\left(P_\nu(\lambda_\nu^{tr})v_\nu + P_\tau(\lambda_\tau^{tr})\cdot v_\tau\right)ds = 0. \quad (3)$$

3 Newton Block Iterative Scheme

Next, the weak form (3) is discretized in space with respect to the overlapping decomposition of the workpiece into the coarse domain Ω_H and the small mobile forming zone $\Omega_{h,n+1}$, using the following FE spaces:

$$\mathbb{V}_h \subset [H^1(\Omega_{h,n+1})]^3, \quad \mathbb{V}_{0h} := \{v_h \in \mathbb{V}_h : v_h = \mathbf{0} \text{ on } \partial\Omega_{h,n+1}\backslash\bar{\Gamma}_C\},$$
$$\mathbb{V}_H \subset [H^1(\Omega_H)]^3, \quad \mathbb{V}_{0H} := \{v_H \in \mathbb{V}_H : v_H = \mathbf{0} \text{ on } \Gamma_D\}.$$

On the coarse space $\mathbb{V}_{0H}$, we only solve a linear elastic problem, yielding the system $\mathbf{A}_H U_H = \mathbf{F}_H^{\text{corr}}$, where $\mathbf{A}_H$ denotes the coarse linear stiffness matrix. The correction forces $\mathbf{F}_H^{\text{corr}}$ due to the coupling with the fine space $\mathbb{V}_h$ are to be specified later. Hence, $\mathbb{V}_h$ is used as a local enrichment of the coarse space $\mathbb{V}_H$ (compare [11]) where additionally the plastification and the contact stresses are taken into account. The overall solution is given by the sum $u_H + u_h$ for $u_H \in \mathbb{V}_{0H}, u_h \in \mathbb{V}_{0h}$. Because of the nonlinear relations on Ω_h, we need to incorporate the coarse displacement vector U_H into the fine problem by means of a suitable interpolation operator $\mathbf{P}$. The coefficient vector of the discrete contact stress is denoted by $\mathbf{\Lambda}_h$, whereas the discretization of the actual plastic strain $\varepsilon_h^{\text{pl}}$ is given by $\mathbf{E}_h^{\text{pl}}$. Then, the discrete version of the equilibrium equation (3) on $\Omega_{h,n+1}$ yields the following nonlinear system:

$$F_h(U_h, \mathbf{\Lambda}_h, \mathbf{E}_h^{\text{pl}}, U_H) := \begin{bmatrix} \mathbf{A}_h(U_h + \mathbf{P}U_H) + \mathbf{B}\mathbf{\Lambda}_h - \mathbf{D}_h\mathbf{E}_h^{\text{pl}} \\ \mathbf{C}_h^{\text{co}}(U_h + \mathbf{P}U_H, \mathbf{\Lambda}_h) \\ \mathbf{C}_h^{\text{pl}}(U_h + \mathbf{P}U_H, \mathbf{E}_h^{\text{pl}}) \end{bmatrix} \stackrel{!}{=} \mathbf{0}. \quad (4)$$

In the following, each row of system (4) is considered in more detail:

1. Equilibrium on $\Omega_{h,n+1}$ with the contact stresses $\mathbf{\Lambda}_h$ and the fine solution U_h as a correction of the coarse displacement U_H. The matrix $\mathbf{B}$ is the coupling matrix between the standard ansatz functions and the shape functions for the dual Lagrange multipliers used for the contact stress [17], whereas $\mathbf{D}_h$ is defined such that $\mathbf{D}_h \mathbf{E}_h^{\text{pl}}$ is the FE assembly of $\int_{\Omega_{h,n+1}} \varepsilon_h^{\text{pl}} : \mathbf{C}^{\text{el}} : \varepsilon(v) \, d\mathbf{x}$. Assuming that $\mathbf{E}_h^{\text{pl}}$ and $\mathbf{\Lambda}_h$ are known, this equation can be used to determine U_h.
2. Nonlinear complementarity function to enforce the contact conditions, stemming from the discretization of the projections on the left side of (2). Here, the normal and tangential parts of the displacement $U_{h,\nu}$, $U_{h,\tau}$ and the contact stress $\mathbf{\Lambda}_{h,\nu}$, $\mathbf{\Lambda}_{h,\tau}$ are employed. Due to the use of dual Lagrange multipliers, the function $\mathbf{C}_h^{\text{co}}$ can be evaluated pointwise for each node $p \subset \Gamma_C$, resulting in the primal-dual active set strategy (see [13] for details):

$$\mathbf{C}_h^{co}(\mathbf{U}_h, \mathbf{\Lambda}_h)[p] := \begin{bmatrix} \mathbf{C}_\nu^{co}(\mathbf{U}_{p,\nu}, \mathbf{\Lambda}_{p,\nu}) \\ \mathbf{C}_\tau^{co}(\mathbf{U}_{p,\tau}, \mathbf{\Lambda}_{p,\nu}, \mathbf{\Lambda}_{p,\tau}) \end{bmatrix}$$

$$:= \begin{bmatrix} \mathbf{\Lambda}_{p,\nu} - \max\left(0, \mathbf{\Lambda}_{p,\nu}^{tr}\right) \\ s_p \left(\mathbf{\Lambda}_{p,\tau} - \min\left(1, \dfrac{g_\tau + \mathfrak{F}\mathbf{\Lambda}_{p,\nu}}{\|\mathbf{\Lambda}_{p,\tau}^{tr}\|}\right) \mathbf{\Lambda}_{p,\tau}^{tr} \right) \end{bmatrix}. \quad (5)$$

The additional factor $s_p := \max(\|\mathbf{\Lambda}_{p,\tau}^{tr}\|, g_\tau + \mathfrak{F}\mathbf{\Lambda}_{p,\nu})$ in $\mathbf{C}_\tau^{co}$ is introduced due to stability reasons.

3. Nonlinear complementarity function to enforce the plasticity complementary conditions, stemming from the discretization of the projection on the right side of (2). (Details will be described in [12].) Similar to (5), this function is evaluated pointwise for each integration point p, using Ξ_p and Ξ_p^{tr} as the discrete nodal values of η and η^{tr}, respectively, which depend on the value of u and ε_h^{pl} at p:

$$\mathbf{C}_h^{pl}(\mathbf{U}_h, \mathbf{E}_h^{pl})[p] := \left[\Xi_p - \min\left(1, \dfrac{\sigma_0 + H\alpha_p}{\|\Xi_p^{tr}\|}\right) \Xi_p^{tr} \right]. \quad (6)$$

After the computation on $\Omega_{h,n+1}$, the globally stored plastic strain $\mathbf{E}_h^{pl}$ is updated, and the equilibrium on the coarse grid Ω_H is computed with the load correction term $F_H^{corr} := \mathbf{P}^\top \mathbf{D}_h \mathbf{E}_h^{pl} - \mathbf{M}^\top \mathbf{\Lambda}_h$. The mortar operator $\mathbf{M}$ links the ansatz functions of the Lagrange multipliers of the contact stress with the coarse standard shape functions. This leads to the overall coupled nonlinear system

$$F(X) := F(X_h, X_H) := \begin{bmatrix} F_h(\mathbf{U}_h, \mathbf{\Lambda}_h, \mathbf{E}_h^{pl}, \mathbf{U}_H) \\ \mathbf{A}_H \mathbf{U}_H - \mathbf{P}^\top \mathbf{D}_h \mathbf{E}_h^{pl} + \mathbf{M}^\top \mathbf{\Lambda}_h \end{bmatrix} \stackrel{!}{=} \mathbf{0}, \quad (7)$$

with the Jacobi matrix

$$\dfrac{\partial F}{\partial X} =: \mathbf{K} = \begin{bmatrix} \mathbf{K}_{hh} & \mathbf{K}_{hH} \\ \mathbf{K}_{Hh} & \mathbf{K}_{HH} \end{bmatrix} = \begin{bmatrix} \mathbf{A}_h & \mathbf{B} & -\mathbf{D}_h & \mathbf{A}_h \mathbf{P} \\ \mathbf{K}_U^{co} & \mathbf{K}_\Lambda^{co} & \mathbf{0} & \mathbf{K}_U^{co}\mathbf{P} \\ \mathbf{K}_U^{pl} & \mathbf{0} & \mathbf{K}_E^{pl} & \mathbf{K}_U^{pl}\mathbf{P} \\ \mathbf{0} & \mathbf{M}^\top & -\mathbf{P}^\top \mathbf{D}_h & \mathbf{A}_H \end{bmatrix} \quad (8)$$

and the abbreviations

$$\mathbf{K}_U^{co} := \dfrac{\partial \mathbf{C}_h^{co}}{\partial \mathbf{U}}, \quad \mathbf{K}_\Lambda^{co} := \dfrac{\partial \mathbf{C}_h^{co}}{\partial \mathbf{\Lambda}_h}, \quad \mathbf{K}_U^{pl} := \dfrac{\partial \mathbf{C}_h^{pl}}{\partial \mathbf{U}}, \quad \mathbf{K}_E^{pl} := \dfrac{\partial \mathbf{C}_h^{co}}{\partial \mathbf{E}_h^{pl}}.$$

Finally, in each Newton step j, the 2×2 block system

$$\begin{bmatrix} \mathbf{K}_{hh} & \mathbf{K}_{hH} \\ \mathbf{K}_{Hh} & \mathbf{K}_{HH} \end{bmatrix}^{(j-1)} \begin{bmatrix} \Delta X_h \\ \Delta X_H \end{bmatrix}^{(j)} = - \begin{bmatrix} F_h \\ F_H \end{bmatrix}^{(j-1)} \quad (9)$$

has to be solved. Thus, the domain decomposition into Ω_H and Ω_h defines in a natural way the block decomposition of (9). It can efficiently be computed by a block iterative scheme such as a block Gauss Seidel with iteration index k, see [16] and the references therein:

Fix j, for $k = 1, \ldots$ solve

$$\mathbf{K}_{hh}^{(j-1)} \Delta X_h^{(j,k)} = -F_h^{(j-1)} - \mathbf{K}_{hH}^{(j-1)} \Delta X_H^{(j,k-1)}, \tag{10}$$

$$\mathbf{K}_{HH}^{(j-1)} \Delta X_H^{(j,k)} = -F_H^{(j-1)} - \mathbf{K}_{Hh}^{(j-1)} \Delta X_h^{(j,k)}. \tag{11}$$

It is possible to treat (11) itself with an iterative solver. In the numerical examples we use the same AMG preconditioned CG solver as in [4], which is denoted by 'AMG solver' for the sake of brevity.

4 Inexact Newton Methods

A very important issue in combining the Newton method for the nonlinear problem (7) with an iterative linear solver as in (10), (11) is to avoid the so-called oversolving.

Algorithm 1 Inexact Newton Method.

1: **input:** $[X^{(j-1)}, F(X^{(j-1)}), \mathbf{K}(X^{(j-1)}), \eta_j]$
2: **for** $k = 1, \ldots$ **do**
3: $\quad \Delta X^{(j,k)} :=$ ITERATIVE_SOLVER$(\Delta X^{(j,k-1)}, \mathbf{K}, F)$
4: $\quad$ **if** $\|F(X^{(j-1)}) + \mathbf{K}(X^{(j-1)}) \Delta X^{(j,k)})\| \leqslant \eta_j \|F(X^{(j-1)})\|$ **break**
5: **end for**
6: **output:** $X^{(j)} := X^{(j-1)} + \Delta X^{(j,k)}$

As illustrated in [9], oversolving occurs if too much accuracy is imposed on the computation of the early Newton steps where the nonlinear function and its local linear model differ much. Several choices for the so-called forcing term η_j which controls the relative error tolerance error of the iterative solver have been proposed to overcome this problem. Algorithm 1 is based on [7] which uses a forcing term η_j computed according to suggestions made in [9]. One of the possible choices is presented here for completeness:

$$\eta_j := \min \left\{ \frac{\|F(X^{(j-1)}) - F(X^{(j-2)}) - \mathbf{K}(X^{(j-2)}) \Delta X^{(j-1)}\|}{\|F(X^{(j-2)})\|}, 1 - \epsilon \right\}.$$

with a small number ϵ. This criterion reflects the agreement between F and its local linear model at the previous Newton step. In the numerical example, the maximum norm $\|\cdot\|_\infty$ is used.

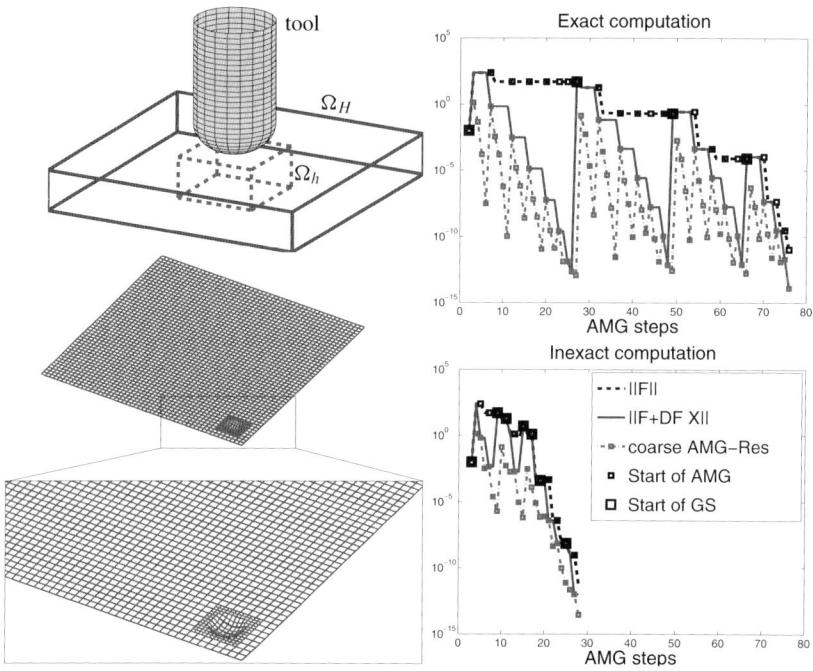

Fig. 1 Left: ODDM (top) and test setting (bottom). Right: Convergence plots of exact (top) and fully inexact (bottom) block Newton scheme.

5 Numerical Results

Two numerical results are presented in this section, both based on the test setting displayed in Figure 1. The domains $\Omega_H = [0, 50] \times [0, 50] \times [0, 1]$ and $\Omega_{h,1} = [2, 8] \times [2, 8] \times [0, 1]$ are discretized with regular hexahedral meshes. The following set of elastic material data is chosen: $E = 69000$, $\nu = 0.33$. For the elastoplastic material behavior, linear isotropic hardening is assumed with the parameters $K = 0$, $H = 2538.930$ and $\sigma_0 = 279.618$. For the first numerical example only one load step is considered, where the tool indents the workpiece by 0.01. The mesh sizes are $H = 1.0$ and $h = 0.5$. The Newton residual $\|F\|$ (dashed line), the block GS residual $\|F + \mathbf{K}\Delta X\|$ (solid line) and the AMG residual (dash-dotted line) are plotted versus the AMG steps in Figure 1. The first computation is an exact one, where block GS and the AMG iterations are performed until convergence in each GS step and in each Newton step, respectively. Here, oversolving occurs as can be seen by the behavior of the norm of the residual. In a second computation, the fully inexact strategy of Algorithm 1 is used. Nearly 64% of the overall number of AMG steps are saved, although the inexact computation needs more Newton iterations.

Fig. 2 Deep-rolling example; 192nd load step; only contact surface plotted; displacements exaggerated.

For a numerical investigation of the error of solving (4) instead of a global nonlinear problem on a fine grid, the interested reader is refered to [3].

In the second example, the mesh sizes are $H = h = 0.5$ and the tool moves horizontally along the workpiece. The deformation of the coarse contact surface grid is shown in Figure 2, the colors indicating the evolution of the isotropic hardening parameter α.

6 Summary

In this contribution, a numerical scheme has been presented for the efficient numerical simulation of cold metal forming. The frictional and plastic complementary conditions are transferred to equality relations by the employment of nonlinear complementary functions. The resulting system is treated by a semi-smooth Newton method, leading to the primal-dual active set strategy. Further, the ODDM proposed in this work defines in a natural way a partitioned solution algorithm which shows a satisfying convergence rate when combined with inexact Newton schemes. As the algorithmic interface between coarse and fine is small, it is possible to work with two different software codes. A generalization of the algorithm to the case where coarse and fine grid are completely nonconforming and not nested would be strongly desirable.

Acknowledgement

The authors express their gratitude for the financial support by the German Research Foundation (DFG) within the framework of the priority program SPP 1146.

References

1. M. Ainsworth and J.T. Oden. *A Posteriori Error Estimation in Finite Element Analysis*. Chichester: Wiley, 2000.
2. P. Alart and A. Curnier. A mixed formulation for frictional contact problems prone to newton like solution methods. *Comput. Methods Appl. Mech. Engrg.*, 92:353–375, 1991.
3. S. Brunssen. *Contact Analysis and Overlapping Domain Decomposition Methods for Dynamic and Nonlinear Problems*. PhD thesis, Universität Stuttgart, 2008.
4. S. Brunssen, F. Schmid, M. Schäfer, and B. Wohlmuth. A fast and robust method for contact problems by combining a primal-dual active set strategy and algebraic multigrid. *Internat. J. Numer. Methods Engrg.*, 69:524–543, 2007.
5. S. Brunssen and B. Wohlmuth. An overlapping domain decomposition method for the simulation of elastoplastic incremental forming processes. Technical report, Universität Stuttgart, SPP 1146, 2007. Internat. J. Numer. Methods Engrg., to appear.
6. P. W. Christensen. A nonsmooth newton method for elastoplastic problems. *Comput. Meth. Appl. Mech. Engrg.*, 191:1189–1219, 2002.
7. R.S. Dembo, S.C. Eisenstat, and T. Steighaug. Inexact Newton methods. *SIAM J. Numer. Anal.*, 19:400–408, 1982.
8. A. Düster, E. Rank, G. Steinl, and W. Wunderlich. A combination of an h- and a p-version of the finite element method for elastic-plastic problems. In *Proceedings of the ECCM 1999*, September 1999.
9. S.C. Eisenstat and H.F. Walker. Choosing the forcing terms in an inexact Newton method. *SIAM J. Sci. Comput.*, 17(1):16–32, 1996.
10. J. Fish. The s-version of the finite element method. *Comput. & Structures*, 43:539–547, 1992.
11. R. Glowinski, J. He, A. Lozinski, J. Rappaz, and J. Wagner. Finite element approximations of multi–scale elliptic problems using patches of elements. *Numer. Math.*, 101:663–687, 2005.
12. C. Hager and B. Wohlmuth. Nonlinear complementary functions for plasticity problems with frictional contact. in preparation.
13. S. Hüeber, G. Stadler, and B. Wohlmuth. A primal-dual active set algorithm for three-dimensional contact problems with Coulomb friction. *SIAM J. Sci. Comput.*, 30(2):572–596, 2008.
14. F. Klocke and S. Mader. Fundamentals of the deep rolling of compressor blades for turbo aircraft engines. In *Proceedings of the 9th International Conference on Shot Peening*, Paris, 2005, pp. 125–130.
15. R. Krause and E. Rank. Multiscale computations with a combination of the h- and p-versions of the finite element method. *Comput. Methods Appl. Mech. Engrg.*, 192:3959–3983, 2003.
16. Y. Saad. *Iterative Methods for Sparse Linear Systems*, 2nd Edition. SIAM, Philadelphia, 2003.
17. B. Wohlmuth. *Discretization Techniques and Iterative Solvers Based on Domain Decomposition*. Lectures Notes in Computational Science and Engineering, Vol. 17. Springer, Berlin/Heidelberg, 2001.
18. Z. Yue and D.H. Robbins. Adaptive superposition of finite element meshes in elastodynamic problems. *Internat. J. Numer. Methods Engrg.*, 63(11):1604–1635, 2005.

Variational Formulation of the Cam-Clay Model

Mohammed Hjiaj[1] and Géry de Saxcé[2]

[1] INSA de Rennes – LGCGM / Structural Engineering Research Group,
20 avenue des Buttes de Coësmes, 35043 Rennes Cedex, France
E-mail: mohammed.hjiaj@insa-rennes.fr
[2] Laboratoire de Mécanique de Lille – UMR 8107, Boulevard Paul Langevin,
59655 Villeneuve d'Ascq Cédex, France
E-mail: gery.desaxce@univ-lille1.fr

Abstract. In this paper, fundamental mathematical concepts for modelling the dissipative behavior of geomaterials are recalled. These concepts are used to revisit the Cam-clay model. A standard version of the model is first discussed. Regarding the non-standard form of the model, the partial normality is exploited and an implicit variational formulation of the modified Cam-clay model is derived. As a result, the solution of boundary value problems can be replaced by seeking stationary points of a functional.

Key words: modified Cam-clay model, internal variables, variational inequality, convex analysis, Fenchel transform, bi-potential.

1 Introduction

The phenomenological approach with internal variables provides a unified framework for developing various models arising in engineering applications. It consists of supplementing the deformation ε by a set of internal (strain-like) variables $\kappa = (\kappa_i, i = 1, \ldots, n)$ which account for the internal restructuring taking place during the dissipative process [1, 8, 9]. The number and the mathematical nature (tensor, vector or scalar) of the internal variables depend on the model under consideration. The notation used here will be one in which symmetric second-order tensors are represented as six-dimensional vectors and denoted by bold letters. More complex operator are capital doubled (e.g. $\mathbb{D}$ for Hooke's tensor). For the sake of a compact representation, internal variables κ_i are grouped together in a unique vector $\kappa \in \mathbb{R}^m$ made by the following ordered n-tuples:

$$\kappa^t = \left[\kappa_1^t, \cdots, \kappa_i^t, \cdots, \kappa_n^t\right] \qquad (1)$$

where "t" stands for the usual transposition, $\mathbb{R}^m$ is a m-dimensional vector space and κ_i^t can be either a vectorial representation of a tensor, a vector or a scalar. The rate of an internal variable, also called velocity, is denoted by a superimposed dot. A set of generalized stresses $\pi = (\pi_i, i = 1, \ldots, n)$, responsible for the internal mod-

B.D. Reddy (ed.), IUTAM Symposium on Theoretical, Modelling and Computational Aspects of Inelastic Media, 165–174.
© Springer Science+Business Media B.V. 2008

ifications, are defined such that (generalized stresses) × (rate of change of internal variables) gives the rate of dissipation. Grouping together the generalized stresses in the vector $\pi \in \mathbb{R}^m$, the rate of dissipation is given as a scalar product in $\mathbb{R}^m$

$$\pi \cdot \dot{k} = \pi_1 \cdot \dot{k}_1 + \cdots + \pi_i \cdot \dot{k}_i + \cdots + \pi_n \cdot \dot{k}_n \qquad (2)$$

where a dot "·" represents the usual scalar product. The m-dimensional linear space $\mathbb{R}^m$ whose elements are the velocities is called the velocity space and denoted by $\mathcal{V}$. The bilinear form generated by the rate of dissipation puts the velocity space $\mathcal{V}$ in duality with the force space $\mathcal{F}$ comprising the generalized stresses π:

$$\forall (\dot{k}, \pi) \in \mathcal{V} \times \mathcal{F} \mapsto \dot{k} \cdot \pi \in \mathbb{R} \qquad (3)$$

It is said that π and $\dot{k}$ are conjugated with respect to the dissipation. While the evolution of the strain can be controlled externally, the internal variables evolve according to some additional laws called *evolution laws* which complement the state laws (e.g. elastic law for an elastic perfectly plastic model). These laws, which describe the evolution of the internal modifications, establish relationships between the rate of change of each k_i and each generalized stress π_i. From a mathematical viewpoint, the global evolution law defines a certain mapping between $\mathcal{V}$ and $\mathcal{F}$, denoted by $\mathcal{A}$, which maps each $\pi \in \mathcal{F}$ to the set, possibly empty, $\mathcal{A}(\pi) \subset \mathcal{V}$.

A class of dissipative materials, interesting from both a mathematical and a computational point of view, are those for which the dissipative operator can be obtained as a gradient or a subgradient[1] of a function for all its elements of its domain. If the operator satisfies the maximal cyclic monotonicity condition, it is proven that this operator can be derived as the subgradient of a *convex* scalar-valued function. The primary advantage of having a potential structure for the constitutive relations is the possibility of applying the calculus of variations.

A broad range of dissipative materials present in engineering have more complex dissipative laws which can not take the convenient form of a potential law. One of most illustrative example is the Coulomb frictional contact law. Other examples are typically those provided by dissipative laws of geomaterials and cyclic (visco)-plasticity models. In this context, the following question arises naturally: how can you preserve all the benefits of a formulation based on the definition of a scalar-valued function, i.e. a potential structure of the dissipative law? An answer to this question is to relax the explicit relation introduced by the potential form by admitting an implicit one.

2 The Cam-Clay Model

It is well known that geomaterials have a very complicated behavior compared to metals even if only monotonic loading is considered. It is therefore a challenge in

---

[1] The subgradient is a generalization of the gradient to non-differentiable functions.

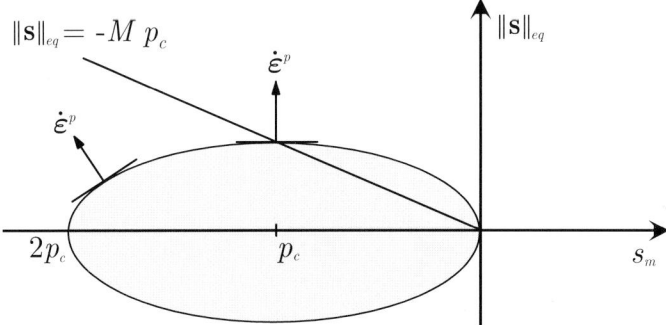

Fig. 1 The Cam-clay model.

geomechanics to develop relatively simple mathematical models able to predict, at least qualitatively, a great number of fundamental aspects of soil behavior. The success of the modified Cam-clay model lies in its ability to capture many of the characteristics of clay behaviour by using only a limited number of material parameters. This model belongs to the class of critical state models which originated from the work of Roscoe and his co-workers at the University of Cambridge [6]. Recent work on the modified Cam-clay model using thermo-dynamics has been carried out by Collins [10]. Commonly observed features such as hardening/softening, contractancy/dilatancy and the tendency to eventually reach a state in which the stress state and the volume change become stationary are all captured by the modified Cam-clay model. Even at present, the modified Cam-clay model remains widely used for computational applications as further evidence of its success. In this section, we first recall the relations governing the dissipation of the model. The elasticity relations are not discussed.

2.1 Classical Formulation

The plastic strain is decomposed into the volumetric plastic strain e_m^p and the plastic strain deviator $\mathbf{e}^p$. The corresponding dual variables are the mean stress s_m and the stress deviator $\mathbf{s}$. The modified Cam-clay yield surface is defined by:

$$f(\mathbf{s}, s_m, p_c) = \|\mathbf{s}\|_{eq}^2 + M^2 s_m^2 - 2M^2 s_m p_c \qquad (4)$$

where $\|\mathbf{s}\|_{eq} = \sqrt{3/2}\,\|\mathbf{s}\|$, p_c is the *"critical state pressure"* and M a material constant defined by

$$M = \frac{6\sin\phi}{3 - \sin\phi}$$

with ϕ the internal friction angle. The solid mechanics convention for strains and stresses is used. In the plane $(s_m, \|\mathbf{s}\|_{eq})$, the yield surface is represented by a family of ellipses passing through the origin, taking a maximum value for $s_m = p_c$ and intercepting the mean stress axis at $s_m = 2\,p_c$. This point corresponds to the elastic limit under a hydrostatic loading and is called the "*preconsolidation pressure*". The plastic flow obeys the normality rule

$$\dot{\mathbf{e}}^p = \dot{\lambda}\frac{\partial f}{\partial \mathbf{s}} = 3\,\dot{\lambda}\,\mathbf{s} \qquad (5)$$

$$\dot{e}_m^p = \dot{\lambda}\frac{\partial f}{\partial s_m} = 2\,\dot{\lambda}\,M^2(s_m - p_c) \qquad (6)$$

and the evolution of the elastic domain is governed by the relation

$$\dot{p}_c = -\upsilon\,p_c\,\dot{e}_m^p \quad \text{with} \quad \upsilon = \frac{1+e}{\eta - \varsigma} \qquad (7)$$

where e is the void ratio of the soil mass, η is the virgin compression index and ς the swell/recompression index. Equation (7) shows that the contractancy leads to a decrease of p_c and therefore the ellipse expands so that the elastic domain is enlarged. On the contrary, dilatancy leads to an increase of p_c (softening phase) which corresponds to a reduction of the elastic domain. When the plastic volumetric strain is zero, p_c becomes constant and the elastic domain stationary. During hardening/softening, the top of the ellipse moves along the straight line $\|\mathbf{s}\|_{eq} = -M\,p_c$ called the "*critical state line*". The critical state line divides the stress space into a contractant and a dilatant region. For a stress state situated on the part of the yield function where $s_m > p_c$, the behavior will be plastically dilatant. On the contrary, for any stress state situated on the part of the yield function where $s_m < p_c$, the behavior will be plastically contractant. The point situated on the yield curve at the intersection with the critical state line strains at constant plastic volume.

2.2 Internal Variable Formulation

As mentioned before, our aim is to discuss the dissipative behaviour of the model. In addition to the plastic strain $\dot{\boldsymbol{\varepsilon}}^p$, a scalar internal variable α is introduced. This variable accounts for hardening/softening. The conjugated variables are $\boldsymbol{\sigma}$ and $\overline{p}_c$ respectively, and the dissipation is given by

$$\boldsymbol{\pi} \cdot \dot{\boldsymbol{\kappa}} = \boldsymbol{\sigma} \cdot \dot{\boldsymbol{\varepsilon}}^p - \overline{p}_c\,\dot{\alpha} \qquad (8)$$

where $\boldsymbol{\pi}, \dot{\boldsymbol{\kappa}}$ and $\overline{p}_c$ are given by

$$\boldsymbol{\pi} = \begin{bmatrix} \boldsymbol{\sigma} \\ \overline{p}_c \end{bmatrix} \;,\quad \dot{\boldsymbol{\kappa}} = \begin{bmatrix} \dot{\boldsymbol{\varepsilon}}^p \\ -\dot{\alpha} \end{bmatrix} \quad \text{and} \quad p_c(\alpha) = p_{c0} + \overline{p}_c(\alpha), \qquad (9)$$

with p_{c0} being the initial size of the elastic domain. The elastic domain K is now defined in the generalized stress space

$$K = \left\{ \pi \in \mathcal{S} \mid \|\mathbf{s}\|_{eq}^2 + M^2(s_m - p_c)^2 \le M^2 p_c^2 \right\} \quad (10)$$

Relation (7), which is referred to as the rate form of the state equation becomes

$$\dot{p}_c = -\upsilon\, p_c\, \dot{\alpha} \quad (11)$$

Comparing (7) and (11), we deduce the so-called hardening rule

$$\dot{e}_m^p - \dot{\alpha} = 0 \quad (12)$$

The evolution law for α does not satisfy the normal rule since we have

$$\dot{\lambda} \frac{\partial f}{\partial p_c} = 2\dot{\lambda} M^2 (p_c - s_m) \ne -\dot{\alpha} \quad (13)$$

Therefore the model is not standard generalized since we do not have generalized normality (normality for each component of $\dot{\kappa}$). Therefore, we need to introduce another scalar function, called plastic potential, from which the evolution laws can be deduced by applying the normality rule. The expression of the plastic potential is

$$g(\mathbf{s}, s_m, p_c) = \|\mathbf{s}\|_{eq}^2 + M^2 (s_m - p_c)^2 \quad (14)$$

and the evolution laws are given by

$$\dot{\mathbf{e}}^p = \dot{\lambda} \frac{\partial g}{\partial \mathbf{s}} = 3\dot{\lambda}\, \mathbf{s}$$

$$\dot{e}_m^p = \dot{\lambda} \frac{\partial g}{\partial s_m} = 2\dot{\lambda} M^2 (s_m - p_c)$$

$$-\dot{\alpha} = \dot{\lambda} \frac{\partial g}{\partial p_c} = 2\dot{\lambda} M^2 (p_c - s_m)$$

Before going further, let us derive an equivalent expression for the yield function which will have the property of being homogeneous of degree one. Taking into account that p_c is always negative, an alternative expression of the elastic domain is given by

$$\sqrt{\|\mathbf{s}\|_{eq}^2 + M^2(s_m - p_c)^2} \le -M\, p_c \quad (15)$$

and the expression of the yield function is now

$$f(\boldsymbol{\sigma}, p_c) = \sqrt{\|\mathbf{s}\|_{eq}^2 + M^2(s_m - p_c)^2} + M\, p_c \quad (16)$$

which is homogenous of degree one:

$$f(\beta\, \boldsymbol{\sigma}, \beta\, p_c) = \beta\, f(\boldsymbol{\sigma}, p_c) \quad (17)$$

By introducing the following notations

$$\boldsymbol{\sigma} = \begin{bmatrix} \mathbf{s} \\ s_m \end{bmatrix} \quad \text{and} \quad \mathbf{X} = \begin{bmatrix} \mathbf{0} \\ p_c \end{bmatrix},$$

the yield function (16) becomes

$$f(\boldsymbol{\sigma}, p_c) = \|\boldsymbol{\sigma} - \mathbf{X}\|_{cc} + M\, p_c$$

where

$$\|\boldsymbol{\sigma} - \mathbf{X}\|_{cc} = \left(\|\mathbf{s}\|_{eq}^2 + M^2 (s_m - p_c)^2 \right)^{\frac{1}{2}} \tag{18}$$

With this new expression of the yield function, the flow rule becomes

$$\dot{\mathbf{e}}^p = \frac{3\,\dot{\lambda}}{2} \frac{\mathbf{s}}{\|\boldsymbol{\sigma} - \mathbf{X}\|_{cc}} \tag{19}$$

$$\dot{e}_m^p = \dot{\lambda}\, M^2 \frac{(s_m - p_c)}{\|\boldsymbol{\sigma} - \mathbf{X}\|_{cc}} \tag{20}$$

which leads to the following expression for the plastic multiplier $\dot{\lambda}$

$$\dot{\lambda} = \|\dot{\boldsymbol{\varepsilon}}^p\|_{cc}^* = \left(\|\dot{\mathbf{e}}^p\|_{eq}^2 + \frac{(\dot{e}_m^p)^2}{M^2} \right)^{\frac{1}{2}} \tag{21}$$

where the norm $\|\bullet\|_{cc}^*$ defined on the velocity space is dual to the norm (18) in the sense that

$$(\boldsymbol{\sigma} - \mathbf{X}) \cdot \dot{\boldsymbol{\varepsilon}}^p \leq \|\boldsymbol{\sigma} - \mathbf{X}\|_{cc}\, \|\dot{\boldsymbol{\varepsilon}}^p\|_{cc}^* \tag{22}$$

3 Standard Version of the Modified Cam-Clay Model

Suppose that we would like to have a generalized standard model [4, 5]. The hardening rule (12) must be different in order to satisfy the generalized normality. Indeed, by applying the normality rule for the internal variable α, we find the following relationship for the hardening rule

$$-\dot{\alpha} = \dot{\lambda} \frac{\partial f}{\partial p_c} = -\dot{e}_m^p + M\, \|\dot{\boldsymbol{\varepsilon}}^p\|_{cc}^* \tag{23}$$

So to have a standard model the relation (23) should be used instead of (12). It is worth mentioning that now $-\alpha$ is a non-decreasing variable and therefore softening cannot occur. With this expression of $-\alpha$, the evolution rule for all internal variables can be written in the following compact relation:

$$\dot{\boldsymbol{\kappa}} \in \partial \psi^*(\boldsymbol{\pi}) \tag{24}$$

where $\psi^*(\pi)$ is the complementary dissipation pseudo-potential which corresponds to the indicator function (see [7]) of the elastic domain K expressed in the generalized stress space

$$\Psi_K(\pi) = \begin{cases} 0 & \text{if } \pi \in K, \\ +\infty & \text{otherwise} \end{cases} \tag{25}$$

A standard model for clay seems not to be appropriate since it does not reproduce the softening behavior observed experimentally. The dissipation function is obtained as follows

$$\psi(\dot{\kappa}) = \sup_{\pi \in K} [\pi \cdot \dot{\kappa}] = \sup_{\pi \in K} \left[\mathbf{s} \cdot \dot{\mathbf{e}}^p + s_m \dot{e}_m^p - \overline{p}_c \dot{\alpha} \right] \tag{26}$$

It is clear that the supremum will be achieved for a vector $\mathbf{s}$ colinear to $\dot{\mathbf{e}}^p$:

$$\mathbf{s} \cdot \dot{\mathbf{e}}^p + s_m \dot{e}_m^p - \overline{p}_c \dot{\alpha} \leq \|\mathbf{s}\| \|\dot{\mathbf{e}}^p\| + s_m \dot{e}_m^p - \overline{p}_c \dot{\alpha} \tag{27}$$

Adding and subtracting $p_c \dot{e}_m^p$, we have

$$\mathbf{s} \cdot \dot{\mathbf{e}}^p + s_m \dot{e}_m^p - \overline{p}_c \dot{\alpha} \leq \|\mathbf{s}\| \|\dot{\mathbf{e}}^p\| + (s_m - p_c) \dot{e}_m^p + \overline{p}_c (\dot{e}_m^p - \dot{\alpha}) + p_{c0} \dot{e}_m^p \tag{28}$$

Using the Cauchy–Schwartz inequality (22), we obtain

$$\mathbf{s} \cdot \dot{\mathbf{e}}^p + s_m \dot{e}_m^p - \overline{p}_c \dot{\alpha} \leq \|\boldsymbol{\sigma} - \mathbf{X}\|_{cc} \|\dot{\mathbf{e}}^p\|_{cc}^* + \overline{p}_c (\dot{e}_m^p - \dot{\alpha}) + p_{c0} \dot{e}_m^p \tag{29}$$

Taking into account that $\|\boldsymbol{\sigma} - \mathbf{X}\|_{cc}$ is bounded by $-M p_c$ and making use of the hardening rule (23), we have

$$\mathbf{s} \cdot \dot{\mathbf{e}}^p + s_m \dot{e}_m^p - \overline{p}_c \dot{\alpha} \leq -M p_c \|\dot{\mathbf{e}}^p\|_{cc}^* + \overline{p}_c \left(M \|\dot{\mathbf{e}}^p\|_{cc}^* \right) + p_{c0} \dot{e}_m^p \tag{30}$$

Therefore, the supremum is equal to

$$-p_{c0} \left(M \|\dot{\mathbf{e}}^p\|_{cc}^* - \dot{e}_m^p \right) \tag{31}$$

and the dissipation pseudo-potential is given by the following expression

$$\psi(\dot{\kappa}) = -p_{c0} \left(M \|\dot{\mathbf{e}}^p\|_{cc}^* - \dot{e}_m^p \right) + \Psi_{(0)} \left(M \|\dot{\mathbf{e}}^p\|_{cc}^* + \dot{\alpha} - \dot{e}_m^p \right) \tag{32}$$

where

$$\Psi_{(0)} = \begin{cases} 0 & \text{if } M \|\dot{\mathbf{e}}^p\|_{cc}^* + \dot{\alpha} - \dot{e}_m^p = 0 \\ +\infty & \text{if } M \|\dot{\mathbf{e}}^p\|_{cc}^* + \dot{\alpha} - \dot{e}_m^p \neq 0 \end{cases} \tag{33}$$

The functions $\psi(\dot{\kappa})$ and $\psi^*(\pi)$ satisfy the following relation:

$$\psi(\dot{\kappa}') + \psi^*(\pi') \geq \pi' \cdot \dot{\kappa}', \quad \forall \left(\pi', \dot{\kappa}' \right) \in \mathcal{V} \times \mathcal{F} \tag{34}$$

A pair $(\pi, \dot{\kappa})$ related by the generalized normality satisfies

$$\dot{\kappa} \in \partial \psi^*(\pi) \Leftrightarrow \pi \in \partial \psi(\dot{\kappa}) \Leftrightarrow \psi(\dot{\kappa}) + \psi^*(\pi) = \pi \cdot \dot{\kappa} \tag{35}$$

4 Implicit Normality Rule

The modified Cam-clay model is non-standard but we will see below that it is still possible to obtain a variational formulation of the evolution law. In the original model, the non-normality is partial and concerns only the internal variable α. Applying the following change of variables

$$-\dot{\vartheta} = -\dot{\alpha} + M \, \|\dot{\boldsymbol{\varepsilon}}^p\|_{cc}^* \tag{36}$$

the normality rule can be recovered

$$\dot{\boldsymbol{\varepsilon}}^p \in \partial_{\boldsymbol{\sigma}} \psi^*(\boldsymbol{\sigma}, \overline{p}_c) \quad \text{and} \quad -\dot{\vartheta} \in \partial_{\overline{p}_c} \psi^*(\boldsymbol{\sigma}, \overline{p}_c) \tag{37}$$

or in more compact form

$$\dot{\boldsymbol{\xi}} \in \partial \psi^*(\boldsymbol{\pi}) \tag{38}$$

where the vector $\dot{\boldsymbol{\xi}}$ is given by

$$\dot{\boldsymbol{\xi}} = \begin{bmatrix} \dot{\boldsymbol{\varepsilon}}^p \\ -\dot{\vartheta} \end{bmatrix} \tag{39}$$

The inverse law is

$$\boldsymbol{\pi} \in \partial \psi(\dot{\boldsymbol{\xi}}) \tag{40}$$

where $\psi(\dot{\boldsymbol{\xi}})$ now depends on $\dot{\boldsymbol{\varepsilon}}^p$ and $-\dot{\vartheta}$

$$\psi(\dot{\boldsymbol{\kappa}}) = p_{c0}\left(M \, \|\dot{\boldsymbol{\varepsilon}}^p\|_{cc}^* - \dot{e}_m^p\right) + \Psi_{(0)}\left(M \, \|\dot{\boldsymbol{\varepsilon}}^p\|_{cc}^* + \dot{\vartheta} - \dot{e}_m^p\right) \tag{41}$$

$\psi(\dot{\boldsymbol{\xi}})$ and $\psi^*(\boldsymbol{\pi})$ satisfy the Fenchel inequality

$$\psi(\dot{\boldsymbol{\xi}}') + \psi^*(\boldsymbol{\pi}') \geq \boldsymbol{\pi}' \cdot \dot{\boldsymbol{\xi}}', \quad \forall (\boldsymbol{\pi}', \dot{\boldsymbol{\xi}}') \in \mathcal{V} \times \mathcal{F} \tag{42}$$

Although this relation provides further insight into such plastic model, additional developments can still be made to establish a relationship between $\dot{\boldsymbol{\kappa}}$ and $\boldsymbol{\pi}$ based on a normality rule. To recover a relation between the dual variables $\dot{\boldsymbol{\kappa}}$ and $\boldsymbol{\pi}$, we add $\boldsymbol{\pi}' \cdot \dot{\boldsymbol{\kappa}}'$ to both sides of (42),

$$\psi(\dot{\boldsymbol{\xi}}') + \psi^*(\boldsymbol{\pi}') + \boldsymbol{\pi}' \cdot (\dot{\boldsymbol{\kappa}}' - \dot{\boldsymbol{\xi}}') \geq \boldsymbol{\pi}' \cdot \dot{\boldsymbol{\kappa}}', \quad \forall (\boldsymbol{\pi}', \dot{\boldsymbol{\xi}}') \in \mathcal{V} \times \mathcal{F} \tag{43}$$

The left-hand side of (43) is a function of both $\dot{\boldsymbol{\kappa}}'$ and $\boldsymbol{\pi}'$, which cannot be represented as the sum of two functions, one of $\dot{\boldsymbol{\kappa}}'$ and another of $\boldsymbol{\pi}'$. We call this function a *bi-potential* and its general expression is given by

$$b_p(\dot{\boldsymbol{\xi}}', \boldsymbol{\pi}') := \psi(\dot{\boldsymbol{\xi}}') + \psi^*(\boldsymbol{\pi}') + \boldsymbol{\pi}' \cdot (\dot{\boldsymbol{\kappa}}' - \dot{\boldsymbol{\xi}}') \tag{44}$$

The right-hand side of (44) is developed using the change of variables (36). Developing the scalar product in (44), we obtain the bi-potential for the modified Cam-clay model:

$$b_p(\dot{\kappa}, \pi) = \Psi_K(\pi) - p_c M \|\dot{\varepsilon}^p\|_{cc}^* + p_{c0}\dot{e}_m^p + \Psi_{(0)}\left(\dot{\alpha} - \dot{e}_m^p\right) \tag{45}$$

The bi-potential is positive function and satisfies the fundamental inequality

$$b_p(\dot{\kappa}', \pi') \geq \dot{\kappa}' \cdot \pi' \tag{46}$$

A strict equality is obtained in (46) for any pair $(\dot{\kappa}, \pi)$ related by the evolution law:

$$b_p(\dot{\kappa}, \pi) = \dot{\kappa} \cdot \pi \tag{47}$$

The relations (46) and (47) can be combined to give

$$\forall \pi' \in \mathcal{F} : \quad b_p(\dot{\kappa}, \pi') - b_p(\dot{\kappa}, \pi) \geq \dot{\kappa} \cdot (\pi' - \pi) \tag{48}$$

$$\forall \dot{\kappa}' \in \mathcal{V} : \quad b_p(\dot{\kappa}', \pi) - b_p(\dot{\kappa}, \pi) \geq \pi \cdot (\dot{\kappa}' - \dot{\kappa}) \tag{49}$$

which means that

- the bi-potential is bi-convex that is $b_p(\dot{\kappa}, \pi)$ is a convex function of $\dot{\kappa} \in \mathcal{V}$ for each $\pi \in \mathcal{F}$ and a convex function of $\pi \in \mathcal{F}$ for each $\dot{\kappa} \in \mathcal{V}$
- the evolution law and its inverse derive from the bi-potential $b_p(\dot{\kappa}, \pi)$

$$\dot{\kappa} \in \partial_\pi b_p(\dot{\kappa}, \pi) \quad \text{and} \quad \pi \in \partial_{\dot{\kappa}} b_p(\dot{\kappa}, \pi) \tag{50}$$

The advantage of the present formulation results in a compact form of the evolution law formulated with one variational inequality. This formulation of the evolution law can be advantageously exploited to derive a robust algorithm. The relations (50) are essential for the derivation of stationary principles involving a functional that depends now on both the velocities and the stresses.

5 Conclusions

The nature of the constitutive operator has a significant influence on convergence of numerical algorithms. It has been recognized that a variational formulation has several advantages, among them, the possibility to associate extremum (or at least stationary) principles to weak formulations of the initial/boundary value problems. However geomaterial models does not exhibit such a strong variational structure. By allowing an implicit form of the evolution rule, a weaker variational formulation can be recovered and the pseudo-potential concept (introduced by Moreau [2, 3]) extended to cover non-standard behaviors. The pseudo-potential is replaced by the bi-potential, which depends on both the generalized stresses and the velocities. The bi-potential is not convex but bi-convex, that means convex with respect to the generalized stresses and the plastic strain rates when considered separately. The partial sub-derivatives of the bi-potential yield the evolution law and its inverse. It has been shown that the evolution law of the modified Cam-clay model can be derived from a bi-potential which serve as a "potential" for both the generalized stresses and the velocities. As a consequence, coupled extremum principles exist. These principles are not as strong as the usual extremum (or stationary) principles since they involve static and kinematic variables, but at least they provide new insights into this difficult problem.

References

1. Martin, J.B. *Plasticity: Fundamentals and General Results*. The MIT Press, 1975.
2. Moreau, J.J. La notion de sur-potentiel et les liaisons en élastostatique. *Comptes Rendus de l'Académie des Sciences*, 267:954–957, 1968.
3. Moreau, J.J. Sur les lois de frottement, de plasticité et de viscosité. *Comptes Rendus de l'Académie des Sciences*, 271:608–611, 1970.
4. Nguyen Quoc Son. Matériaux élasto-visco-plastiques à potentiel généralisé. *Comptes Rendus de l'Académie des Sciences*, 277:915–918, 1973.
5. Halphen, B. and Nguyen Quoc Son. Sur les matériaux standard généralisés. *Journal de Mécanique*, 14:39–63, 1975.
6. Roscoe, K.H. and Burland, J.B. On the generalized stress-strain behaviour of "wet" clays. In: Heyman, J. and Leckie, F.A. (Eds.), *Engineering Plasticity*, Cambridge University Press, pp. 535–609, 1968.
7. Rockafellar, R.T. and Wets, J.B. *Variational Analysis*. Springer, Berlin, 1998.
8. Mróz, Z. *Mathematical Models of Inelastic Behavior*. Solid Mechanics Division, University of Waterloo, Canada, 1973.
9. Han, W. and Reddy, B.D. *Plasticity. Mathematical Theory and Numerical Analysis*. Springer, Berlin, 1999.
10. Collins, I.F. A systematic procedure for constructing critical state models in three dimensions. *International Journal of Solids and Structures*, 40:4379–4397, 2003.

Anisotropic Modelling of Metals in Forming Processes

S. Reese and I.N. Vladimirov

*Institute of Solid Mechanics, Technische Universität Braunschweig,
D-38106 Braunschweig, Germany
E-mail: {s.reese, ivaylo.vladimirov}@tu-bs.de*

Abstract. The accuracy of sheet metal forming simulations crucially relies on the quality of the material description which involves the modelling of large strains, elastic and plastic anisotropy as well as isotropic and kinematic hardening. In the present paper we propose a hyperelasticity-based concept using structural tensors whose evolution is in contrast to previous works explicitly taken into account. In this way we consider the evolution of the elastic anisotropy (through the dependence of the Helmholtz free energy function on the evolving structural tensors) as well as of the plastic anisotropy (through the dependence of the yield locus on the structural tensors). Exploiting the dissipation inequality leads to the important result that the model includes only symmetric tensor-valued internal variables the integration of which can be performed very efficiently by means of the exponential map. Numerical examples are shown to qualitatively validate the model.

Key words: structural tensors, large deformations, deep drawing.

1 Introduction

The best possible agreement between the finally manufactured product and the expected, i.e. desired and designed shape is very important in modern sheet metal forming processes. In order to investigate in advance whether a newly conceived process chain will be successful, finite element simulation can be very helpful, on the condition that it is based on model modules which capture the realistic behaviour sufficiently well. There are multiple sources for modelling errors: e.g. unrealistic material modelling, inappropiate finite element technology, inaccurate geometric boundary conditions or loading functions. The present paper focuses on the first aspect.

Today many authors incorporate the micro structural behaviour into their constitutive concept. Among these one differentiates between approaches based on crystallographic homogenization theory [2, 18] numerical coupling of multiple scales to predict crystallographic texture and mechanical anisotropy [8, 11, 12] and purely phenomenological models [3, 6, 20] which are motivated by the micro mechanical behaviour of polycrystals. Purely continuum mechanical approaches using structural tensors (see the theoretical contributions of [1, 16, 21]) can be found e.g. in [9, 10, 13–15, 17]. These papers include different concepts to model anisotropic plasticity of various materials, mainly soft materials (polymers, soft tissue). Further, efficient finite element implementations are discussed. However, the use of such models in forming simulations has not been fully exploited. In this context also the link to experiments is missing.

In the present paper we go beyond previous works by taking the evolution of the structural tensors explicitly into account. In this way we consider the evolution of elastic anisotropy and plastic anisotropy.

2 Motivation by a Rheological Model

Figure 1 shows a typical rheological combination to model the Bauschinger effect which will serve as motivation for our continuum mechanical modelling. A spring (Young's modulus E) is connected in series to another part consisting of a friction element (yield stress σ_y) and a Maxwell element (pseudo-viscosity $c/(\dot\lambda\, b)$, Young's modulus c). Here, b and c are material parameters, λ represents the plastic multiplier. It becomes further visible that the total strain ε is decomposed into three parts: $\varepsilon = \varepsilon_e + \varepsilon_{p_e} + \varepsilon_{p_i}$ where the first summand represents the elastic strain and the sum of the two other terms the total plastic strain which can be again additively decomposed into the strain of the second (hardening) spring and the strain in the pseudo-dashpot. We use here the notions pseudo-viscosity and pseudo-dashpot to indicate that we restrict ourselves to rate-independent material behaviour where the viscosity $c/(\dot\lambda\, b)$ of the dashpot is not a true material parameter but depends in addition to c and b on the plastic multiplier λ. The inelastic part of the plastic strain ε obeys then the evolution law $\dot\varepsilon_{p_i} = \dot\lambda\, b\, \varepsilon_{p_e}$. For the total plastic strain we can state the classical flow rule $\dot\varepsilon_p = \dot\lambda\, (\partial\Phi/\partial\sigma)$ where the yield function Φ is a function of the effective stress, i.e. the difference between the total stress and the back stress $c\,\varepsilon_{p_e}$. The Helmholtz free energy density ψ consists of the two Helmholtz free energy densities measured in the springs, i.e. $\psi = \psi_e + \psi_{\text{kin}}$ with $\psi_e = E\,\varepsilon^2/2$ and $\psi_{\text{kin}} = c\,\varepsilon_{p_e}^2/2$.

3 Continuum Mechanical Model

In order to achieve the transition to continuum mechanics it is important to differentiate between two physical phenomena. First of all dislocations in the material cause isotropic and kinematic hardening. Secondly, due to its crystal structure (texture) the material shows elastic (initial) as well as flow anisotropy. Both, elastic and flow anisotropy, are influenced by texture evolution and as such vary during the deformation process.

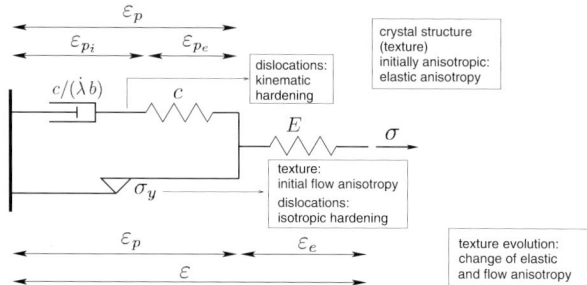

Fig. 1 Rheological model to motivate the present continuum mechanical approach.

3.1 Summary of the Constitutive Model

The complete constitutive model is summarized in Table 1 where also the symbols used in the equations are defined.

3.2 Basic Assumptions

Analogously to the additive split of the total linearized strain used in Section 3 we assume the triple multiplicative split $\mathbf{F} = \mathbf{F}_e \mathbf{F}_{p_e} \mathbf{F}_{p_i}$ (1) where the product (2) yields the plastic deformation gradient $\mathbf{F}_p$ (see also [5, 7, 19]).

Taking into account the information given in Section 3 and at the beginning of Section 4, the Helmholtz free energy density (3) is additively decomposed into two parts both of which are isotropic functions of its arguments.

To model anisotropic material behaviour the Helmholtz free energy density (3) is formulated in terms of so-called structural tensors $\mathbf{M}_i$ (4) with $\mathbf{N}_i$ having the absolute value $|\mathbf{N}_i| = 1$. The structural vectors $\mathbf{N}_i$ determine in general the anisotropy directions of the material. Here we focus on metallic materials which are characterized by a crystallographic structure. In this case the structural vectors $\mathbf{N}_i$ correspond to the reference lattice vectors and are as such defined in the intermediate configuration. Thereby $\mathbf{N}_3$ is assumed to point in thickness direction whereas $\mathbf{N}_1$ and $\mathbf{N}_2$ span the plane of the sheet metal. In this way $\mathbf{N}_j \cdot \mathbf{N}_3 = 0$ ($j = 1, 2$) holds throughout the analysis. For the derivation of the constitutive model we further need a relation between the structural tensors $\mathbf{M}_i$ ($i = 1, 2, 3$) in the intermediate configuration and the corresponding quantities $\bar{\mathbf{M}}_i$ ($i = 1, 2, 3$) in the undeformed (reference) configuration. As suggested e.g. by [13] we use for this purpose the relation $(4)_2$.

Table 1 Constitutive model.

Description	Equation	Number
multiplicative split of the deformation gradient $\mathbf{F}$ into elastic ($\mathbf{F}_e$) and and plastic ($\mathbf{F}_p$) parts	$\mathbf{F} = \mathbf{F}_e \mathbf{F}_p$	(1)
multiplicative split of the plastic deformation gradient $\mathbf{F}_p$ into elastic ($\mathbf{F}_{p_e}$) and inelastic ($\mathbf{F}_{p_i}$) parts	$\mathbf{F}_p = \mathbf{F}_{p_e} \mathbf{F}_{p_i}$	(2)
Helmholtz free energy density	$\psi = \psi_e(\mathbf{C}_e, \mathbf{M}_1, \mathbf{M}_2, \mathbf{M}_3)$ $+ \psi_{\text{kin}}(\mathbf{C}_{p_e})$	(3)
right Cauchy–Green tensor	$\mathbf{C}$	
elastic part of $\mathbf{C}$	$\mathbf{C}_e = \mathbf{F}_p^{-T} \mathbf{C} \mathbf{F}_p^{-1}$	
plastic right Cauchy–Green tensor	$\mathbf{C}_p$	
elastic part of $\mathbf{C}_p$	$\mathbf{C}_{p_e} = \mathbf{F}_{p_i}^{-T} \mathbf{C}_p \mathbf{F}_{p_i}^{-1}$	
structural tensor in the undeformed configuration	$\bar{\mathbf{M}}_i = \bar{\mathbf{N}}_i \otimes \bar{\mathbf{N}}_i \ (i = 1, 2, 3)$	
corresponding structural vector	$\bar{\mathbf{N}}_i$	
structural tensor in the intermediate configuration	$\mathbf{M}_i = \mathbf{N}_i \otimes \mathbf{N}_i$ $= \mathbf{F}_p \bar{\mathbf{M}}_i \mathbf{F}_p^T / (\mathbf{C}_p \cdot \bar{\mathbf{M}}_i)$	(4)
corresponding structural vector	$\mathbf{N}_i$	
second Piola–Kirchhoff stress tensor	$\mathbf{S} = 2 \mathbf{F}_p^{-1} (\partial \psi_e / \partial \mathbf{C}_e) \mathbf{F}_p^{-T}$	(5)
"Mandel"-type effective stress tensor	$\mathbf{\Sigma} = 2 \mathbf{C}_e (\partial \psi_e / \partial \mathbf{C}_e)$ $- \sum_{i=1}^{3} 2 [(\partial \psi_e / \partial \mathbf{M}_i) \mathbf{M}_i$ $+ \text{tr}((\partial \psi_e / \partial \mathbf{M}_i) \mathbf{M}_i) \mathbf{I}]$ $- 2 \mathbf{F}_{p_e} (\partial \psi_{\text{kin}} / \partial \mathbf{C}_{p_e}) \mathbf{F}_{p_e}^T$	(6)
back stress due to kinematic hardening	$\mathbf{\Sigma}_{\text{kin}} = 2 \mathbf{C}_{p_e} (\partial \psi_{\text{kin}} / \partial \mathbf{C}_{p_e})$	(7)
"anisotropy" stress tensor in the intermediate configuration	$\mathbf{\Omega}_i = -\partial \psi_e / \partial \mathbf{M}_i$ $+ \text{tr}((\partial \psi_e / \partial \mathbf{M}_i) \mathbf{M}_i) \mathbf{I}$	(8)
"anisotropy" stress tensor in the undeformed configuration	$\bar{\mathbf{\Omega}}_i = \mathbf{F}_p^T \mathbf{\Omega}_i \mathbf{F}_p$	(9)
evolution of anisotropy directions	$(\bar{\mathbf{N}}_i)^{\cdot} = \dot{\lambda} \, d_i \, \bar{\mathbf{\Omega}}_i [\bar{\mathbf{N}}_i]$	(10)
plastic multiplier	$\dot{\lambda}$	
evolution of plastic deformation	$\mathbf{d}_p = \text{sym} \, \mathbf{l}_p = \dot{\lambda} \, \partial \Phi / \partial \mathbf{\Sigma}^D$	(11)
evolution of the inelastic part of the plastic deformation	$\mathbf{d}_{p_i} = \dot{\lambda} \, \frac{b}{c} \mathbf{\Sigma}_{\text{kin}}^D$	(12)
yield function $((\ldots)^D$ deviator)	$\Phi = (\mathbf{\Sigma}^D \cdot (\mathcal{Y}[\mathbf{\Sigma}^D]))^{1/2} - \sigma_y$	(13)
orthotropic fourth order tensor	$\mathcal{Y}$	
Kuhn–Tucker conditions	$\dot{\lambda} \geq 0, \Phi \leq 0, \dot{\lambda} \Phi = 0$	(14)

3.3 Exploitation of the Clausius–Duhem Inequality

At this point we exploit the Clausius–Duhem inequality $-\dot{\psi} + \mathbf{S} \cdot (\dot{\mathbf{C}}/2) \geq 0$ which is especially important to derive thermodynamically consistent equations for the evolution of the anisotropy directions and to incorporate the influence of anisotropy into the flow rules. We obtain after some steps of tensor analysis the requirement

$$\left(\mathbf{S} - 2\mathbf{F}_p^{-1} \frac{\partial \psi_e}{\partial \mathbf{C}_e} \mathbf{F}_p^{-T}\right) \cdot \frac{1}{2}\dot{\mathbf{C}} + \mathbf{\Sigma} \cdot \mathbf{l}_p + \mathbf{\Sigma}_{\text{kin}} \cdot \mathbf{d}_{p_i} + \sum_{i=1}^{3} \mathbf{\Omega}_i \cdot \mathbf{m}_i \geq 0 \quad (15)$$

The variables $\mathbf{S}$, $\mathbf{\Sigma}$, $\mathbf{\Sigma}_{\text{kin}}$ and $\mathbf{\Omega}_i$ ($i = 1, 2, 3$) are defined in Table 1 (5–8). The so-called "anisotropy" stress tensor $\mathbf{\Omega}_i$ is a stress quantity thermodynamically conjugate to the rate

$$\mathbf{m}_i := \frac{\mathbf{F}_p (\bar{\mathbf{M}}_i)^{\cdot} \mathbf{F}_p^T}{\mathbf{C}_p \cdot \bar{\mathbf{M}}_i} \quad (16)$$

It is only present when the Helmholtz free energy density depends on the structural tensors, i.e. when the so-called elastic anisotropy is incorporated. The tensor $\mathbf{l}_p = \dot{\mathbf{F}}_p \mathbf{F}_p^{-1}$ refers to the plastic velocity gradient. Analogously we define $\mathbf{l}_{p_i} = \dot{\mathbf{F}}_{p_i} \mathbf{F}_{p_i}^{-1}$ and $\mathbf{d}_{p_i} = \text{sym}\, \mathbf{l}_{p_i}$.

At the first sight one would assume that the Mandel-type stress tensor $\mathbf{\Sigma}$ is non-symmetric and seek to find a physically reasonable equation for $\mathbf{l}_p$. However, it is one of the crucial points of the present paper to show that $\mathbf{\Sigma}$ is in fact symmetric. For this purpose we exploit the isotropy of ψ_e by stating

$$\psi_e = \hat{\psi}(\mathbf{C}_e, \mathbf{M}_1, \mathbf{M}_2, \mathbf{M}_3) \quad (17)$$
$$= \hat{\psi}(\mathbf{Q}\,\mathbf{C}_e\,\mathbf{Q}^T, \mathbf{Q}\,\mathbf{M}_1\,\mathbf{Q}^T, \mathbf{Q}\,\mathbf{M}_2\,\mathbf{Q}^T, \mathbf{Q}\,\mathbf{M}_3\,\mathbf{Q}^T) \quad (18)$$

for all rotation tensors $\mathbf{Q}$. See for more details in this context [17]. The total time derivative of (17) must equal the total time derivative of (18). This leads us to the equation

$$\frac{\partial \hat{\psi}(\mathbf{Q}\,\mathbf{C}_e\,\mathbf{Q}^T, \mathbf{Q}\,\mathbf{M}_1\,\mathbf{Q}^T, \mathbf{Q}\,\mathbf{M}_2\,\mathbf{Q}^T, \mathbf{Q}\,\mathbf{M}_3\,\mathbf{Q}^T)}{\partial \mathbf{Q}} \cdot \dot{\mathbf{Q}} = 0 \quad (19)$$

which alternatively reads $\mathbf{\Sigma} \cdot \mathbf{W} = 0$. Due to the fact that $\mathbf{W} := \dot{\mathbf{Q}}^T \mathbf{Q}$ is a skew-symmetric tensor the latter relation can only be fulfilled if $\mathbf{\Sigma}$ is symmetric. This result is also tacitly included e.g. in [9] and has already been numerically exploited by Reese [13] for the multi-scale modelling of pneumatic membranes. An important consequence of the symmetry of $\mathbf{\Sigma}$ is the fact that the second term in the Clausius–Duhem inequality (15) reduces to

$$\mathbf{\Sigma} \cdot \mathbf{l}_p = (\mathbf{F}_p^{-1} \mathbf{\Sigma} \mathbf{F}_p^{-T}) \cdot \dot{\mathbf{C}}_p \frac{1}{2} = \bar{\mathbf{\Sigma}} \cdot \dot{\mathbf{C}}_p \frac{1}{2} \quad (20)$$

It can be shown that the referential stress measure $\bar{\mathbf{\Sigma}}$ is a function of $\mathbf{C}$, $\mathbf{C}_p$, $\mathbf{C}_{p_i} = \mathbf{F}_{p_i}^T \mathbf{F}_{p_i}$ and $\bar{\mathbf{M}}_i$ ($i = 1, 2, 3$). The quantities $\mathbf{C}_p$, $\mathbf{C}_{p_i}$ and $\bar{\mathbf{M}}_i$ ($i = 1, 2, 3$) take the function of internal variables. They are all symmetric. Hence, the tensors $\mathbf{F}_p$ and $\mathbf{F}_{p_i}$ are only determined up to their stretch parts $\mathbf{U}_p = \sqrt{\mathbf{C}_p}$ and $\mathbf{U}_{p_i} = \sqrt{\mathbf{C}_{p_i}}$, respectively. The rotational parts $\mathbf{R}_p$ and $\mathbf{R}_{p_i}$ remain *unknown*. Likewise it can be said that the spins $\mathbf{w}_p = \text{skew}\,\mathbf{l}_p$ and $\mathbf{w}_{p_i} = \text{skew}\,\mathbf{l}_{p_i}$ remain unknown. It should be clearly emphasized that the fact that a quantity remains unknown does *not* imply that it is set equal to zero. We simply never need and never obtain any information about $\mathbf{R}_p$ or $\mathbf{R}_{p_i}$. Consequently $\mathbf{F}_p$ and $\mathbf{F}_{p_i}$ remain undetermined.

Concerning the modelling of kinematic hardening we do not go beyond the scope of [19]. For this reason we will not discuss this issue further. Important is, however, the evolution of the structural tensors because the change of the anisotropy axes has to be realistically described.

3.4 Evolution of Structural Tensors

Assuming that $\mathbf{m}_i$ is independent of $\mathbf{d}_p$ and $\mathbf{d}_{p_i}$ we need to guarantee that the term $\sum_{i=1}^{3} \mathbf{\Omega}_i \cdot \mathbf{m}_i$ is zero or positive at all times. Further we need to take into account that $\mathbf{M}_i$ as well as $\bar{\mathbf{M}}_i$ represent dyadic products of the structural vectors $\mathbf{N}_i$ and $\bar{\mathbf{N}}_i$, respectively. As already mentioned the vectors $\mathbf{N}_i$ and $\bar{\mathbf{N}}_i$ ($i = 1, 2, 3$) have the length 1. Both properties must not be violated by the evolution equations if the physical meaning of the structural tensors should be maintained. Therefore it is more natural to formulate an evolution equation for the structural vector $\bar{\mathbf{N}}_i$. The scalar product $\mathbf{\Omega}_i \cdot \mathbf{m}_i$ can be alternatively represented by $(\bar{\mathbf{N}}_i)^{\cdot} \cdot \bar{\mathbf{\Omega}}_i [\bar{\mathbf{N}}_i]$ where the referential stress $\bar{\mathbf{\Omega}}_i$ is given by (9). In the context of a rate-independent model this suggests the evolution equation (10) where $d_i \geq 0$ ($i = 1, 2, 3$) represent material-dependent factors which are additionally used to preserve the length $|\bar{\mathbf{N}}_i| = 1$.

3.5 Further Evolution Equations

In the following we have to formulate evolution equations for the other internal variables (i.e. $\mathbf{C}_p$ and $\mathbf{C}_{p_i}$) as well. This is done in the classical way by stating (11) and (12). Further we have to fulfill the Kuhn–Tucker conditions (14). The yield function Φ is e.g. the one suggested by Hill (13).

4 Integration Algorithm

The next point is to discuss the implementation of the evolution equation into a finite element formulation. A main problem in large strain inelasticity is in general the fact that metallic materials usually exhibit approximately plastically incompressible behaviour. Therefore, it is important to establish an integration algorithm which properly accounts for this property also in the numerical analysis.

It can be easily shown that using above evolution equations for $\mathbf{d}_p$ and $\mathbf{d}_{p_i}$, respectively, the referential rates $\dot{\mathbf{C}}_p = 2\mathbf{F}_p^T \mathbf{d}_p \mathbf{F}_p$ and $\dot{\mathbf{C}}_{p_i} = 2\mathbf{F}_{p_i}^T \mathbf{d}_{p_i} \mathbf{F}_{p_i}$ are functions of $\mathbf{C}$, $\mathbf{C}_p$, $\mathbf{C}_{p_i}$ and $\bar{\mathbf{M}}_i$ ($i = 1, 2, 3$) only. Note that in contrast to crystal plasticity models where usually an evolution equation for $\dot{\mathbf{F}}_p$ has to be integrated we deal here with six-dimensional internal variables (in Voigt notation). In recent papers (see [5], [19]) we have developed efficient integration algorithms for the corresponding isotropic situation which are both based on the exponential map and work very robustly and efficiently. It is another advantage of the present model of anisotropic plasticity that these concepts can be taken over with hardly any changes.

5 Numerical Examples

To carry out numerical simulations the material model has to be further specified. This concerns at first the choice of the Helmholtz free energy density parts ψ_e and ψ_{kin}. For ψ_e we choose as first summand a Neo-Hooke term. It depends on the three invariants of $\mathbf{C}_e$ and includes the two material parameters μ and Λ (shear modulus and Lamé constant, respectively). The second summand of ψ_e reads $K_1 (I_4 - 1)^{\alpha_1} + K_2 (I_6 - 1)^{\alpha_2}$ (see also [13]). It is a function of $I_4 = \mathbf{C}_e \cdot \mathbf{M}_1$ and $I_6 = \mathbf{C}_e \cdot \mathbf{M}_2$. K_i and α_i ($i = 1, 2$) are material parameters.

In principal the model is general enough to allow large elastic deformation, although this is usually not needed in the context of metals. However, the concept is in this way applicable to other material classes as e.g. polymers which also show a high amount of anisotropic plastification. The form of ψ_{kin} is similar to the Neo-Hooke form. But as it is common in the context of kinematic hardening we neglect the volumetric part. The Bauschinger effect is here modelled in a large strain version of the Armstrong-Frederick form. For the yield potential we choose the Hill function. The following parameters are used: $\mu = 80000$ MPa, $\Lambda = 120000$ MPa, $K_1 = 30000$ MPa, $K_2 = 20000$ MPa, $\alpha_1 = \alpha_2 = 2$, $c = 600$ MPa, $b = 4$, Hill parameters: f = 0.25, g = h = 1, l = m = n = 1.5 (if nothing said otherwise).

5.1 Evolution of Structural Vectors

The above choice for ψ_e (with $\alpha_1 = \alpha_2 = 2$) lets us further specify the stresses $\bar{\boldsymbol{\Omega}}_i$ ($i = 1, 2, 3$) which read

$$\bar{\boldsymbol{\Omega}}_i = 2 K_i \frac{(\mathbf{C} - \mathbf{C}_p) \cdot \bar{\mathbf{M}}_i}{\mathbf{C}_p \cdot \bar{\mathbf{M}}_i} \left((\mathbf{C} \cdot \bar{\mathbf{M}}_i) \mathbf{C}_p - (\mathbf{C}_p \cdot \bar{\mathbf{M}}_i) \mathbf{C} \right) \qquad (21)$$

Let us look at two special cases.

Biaxial tension. If the anisotropy directions coincide with the loading directions the structural vectors are given by $\bar{\mathbf{N}}_1^T = \{1, 0, 0\}$, $\bar{\mathbf{N}}_2^T = \{0, 1, 0\}$ and $\bar{\mathbf{N}}_3^T = \{0, 0, 1\}$. We choose $d_1 = d_2 = d$ and $d_3 = 0$. The coefficients C_{jk} and $(C_p)_{jk}$ with $j \neq k$ and $j, k = 1, 2, 3$ are zero. Consequently, also $\bar{\boldsymbol{\Omega}}_i$ ($i = 1, 2, 3$) has diagonal form. We obtain for the 11- and the 22-

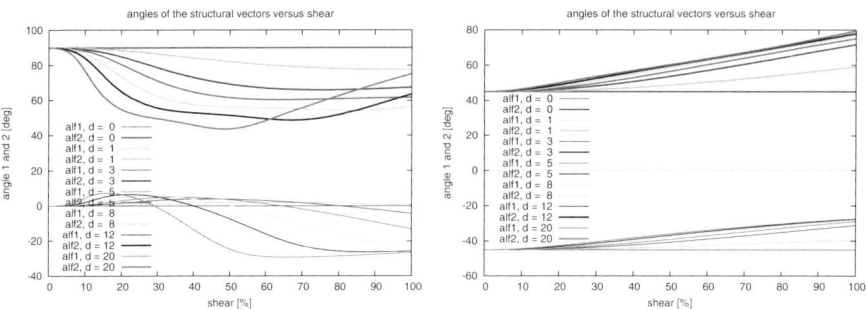

Fig. 2 Orientation of the structural vectors dependent on the shear strain. Initial orientations: (a) $\alpha_1 = 0°, \alpha_2 = 90°$, (b) $\alpha_1 = -45°, \alpha_2 = 45°$.

coefficients of the tensor $\mathbf{X}_i := (\mathbf{C} \cdot \bar{\mathbf{M}}_i) \mathbf{C}_p - (\mathbf{C}_p \cdot \bar{\mathbf{M}}_i) \mathbf{C}$ the expressions $(X_1)_{11} = C_{11}(C_p)_{11} - (C_p)_{11} C_{11} = 0$ and $(X_1)_{22} = C_{11}(C_p)_{22} - (C_p)_{11} C_{22} \neq 0$. Analogously we write $(X_2)_{11} = C_{22}(C_p)_{11} - (C_p)_{22} C_{11} \neq 0$ and $(X_2)_{22} = C_{22}(C_p)_{22} - (C_p)_{22} C_{22} = 0$. Thus, for biaxial tension the statement $(\bar{\Omega}_1)_{11} = (\bar{\Omega}_2)_{22} = 0$ holds. The products $\bar{\Omega}_1 [\bar{\mathbf{N}}_1]$ and $\bar{\Omega}_2 [\bar{\mathbf{N}}_2]$ vanish which means that no evolution of $\bar{\mathbf{N}}_1$ and $\bar{\mathbf{N}}_2$ takes place. This is the expected and physically reasonable result.

Simple shear. In the second place we consider a homogeneous simple shear experiment where the displacement in horizontal direction depends linearly on the coordinate in vertical direction. The vertical displacement is zero. The structural vectors $\bar{\mathbf{N}}_1$ and $\bar{\mathbf{N}}_2$ are given by $\bar{\mathbf{N}}_1^T = \{\cos\alpha_1, \sin\alpha_1, 0\}$ and $\bar{\mathbf{N}}_2^T = \{\cos\alpha_2, \sin\alpha_2, 0\}$. The dependence of the angles α_1 ("alf1") and α_2 ("alf2") on the shear strain (given in percent) for different values $d_1 = d_2 = d$ and two initial orientations is plotted in Figures 2a and 2b. In the moderate shear regime (Figure 2a, initial orientation: $\alpha_1 = 0°, \alpha_2 = 90°$) the angle α_1 decreases whereas the angle α_2 increases. The anisotropy directions orient themselves towards the directions of maximum strain. This observation agrees well with physical intuition. However, the situation changes for rather large shear angles. Then α_1 takes on negative values and α_2 exceeds 90°. Whether such a phenomenon is also seen in reality needs to be clarified by means of experimental investigations. It could be also a consequence of working with an anisotropic yield potential which is influenced by the rotation of the axes as well. Certainly experimental data are in general needed to determine the material parameter d. In Figure 2b (initial orientation: $\alpha_1 = -45°, \alpha_2 = 45°$) both angles monotonically increase, i.e. the anisotropy axes rotate counterclockwise.

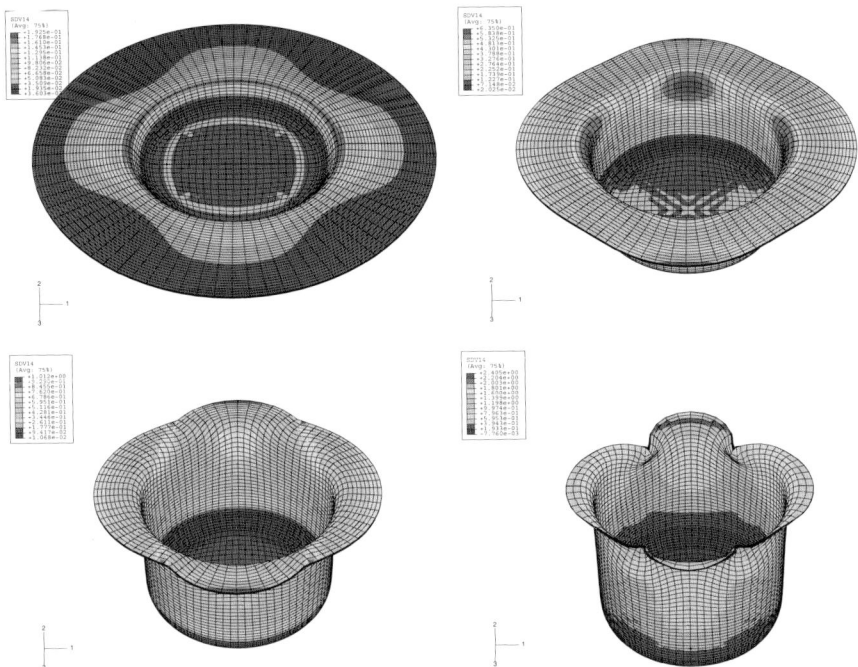

Fig. 3 Simulation of a deep-drawing process.

5.2 Forming Processes

The final task is to apply the model to the simulation of forming processes. Implementing the new constitutive concept as UMAT subroutine into the finite element software package ABAQUS lets us numerically investigate complex processes as e.g. the deep-drawing of a cup. For this purpose the present model has been adapted to results of Choi et al. [4]. In Figure 3 different stages of the deep-drawing process are depicted. The expected deviation from axial symmetry and the occurrence of 'earings' are obvious.

References

1. Boehler JP (1979) *Zeitschrift für Angewandte Mathematik und Mechanik* 59:157–167.
2. Böhlke T, Risy G, Bertram A (2006) *Modelling and Simulation in Materials Science and Engineering* 14:365–387.
3. Cazacu O, Plunkett B, Barlat F (2006) *International Journal of Plasticity* 22:1171–1194.
4. Choi Y, Han CS, Lee JK, Wagoner RH (2006) *International Journal of Plasticity* 22:1765–1783.

5. Dettmer W, Reese S (2004) *Computer Methods in Applied Mechanics and Engineering* 193:87–116.
6. Li S, Hoferlin E, Van Bael A, Van Houtte P, Teodosiu C (2003) *International Journal of Plasticity* 19:647–674.
7. Lion A (2000) *International Journal of Plasticity* 16:469–494.
8. Logé RE, Chastel YB (2006) *Computer Methods in Applied Mechanics and Engineering* 195:6843–6857.
9. Menzel A, Steinmann P (2003) *Computer Methods in Applied Mechanics and Engineering* 192:3431–3470.
10. Papadopoulos P, Lu J (2001) *Computer Methods in Applied Mechanics and Engineering* 190:4889–4910.
11. Plunkett B, Lebensohn RA, Cazacu O, Barlat F (2006) *Acta Materialia* 54:4159–4169.
12. Raabe D, Wang Y, Roters F (2005) *Computational Materials Science* 34:221–234.
13. Reese S (2003) *International Journal of Solids and Structures* 40:951–980.
14. Sansour C, Karsaj I, Soric J (2006) *International Journal of Plasticity* 22:2246–2365.
15. Schröder J, Gruttmann F, Löblein J (2002) *Computational Mechanics* 30:48–64.
16. Spencer AJM (1984) *Continuum Theory of the Mechanics of Fibre-Reinforced Composites*. Springer, Wien New York.
17. Svendsen B (2001) *International Journal of Solids and Structures* 38:9579–9599.
18. Van Houtte P, Kanjarla AK, Van Bael A, Seefeldt M, Delannay L (2006) *European Journal of Mechanics A/Solids* 25:634–648.
19. Vladimirov IN, Pietryga M, Reese S (2008) *International Journal for Numerical Methods in Engineering* (in press) doi:10.1002/nme.2234.
20. Wang J, Levkovitch V, Reusch F, Svendsen B, Huétink J, van Riel M (2008) *International Journal of Plasticity* (in press) doi:10.1016/j.ijplas.2007.08.009.
21. Zhang JM, Rychlewski J (1990) *Archives of Mechanics* 42:267–277.

Inelastic Media under Uncertainty: Stochastic Models and Computational Approaches

Hermann G. Matthies and Bojana V. Rosić

Institute of Scientific Computing, Technische Universität Braunschweig, D-38092 Brunswick, Germany
E-mail: wire@tu-bs.de

Abstract. We discuss inelastic media under uncertainty modelled by probabilistic methods. As a prototype inelastic material we consider perfect plasticity. We propose a mathematical formulation as a stochastic variational inequality. The new element vis à vis a stochastic elastic medium is the so-called return map at each Gauss-point. We concentrate on a stochastic version of this, showing how it may be solved via (generalised) polynomial expansion.

Explicit formulas are provided for plane strain, together with results from example computations, giving a stochastic inelastic or irreversible time-step procedure. This may serve as a prototype example for any other irreversible behaviour, especially of the rate-independent kind.

Key words: inelastic irreversible behaviour, uncertainty, stochastic model, stochastic variational inequality, stochastic generalised return map, chaos expansion.

1 Introduction

In many cases in systems with inelastic materials some details are not precisely known. This is termed uncertainty. It arises from different sources: loading of the material (due to wind, waves, etc.), or more generally the action from the surrounding environment; but the system itself may also contain only incompletely known parameters, processes or fields (not possible or too costly to measure), or there may be small, unresolved scales in the model, which act as a kind of background noise. All these items introduce some uncertainty in the model. We treat such uncertainties with stochastic models, which allow a rich theoretical and computational structure.

Material parameters describing both the elastic/reversible behaviour as well as those pertaining to the inelastic/irreversible behaviour are modelled as spatial random fields. The elastic as well as the inelastic behaviour has to be derived in terms of these random descriptors. We here focus on the simplest such situation, which nonetheless is completely typical of irreversible material behaviour and easily extendible to any kind of inelastic description, namely linear elastic perfectly plastic material. The model is described by partial differential equations with random coefficients. In general, there are several mathematical/numerical methods to deal with the time evolution of such stochastic systems. They each involve very different views of the underlying randomness [21, 25]. Some of them are perturbation techniques [18], theories of probability distributions (Master-Equation, Fokker–Planck equation [11]), methods based on direct integration (Monte Carlo [4], Smolyak [24, 28]), methods based on direct approximation (traditional response surface methods, White Noise Analysis [9, 10, 17], stochastic Galerkin, stochastic collocation [3, 7, 12, 13, 19, 20, 22, 23, 26, 29]) and approximate descriptions for plasticity [2]). While the reversible behaviour has been addressed in the stochastic setting already numerous times, e.g. [23], for the inelastic behaviour there are only few attempts. We give stochastic evolution laws for the internal variables, complementing the stochastic description of the elastic behaviour. When attempting a computational treatment of such stochastic inelastic materials, it becomes a necessity to employ as few random variables as possible. We use the Karhunen–Loève expansion and sparse tensor product representations to approach this problem [20, 23]. Random variables are represented through polynomial chaos type expansions à la Wiener, the truncation of which finally gives a completely discretised formulation which may be implemented on a computer. The result is a description of the evolution of the response variables and internal parameters in terms of those simpler random variables alluded to before. This in turn then enables us to compute various quantities of interest which are functionals of the solution.

2 Model Problem

The first Section 2.1 will mainly define notation and recall the deterministic setting, so it can be contrasted with the stochastic formulation in Section 2.3.

2.1 Deterministic Problem of Elasto-Plastic Flow

The object of our investigation is a quasistatic linear elastic material with perfect plasticity occupying a domain $\mathcal{G} \subset \mathbb{R}^d$ [6, 8, 14, 27]. Let $I = [0, T]$ be the interval of of interest, and let Σ and $\mathcal{U}$ be the space of stresses resp. displacements. Denote by $(\sigma, u) : I \to \Sigma \times \mathcal{U}$ the stress and displacement evolution. The data of the problem are the elastic domain, a non-empty convex closed set $\mathcal{K} \subset \Sigma$, the boundary $\partial \mathcal{K}$ of which denotes the yield surface, and a loading function $f : I \to \mathcal{F}$, where $\mathcal{F}$ is a space of forces, in duality with $\mathcal{U}$. For simplicity we assume homogeneous boundary conditions. The equilibrium equation for each time $t \in I$ are given by

$$\langle D^*\sigma, v \rangle_\mathcal{U} = \langle f, v \rangle_\mathcal{U} := \int_\mathcal{G} f(x, t) \cdot v(x) \, dx \quad \forall v \in \mathcal{U}, \qquad (1)$$

where D is a linear operator $D : \mathcal{U} \to \mathcal{E}$ mapping displacements to strains $\varepsilon = Du \in \mathcal{E}$, D^* is the dual operator, and $\langle .,.\rangle_\mathcal{U}$ denotes the duality paring. In the case of small deformations the total deformation consist of two parts: the plastic deformation ε_p and the elastic deformation ε_e i.e. $\varepsilon = \varepsilon_e + \varepsilon_p$. In that case one can write the elastic constitutive equation $\langle \epsilon, A\varepsilon_e\rangle_\Sigma = \langle \epsilon, \sigma\rangle_\Sigma$ for all $\epsilon \in \mathcal{E}$, defined by Hooke's law with elastic constants in A. The material behaves elastic while the stress is in the interior int $\mathcal{K}$ of the elastic domain, whereas it becomes perfectly plastic when the stress lies on the boundary $\partial \mathcal{K}$. This means that the following variational inequality holds

$$\langle \dot{\varepsilon}_p, \sigma - \tau\rangle_\Sigma \leq 0 \qquad \forall \tau \in \mathcal{K}, \tag{2}$$

i.e. that the plastic flow $\dot{\varepsilon}_p$ must be in the normal cone to $\mathcal{K}$ at the point σ.

2.2 Random Variables and Random Fields

Assume that $(\Omega, \mathfrak{A}, \mathbb{P})$ is a probability space with probablity measure $\mathbb{P}$ and a σ-algebra of events $\mathfrak{A}$. A random variable (RV) r we will consider to be a measurable map $r : \Omega \to \mathcal{V}$ into a vector space $\mathcal{V}$, which is often equal to $\mathbb{R}$ [1, 15, 16].

The so-called second order information of a random variable involves the mean value $\bar{r} := \mathbb{E}(r(\cdot)) := \int_\Omega r(\omega)\, \mathbb{P}(d\omega)$ with the expectation $\mathbb{E}(\cdot)$, and its auto-covariance $\mathbf{C}_r := \mathbb{E}(\tilde{r} \otimes \tilde{r})$, where we have a splitting into the mean and the fluctuating part $\tilde{r}(\omega) = r(\omega) - \bar{r}$, with $\mathbb{E}(\tilde{r}) = 0$.

If the vector space $\mathcal{V}$ where the RV takes its values is a space of functions, say $C(\Phi, \mathbb{R})$, the space of continous functions on Φ, we call such a RV a random process if Φ is interpreted as time, or a random field if Φ is interpreted as space [1, 15]. Similarly to a "simple" random variable we may define the mean value of a random field $\bar{\kappa}(x) := \mathbb{E}(\kappa_\omega(x))$, and also the fluctuating part $\tilde{\kappa}(x, \omega)$. The covariance can consider different positions, it is defined as $\mathbf{C}_\kappa(x_1, x_2) := \mathbb{E}(\tilde{\kappa}(x_1, \cdot) \otimes \tilde{\kappa}(x_2, \cdot))$. The covariance function will be crucial later on in Section 3.1 for the dicretisation of the input random fields.

2.3 Stochastic Model Problem

Heterogeneities at the micro-structural level are usually subject to a number of uncertainties. Material properties of a heterogeneous material usually are not known with certainty at each point. So the behaviour of these materials must be derived in terms of random descriptors, which are modeled as random fields.

Assume that the loading f, elastic law A, and elastic domain $\mathcal{K}$ (plastic flow rule) are uncertain, given by a probabilistic model with realisations $\omega \in \Omega$. The formulation of the stochastic model problem is based on introducing the uncertainty as random processes and fields, i.e. we are given a random loading history $f(t, \omega) = f(x, t, \omega)$ a random field $A(x, \omega)$ of elastic coefficients, and random elastic domains $\mathcal{K}(x, \omega)$, and the equilibrium equations are required to hold for (almost) all ω (almost surely), which may be put in a variational statement:

$$\langle\!\langle D^*\sigma, v \rangle\!\rangle_\mathcal{V} = \langle\!\langle f, v \rangle\!\rangle_\mathcal{V} := \mathbb{E}(\langle f(\omega), v(\omega) \rangle_\mathcal{V}) \qquad \forall v \in \mathcal{V}, \tag{3}$$

as well as the stochastic elastic constitutive model

$$\langle\!\langle \epsilon, A\varepsilon_e \rangle\!\rangle_\Sigma = \langle\!\langle \epsilon, \sigma \rangle\!\rangle_\Sigma := \mathbb{E}(\langle \epsilon, \sigma \rangle_\Sigma) \qquad \forall \epsilon \in \mathcal{E}. \tag{4}$$

The stresses must be in the elastic domain $\sigma(t, \omega) \in \mathcal{K}(\omega)$ almost surely, where the plastic flow rule is given by the stochastic Prandtl law

$$\langle\!\langle \dot{\varepsilon}_p, \sigma - \tau \rangle\!\rangle_\Sigma \leq 0 \qquad \forall \tau \in \mathcal{K}. \tag{5}$$

We will consider isotropic material for the sake of demonstration and simplicity, although this is an unrealistic representation of heterogeneous realisations. The yield surface will be described with von Mises yield criteria, making this model the simplest case of an irreversible material behaviour. The properties of the material (bulk modulus $K(x, \omega)$, shear modulus $G(x, \omega)$ and yield stress $\sigma_y(x, \omega)$) are supposed to be random fields. Randomness will appear also in solutions like quantities $u(x, t, \omega)$, $\varepsilon_e(x, t, \omega)$, and $\varepsilon_p(x, t, \omega)$. Properties of the material like bulk and shear modulus as well as yield stress must be positive, so they are assumed to be log-normal random fields. Let κ be a general symbol for one of these random fields, then we assume $\kappa(x, \omega) \geq \kappa_0(x) > 0$, and also assume a modified log-normal distribution of the form $\kappa(x, \omega) = \kappa_0(x) + \kappa_1(x) \exp(\gamma(x, \omega))$, where $\gamma(x, \omega)$ is a Gaussian random field.

3 Computational Approach

The goal of the stochastic analysis is to compute functionals of the solution in the form

$$\Psi_u = \langle \Psi(u) \rangle := \mathbb{E}(\Psi(u)) := \int_\Omega \int_\mathcal{G} \Psi(u(x, \omega), x, \omega) \, dx \, \mathbb{P}(d\omega). \tag{6}$$

In the simplest case this could be the mean value of $\bar{u} = \mathbb{E}(u)$, or the variance of the fluctuacting part $\text{var}_u = \mathbb{E}((\tilde{u})^2)$, where $\tilde{u} = u - \bar{u}$, or the probability $\mathbb{P}\{u \leq u_0\} = \mathbb{P}(\{\omega \in \Omega | u(\omega) \leq u_0\}) = \mathbb{E}(\chi_{\{u \leq u_0\}})$. All such desirables are usually expected values of some functional, to be computed via (high dimensional) integration over Ω. Methods for calculating it are distinguished by how expensive the evaluation of $u(\omega_z)$ is in a sum approximating the above integrals, and how many evaluation points ω_z are needed.

3.1 Karhunen–Loève Expansion

For computing the desired output functionals an effective computational representation of random fields is needed. This is furnished using the Karhunen–Loève expansion (KLE) for discretising the spatial aspect, and Wiener polynomial chaos expansion (PCE) for discretising random variables in terms of simpler – in this case Gaussian RVs. The KLE is also known as a proper orthogonal decomposition, and may also be seen as a singular value decomposition of the random field [20]. Given the covariance function $\mathbf{C}_\kappa(x, y)$ one considers the Fredholm eigenproblem [5]:

$$\int_\mathcal{G} \mathbf{C}_\kappa(x, y) g_j(y) \, dy = \varkappa_j^2 g_j(x) \quad \text{with} \quad \int_\mathcal{G} g_j(x) g_k(x) \, dx = \delta_{jk}. \qquad (7)$$

As the covariance function is symmetric and positive semi-definite, (7) yields positive decreasingly ordered eigenvalues $\varkappa_j^2$ with only accumulation point zero, and a complete orthonormal set of eigenfunctions, such that the spectral decomposition will have the form

$$\mathbf{C}_\kappa(x, y) = \sum_{j=1}^\infty \varkappa_j^2 g_j(x) g_j(y). \qquad (8)$$

This gives a possibility to synthesise the random field κ through its KLE:

$$\kappa(x, \omega) = \bar{\kappa}(x) + \sum_{j=1}^\infty \varkappa_j \, g_j(x) \xi_j(\omega) =: \sum_{j=0}^\infty \varkappa_j \, g_j(x) \xi_j(\omega), \qquad (9)$$

where $\xi_j(\omega)$ are centred, normalised, uncorrelated random variables, i.e. $\mathbb{E}\left(\xi_j\right) = 0$, $\mathbb{E}\left(\xi_j \xi_k\right) =: \langle \xi_j, \xi_k \rangle_{L_2(\Omega)} = \delta_{jk}$. In actual computations the sum in (9) is truncated to a finite number of terms.

The RVs in the KLE (9) can be represented as functions of other simpler RVs, i.e. Gaussian RVs (classical Wiener Chaos). Other possibilities exist also [3, 29]. Best is to use orthogonal polynomials in the underlying measure, in this case Hermite polynomials for Gaussian random variables.

3.2 Polynomial Chaos Expansion

Each random variable $\xi_j(\omega)$ from the KLE (9) may be expanded in a polynomial chaos expansion (PCE) [20, 23]:

$$\xi_j(\omega) = \sum_{\alpha \in \mathcal{J}} \xi_j^{(\alpha)} H_\alpha(\boldsymbol{\theta}(\omega)), \qquad (10)$$

with Hermite polynomials in Gaussian random variables [10, 17] $\{\theta_m(\omega)\}_{m=1}^\infty =: \boldsymbol{\theta}(\omega)$:

$$H_\alpha(\boldsymbol{\theta}(\omega)) = \prod_{j=1}^{\infty} h_{\alpha_j}(\theta_j(\omega)), \tag{11}$$

where $h_\ell(\vartheta)$ are the usual Hermite polynomials, and

$$\mathcal{J} := \left\{ \alpha = (\alpha_1, \ldots, \alpha_j, \ldots), \alpha_j \in \mathbb{N}_0, |\alpha| := \sum_{j=1}^{\infty} \alpha_j < \infty \right\} \tag{12}$$

are multi-indices, only finitely many of which are non-zero. The Hermite polynomials satisfy: $\langle H_\alpha, H_\beta \rangle_{L_2(\Omega)} = \mathbb{E}\left(H_\alpha H_\beta\right) = \alpha! \delta_{\alpha\beta}$, where $\alpha! := \prod_{j=1}^{\infty}(\alpha_j!)$. Again in a practical calculation, only finitely many RVs θ_m will be used, and the sum in (10) will be truncated to a finite number of terms.

3.3 Time Discrete Computation

Let quantities at the initial time t_0 be denoted with $[.]_0$, those at time t_1 with $[.]_1$ and increments with $\Delta[.]$, where $t_1 := t_0 + \Delta t$. The computation is based on the so-called return algorithm [27].

We will take into consideration just one material point, assuming that the elastic-plastic evolution is independent from all other material points. In one point we define one typical discrete time-step for plasticity computation, and give a stochastic version of the radial return algorithm. All random fields $\kappa(x, \omega)$ at a material point (Gauss point in FEM) x_0 are simple random variables $\varkappa(\omega) := \kappa(x_0, \omega)$ and have a PCE approximation like (10). The stochastic radial return map is given in Table 1, and it is based on the radial return map for classical plasticity.

Table 1 Stochastic radial return algorithm.

1. Lognormal random variables $K(\theta), G(\theta), \sigma_y(\theta)$.
2. Starting stress $\sigma_0(\theta) = -p_0(\theta)I + \sigma_0^D(\theta)$, and strain increment $\Delta\varepsilon(\theta)$, with initial pressure $p_0(\theta)$ and initial deviatoric part of the stress $\sigma_0^D(\theta)$.
3. Trial deviatoric elastic stress $\sigma_e^D(\theta) = \sigma_0^D(\theta) + G(\theta)(\Delta\varepsilon(\theta) - (\text{tr } \Delta\varepsilon(\theta)/3)I)$.
4. Pressure $p_1(\theta) = p_0(\theta) - (K(\theta) \text{ tr } \Delta\varepsilon(\theta))/3$.
5. Stress at the end of the step $\sigma_1(\theta) = -p_1(\theta)I + \sigma_1^D(\theta)$.
6. Yield criterion requires $\forall \theta$ that $\sigma_1(\theta)$ is inside the yield surface:
  $\sigma_1^D(\theta) = \sigma_y(\theta)\sigma_e^D(\theta)/\|\sigma_e^D(\theta)\|_{J_2}$.
7. Check yield criterion $\phi = \|\sigma_e^D(\theta)\|_{J_2} - \sigma_y$,
  where $\|\sigma_e^D(\theta)\|_{J_2} = \|\sigma_e - \text{tr}(\sigma_e)I\|/\sqrt{6}$.
  Elastic step: set variables equal to trial one and exit.
  Plastic step: Proceed to step 8.
8. Elastic part of the strain increment
  $\Delta\varepsilon_e(\theta) = -(p_1(\theta) - p_0(\omega))/K(\theta)I + 1/G(\theta)(\sigma_1^D(\theta) - \sigma_0^D(\theta))$.
  Plastic part of the strain increment $\Delta\varepsilon_p(\theta) = \Delta\varepsilon(\theta) - \Delta\varepsilon_e(\theta)$.

All linear operations with RVs in PCE representation are easily performed, and for the algebraically non-linear ones – like products – we can use the Hermite algebra as shown in the Appendix. In the yield criterion one can notice the norm of the deviatoric elastic stress, which involves a square root. The problem is to find the polynomial chaos expansion of this. Here it is computed using Monte Carlo simulation.

4 Model Results

Here we consider a plane strain problem defined with a stress tensor involving only the components $\{\sigma_{xx}, \sigma_{yy}, \sigma_{xy}, \sigma_{zz}\}$, and a strain tensor with only non-zero components $\{\varepsilon_{xx}, \varepsilon_{yy}, \varepsilon_{xy}\}$. For outputs we observe the pressure $p = -\frac{1}{3}\mathrm{tr}(\sigma)$, the in-plane shear stresses σ_{xy} and $\sigma_\delta = \frac{1}{2}(\sigma_{xx} - \sigma_{yy})$, and the out-of-plane shear stress $\sigma_\Delta = \frac{1}{4}(2\sigma_{zz} - \sigma_{xx} - \sigma_{yy})$.

Results of a polynomial chaos expansion of a forth order are compared with the results of Monte Carlo (MC) simulations for 10^5 samples. We may assume that the results of the MC simulation are exact. The bulk modulus is given with the function $\exp(\mu + \nu * \xi)$, where $\mu = 1, \nu = 0.2$, similarly the shear modulus, while the yield stress is given with $\mu = 1, \nu = 0.3$. The approximation for the bulk modulus is shown in Figure 1, both the difference between PCE and MC simulation and its probability density function (PDF). The difference between the fourth order PCE and the MC simulation is small. The bulk modulus has mean value 3.6755 and a standard deviation equal to 1.6505.

Similar results are shown for the pressure, its probability density function is in Figure 2, with mean value -4.0417 and standard deviation 4.1776, where again the difference between the fourth order PCE and the MC simulation is small.

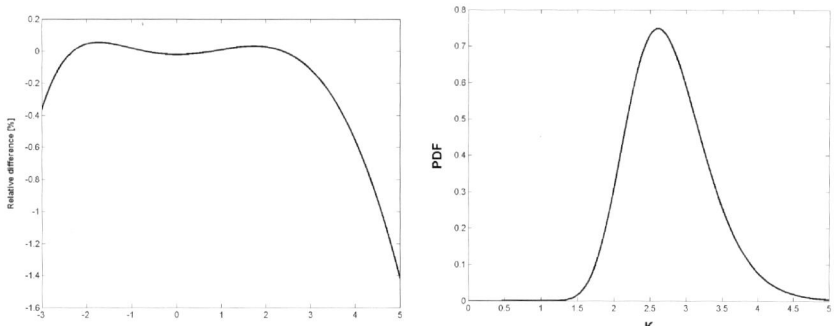

Fig. 1 Realisation of K (bulk modulus) and its PDF.

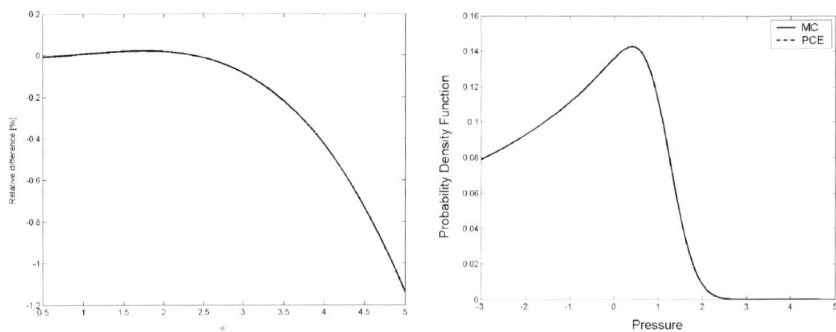

Fig. 2 Difference PCE-MC and PDF of the pressure.

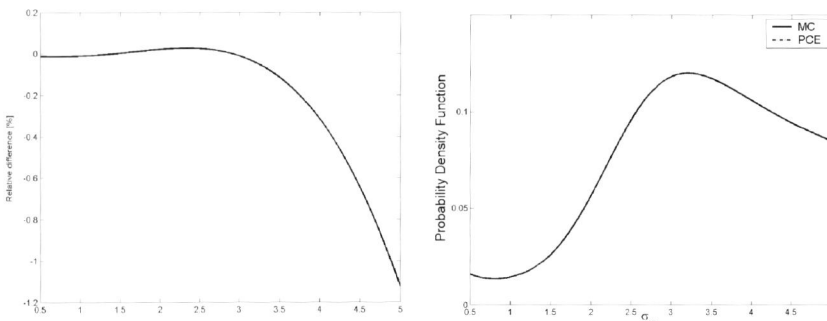

Fig. 3 Difference PCE-MC and PDF of stress σ_{xy}.

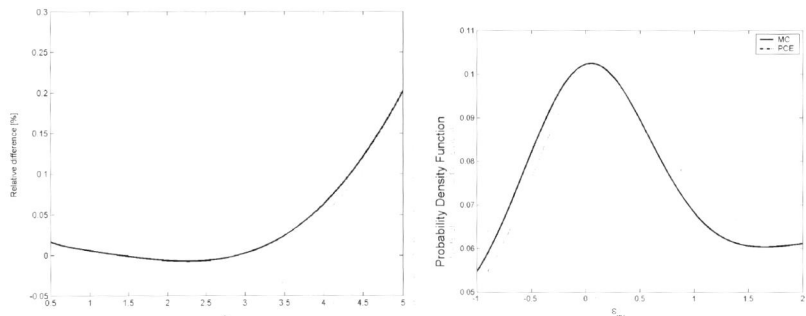

Fig. 4 Difference PCE-MC and PDF of strain ε_{xy}.

For the in-plane shear stress σ_{xy} the results are shown in Figure 3, with mean value 2.6884 and standard deviation 3.5910. In Figure 4 we show analogous results for the strain ε_{xy}, with mean value 1.2041 and standard deviation 4.7824.

5 Conclusion

This article has tried to iluminate the white noise setting for a stochastic plasticity formulation. The idea of random variables as functions in an infinite dimensional space which have to be approximated by elements of finite dimensional spaces has brought a new view to the field. The stochastic Galerkin method (or collocation method) is useful in the approximation of partial differential and integral equations, and it has been applied in the field of elasticity very successfuly. Here we apply it also on irreversible materials, e.g. plasticity. This new method is a contrast to traditional Monte Carlo methods. It remains yet to be seen which of these approaches is more effective in which situation.

Appendix: Hermite Algebra

Univariate Hermite polynomials are a basis for the polynomial algebra [10,17], their products are defined by

$$h_k(\vartheta) h_\ell(\vartheta) = \sum_{m=0}^{k+\ell} c_{k\ell}^{(m)} h_m(\vartheta), \qquad c_{k\ell}^{(m)} = \frac{k!\,\ell!}{(g-k)!\,(g-\ell)!\,(g-m)!}, \qquad (13)$$

where the coefficients (structure constants of the algebra) are nonzero only for integer $g = (k + \ell + m)/2$ and $g \geq k, \ell, m$. For the multivariate Hermite algebra, analogous statements hold:

$$H_\alpha(\theta) H_\beta(\theta) = \sum_\gamma c_{\alpha\beta}^{(\gamma)} H_\gamma(\theta), \qquad c_{\alpha\beta}^{(\gamma)} = \prod_J c_{\alpha_J \beta_J}^{(\gamma_J)}. \qquad (14)$$

The product two polynomial chaos expansions is totaly defined by the previous equations, and the result is again a polynomial chaos. A division or the inverse of a polynomial chaos expansion require more work.

References

1. R. J. Adler. *The Geometry of Random Fields.* John Wiley & Sons, Chichester, 1981.
2. M. Anders and M. Hori. Stochastic finite element methods for elasto-plastic body. *Int. J. Num. Meth. Engnrng.*, 46:1897–1916, 1999.
3. I. Babuška, R. Tempone, and G. E. Zouraris. Galerkin finite element approximations of stochastic elliptic partial differential equations. *SIAM J. Numer. Anal.*, 42:800–825, 2004.
4. R. E. Caflisch. Monte Carlo and Quasi-Monte-Carlo methods. *Acta Numerica*, 7:1–49, 1998.
5. R. Courant and D. Hilbert. *Methods of Mathematical Physics.* John Wiley & Sons, Chichester, 1989.
6. G. Duvaut and J. L. Lions. *Inequalties in Mechanics and Physics.* Springer Verlag, Berlin, 1976.

7. R. Ghanem and P. Spanos. *Stochastic Finite Elements – A Spectral Approach*. Springer Verlag, Berlin, 1991.
8. W. Han and B. D. Reddy. *Plasticity: Mathematical Theory and Numerical Analysis*. Springer Verlag, Berlin, 1999.
9. T. Hida, H. H. Kuo, J. Potthoff, and L. Streit. *White Noise – An Infinite Dimensional Calculus*. Kluwer, Dordrecht, 1993.
10. H. Holden, B. Øksendal, J. Uboe, and T. Zhang. *Stochastic Partial Differential Equations. A Modeling, White Noise, Functional Approach*. Birkhäuser, Basel, 1996.
11. B. Jeremić, K. Sett, and M. Levent Kavvas. Probabilistic elasto-plasticity: Formulation in 1d. *Acta Geotechnica*, 2:197–210, 2007.
12. A. Keese. *Numerical Solution of Systems with Stochastic Uncertainties. A General Purpose Framework for Stochastic Finite Elements*. PhD thesis, TU Braunschweig, Brunswick, 2003.
    url: http://www.digibib.tu-bs.de/?docid=00001595
13. A. Keese. A review of recent develpoments in the numerical solution of stochastic partial differential equations. Informatikbericht 2003-6, Institute of Scientific Computing, TU Braunschweig, Brunswick, 2003.
    url: http://www.digibib.tu-bs.de/?docid==00001504
14. M. Kojić and K.-J. Bathe. *Inelastic Analysis of Solids and Structures*. Springer Verlag, Berlin, 2004.
15. P. Krée and C. Soize. *Mathematics of Random Phenomena-Random Vibrations of Mechanical Structures*. D. Reidel, Dordrecht, 1986.
16. M. Loève. *Probability Theory*. Springer-Verlag, Berlin, 1977.
17. P. Malliavin. *Stochastic Analysis*. Springer-Verlag, Berlin, 1997.
18. K. Z. Markov. Application of Volterra-Wiener series for bounding the overall conductivity of heterogeneous media, I. general procedure. *SIAM J. Appl. Math.*, 4:831–849, 1987.
19. H. G. Matthies. Quantifying uncertainty: Modern computational representation of probability and applications. In A. Ibrahimbegović and I. Kožar (Eds.), *Extreme Man-Made and Natural Hazards in Dynamics of Structures*, NATO-ARW, Springer Verlag, Berlin, 2007.
20. H. G. Matthies. Uncertainty quantification with stochastic finite elements. In E. Stein, R. de Borst, and T. J. R. Hughes (Eds.), *Encyclopedia of Computational Mechanics*. John Wiley & Sons, Chichester, 2007.
21. H. G. Matthies, C. E. Brenner, C. G. Bucher, and C. Guedes Soares. Uncertainties in probabilistic numerical analysis of structures and solids – Stochastic finite elements. *J. Stuctural Safety*, 19:283–336, 1997.
22. H. G. Matthies and C. Bucher. Finite elements for stochastic media problems. *Comp. Meth. Appl. Mech. Engnrng.*, 168:3–17, 1999.
23. H. G. Matthies and A. Keese. Galerkin methods for linear and nonlinear elliptic stochastic partial differential equations. *Comp. Meth. Appl. Mech. Engnrng.*, 194:1295–1331, 2005.
24. E. Novak and K. Ritter. Simple cubature formulas with high polynomial exactness. *Constructive Approximation*, 15:499–522, 1999.
25. G. Schuëller. A state of the art report on computational stochastic mechanics. *Prob. Engnrng. Mech.*, 12:197–321, 1997.
26. C. Schwab and R. A. Todor. Sparse finite elements for elliptic problems with stochastic loading. *Numer. Math.*, 95:707–734, 2003.
27. J. C. Simo and T. J. R. Hughes. *Computational Inelasticity*. Springer Verlag, Berlin, 1998.
28. S. A. Smolyak. Quadrature and interpolation formulas for tensor products of certain classes of functions. *Dokl. Akad. Nauk SSSR (Soviet Math. Dokl.)*, 4:240–243, 1963.
29. D. Xiu and G. E. Karniadakis. Modeling uncertainty in steady state diffusion problems via generalized polynomial chaos. *Comp. Meth. Appl. Mech. Engnrng.*, 191:4927–4948, 2002.

Automated Computational Modelling for Solid Mechanics

Kristian B. Ølgaard[1], Garth N. Wells[2] and Anders Logg[3]

[1] *Faculty of Civil Engineering and Geosciences, Delft University of Technology, Stevinweg 1, 2628 CN Delft, The Netherlands*
E-mail: k.b.oelgaard@tudelft.nl
[2] *Department of Engineering, University of Cambridge, Trumpington Street, Cambridge CB2 1PZ, United Kingdom*
E-mail: gnw20@cam.ac.uk
[3] *Center for Biomedical Computing, Simula Research Laboratory/Department of Informatics, University of Oslo, P.O. Box 134, 1325 Lysaker, Norway*
E-mail: logg@simula.no

Abstract. We present some recent developments in automated computational modelling with an emphasis on solid mechanics applications. The automation process permits an abstract mathematical model of a physical problem to be translated into computer code rapidly and trivially, and can lead to computer code which is faster than hand-written and optimised code. Crucial to the approach is ensuring that mathematical abstractions inherent in the mathematical model are inherited by the software library.

Key words: automated modelling, variational methods, incompressible elasticity, plasticity, spinodal decomposition.

1 Introduction

As increasingly sophisticated and diverse models are developed for problems in solid mechanics, the burden of translating an abstract mathematical representation into efficient computer code grows. The process of creating computer code for a given model becomes increasingly time consuming and error prone, and efficiency can become elusive.

In the context of automated computer modelling, we present some recent developments within the FEniCS Project [1], with an emphasis on solid mechanics applications. Central to this is the FEniCS Form Compiler (FFC) [2, 3] which transforms high-level mathematical input into low-level computer code. Complex models can be developed rapidly, and the time required for testing and debugging is dramatically reduced. The abstract computer representation of a variational problem also presents various opportunities for optimisation of the generated code. The concepts which we present are generic in that they apply to partial differential equations, although we place an emphasis in this work on some developments that have been driven by challenges typically arising in solving solid mechanics problems.

B.D. Reddy (ed.), IUTAM Symposium on Theoretical, Modelling and Computational Aspects of Inelastic Media, 195–204.
© *Springer Science+Business Media B.V. 2008*

The rest of this work is organised as follows. An overview of concepts for automated modelling and code generation is presented in Section 2. The approach is laid down concretely in Section 3 through several example problems and methods, including plasticity, discontinuous Galerkin methods and spinodal decomposition. This is followed by conclusions in Section 4.

2 Automated Modelling Concepts

2.1 Overview

We focus here on the automation of the step from abstract variational formulation to computer code for evaluating finite element tensors (functionals, the 'right-hand side vector' and the 'stiffness matrix'). This is performed via a compiler [2] which translates an abstract mathematical model into low-level code. A function of the compiler approach is to bridge the traditional performance versus generality paradox. The compiler approach for variational problems is an example of meta-programming (programs which write programs) and allows *a priori* optimisations to be performed which are not feasible at run-time. Furthermore, the output from the compiler can be tailored as desired. For the examples which we show in Section 3, C++ code consistent with the Unified Form-assembly Code format (UFC) [4] has been produced. Any finite element assembly library which supports the UFC format could be used. We have used the FEniCS component DOLFIN [5], which is a C++ library that handles the assembly and solution of a global system of equations to solve the example problems presented in this work.

The form compiler FFC and the library DOLFIN are freely available under the GNU General Public License and the Lesser GNU General Public License, respectively, at http://www.fenics.org.

2.2 Mathematical Abstractions

Four key components are needed in order to implement variational forms in a high-level language that mirrors conventional mathematical syntax. These components are: the definition of function spaces and the ability to define functions on these spaces; differential operators; linear algebra operators; and domain integrals.

Function spaces in FFC are defined in terms of finite element spaces. For example,

```
element = FiniteElement("Lagrange", "tetrahedron", 3)
```

which defines a cubic Lagrange basis on tetrahedra. Various element families are available and can be defined in $\mathbb{R}^1$, $\mathbb{R}^2$ and $\mathbb{R}^3$ of arbitrary polynomial degree. Test and trial functions, as well as ordinary functions, can be defined on elements in a straightforward fashion. For example,

```
v = TestFunction(element)
```

defines a test function based on the previously declared element space. A range of standard and more esoteric finite element bases are provided to the form compiler by the library FIAT [6], which is also a component of the FEniCS Project.

A basic set of differential and linear algebra operators, defined in FFC, can act on the various functions. Such operators include partial derivatives, `v.dx(i)`; the gradient, `grad(v)`; the divergence, `div(v)`; the curl, `curl(v)`; inner products `dot(v, w)` and matrix-vector multiplications `mult(A, v)`.

The form compiler FFC also supports operators for discontinuous Galerkin finite element methods. When working with discontinuous Galerkin methods it is common to 'restrict' some functions to either side of an element facet. The syntax for evaluating a function v on one side of a facet is `v('+')`, and the syntax for evaluating the same function on the other side is `v('-')`. By using restricted functions it is possible to construct discontinuous Galerkin operators, and for convenience various jump operators and average operators are available in FFC. The integrals available in FFC are integration over cells which is denoted by `*dx`, integration over exterior facets denoted by `*ds` and integration over interior facets denoted by `*dS`.

When solving many differential equations it is possible to work exclusively with functions from a finite element space or with functions interpolated in a finite space. This approach can however break down for problems such as plasticity where despite the displacement field coming from a finite element space, the stress field inevitably will not. To deal with this problem without disturbing the mathematical structure of the form compiler, we have introduced the concept of a 'quadrature function'. A quadrature function can be evaluated at discrete points and can be integrated approximately, but cannot be differentiated. A quadrature function is defined via a quadrature element. For example,

```
element = QuadratureElement("tetrahedron", 3)
```

This element represents discrete function values for a $3 \times 3 \times 3$ quadrature scheme on a tetrahedron. A function u can then be defined on this element by `u = Function(element)` and used to construct variational forms just like any other function.

Finally, FFC is implemented in Python and inherits the Python syntax. This makes the addition of user-defined operators simple, as will be demonstrated in the following section.

```
P2 = VectorElement("Lagrange", "tetrahedron", 2)
P1 = FiniteElement("Lagrange", "tetrahedron", 1)
TH = P2 + P1                  # Taylor-Hood element

(v, q) = TestFunctions(TH)
(u, p) = TrialFunctions(TH)   # displacement, pressure
f = Function(P2)              # source term

nu = 5.0

a = nu*dot(grad(v), grad(u))*dx - div(v)*p + q*div(u))*dx
L = dot(v, f)*dx
```

Fig. 1 FFC input for incompressible elasticity in three dimensions using a Taylor–Hood element.

3 Examples

3.1 Incompressible Elasticity

Consider the following function spaces on $\Omega \subset \mathbb{R}^d$,

$$V = \left\{ v \in \left(H_0^1(\Omega) \right)^d : v_i \in P^2(K) \ \forall K \in \mathcal{T}, \ 1 \leq i \leq d \right\}, \quad (1)$$

$$Q = \left\{ q \in H^1(\Omega) : q \in P^1(K) \ \forall K \in \mathcal{T} \right\}, \quad (2)$$

where K are finite element cells and $\mathcal{T}$ is a standard finite element partition of the domain Ω into element cells. A variational problem for incompressible elasticity using a Taylor–Hood element reads: find $(u; p) \in V \times Q$ such that

$$a(v; q, u; p) = L(v; q) \quad \forall (v; q) \in V \times Q, \quad (3)$$

where the bilinear and linear forms are given by

$$a(v; q, u; p) = \int_\Omega \nu \nabla v : \nabla u - (\nabla \cdot v) p + q \nabla \cdot u \, dx, \quad (4)$$

$$L(v; q) = \int_\Omega v \cdot f \, dx, \quad (5)$$

where u and p denotes the displacement and pressure fields, respectively, $\nu > 0$ is a parameter and f is a body force. The FFC input for this problem is shown in Figure 1 for the case of tetrahedral elements. Note how closely the FFC input resembles the mathematical notation of (4) and (5). This example also demonstrates the ease with which mixed function spaces of different orders can be used.

3.2 Plasticity

Linearisation of the balance of momentum equation for a plasticity problem yields a bilinear form and linear form which are the basic ingredients in a Newton method. In the absence of inertia terms, the variational forms for linearised standard plasticity read:

$$a(v, u) = \int_\Omega \nabla v : \mathcal{C} : \nabla^s du \, dx, \tag{6}$$

$$L(v) = \int_\Omega v \cdot f \, dx + \int_\Gamma v \cdot h \, ds - \int_\Omega \nabla v : \sigma \, dx, \tag{7}$$

where v and du are the test function and displacement field increment, respectively, and come from an appropriate finite element, $\mathcal{C}$ is the consistent tangent operator, f is a body force, h is the prescribed traction on part of the boundary and σ is the stress from the previous iteration. Functions such as the displacement u will come from a finite element space, and functions such as the body force can be interpolated without impacting the convergence properties of the scheme. However, components of the stress and the tangent operator will not in general come from a finite element space. Therefore, when generating computer code automatically for these forms, quadrature functions are used for the stress and tangent terms. The FFC code for plasticity using quadratic elements can be seen in Figure 2. The format of the FFC input for this case is more complex in this case due to the presence of tensor-valued functions. Note however that the Python interface to FFC makes it straightforward to define functions by combining existing operators. Manipulating tensor-valued functions in a computer code is highly error prone and developments in better supporting such functions in an automated fashion are ongoing.

3.3 Discontinuous Galerkin Methods

In Section 2.2, the restriction of functions to either the plus or minus side of an interior facet was introduced. Some discontinuous Galerkin methods also involve the evaluation of derivatives of functions restricted to either side of a facet. We present here one such method for the biharmonic equation.

Classically, Galerkin methods for the biharmonic equation seek approximate solutions in a subspace of $H^2(\Omega)$. However, such functions are difficult to construct in a finite element context. Based on discontinuous Galerkin principles, methods have been developed which utilise functions from $H^1(\Omega)$ [7, 8]. Rather than considering jumps in functions across element boundaries, terms involving the jump in the normal derivative across element boundaries are introduced. Unlike fully discontinuous approaches, this method does not involve double-degrees of freedom on element edges and therefore does not lead to the significant increase in the number of degrees of freedom relative to conventional methods. Consider the continuous

```
element  = VectorElement("Lagrange", "tetrahedron", 2)
elementT = VectorQuadratureElement("tetrahedron", 2, 36)
elementS = VectorQuadratureElement("tetrahedron", 2, 6)

v = TestFunction(element)
u = TrialFunction(element)
f = Function(element)
h = Function(element)
t = Function(elementT) # Consistent tangent coefficients
s = Function(elementS) # Stresses from previous iteration

# Strain vector
def epsilon(u):
return [u[0].dx(0), u[1].dx(1), u[2].dx(2),
        u[0].dx(1) + u[1].dx(0),
        u[0].dx(2) + u[2].dx(0),
        u[1].dx(2) + u[2].dx(1)]

# Stress tensor from previous iteration
def sigma(s):
return [[s[0], s[3], s[4]],
        [s[3], s[1], s[5]],
        [s[4], s[5], s[2]]]

# Consistent tangent tensor
def tangent(t):
  return [[t[i*6 + j] for j in range(6)] for i in range(6)]

# Incremental stresses
def dsigma(t, u):
  return mult(tangent(t), epsilon(u))

a = dot(epsilon(v), dsigma(t, u) )*dx
L = dot(v, f)*dx + dot(v, h)*ds - dot(grad(v), sigma(s))*dx
```

Fig. 2 FFC input for the linearised plasticity problem using $k = 2$ in three dimensions.

function space
$$V = \left\{ v \in H_0^1(\Omega) : v \in P^k(K) \, \forall K \in \mathcal{T} \right\}. \tag{8}$$

The bilinear and linear forms for the biharmonic equation, with the boundary conditions $u = 0$ on $\partial\Omega$ and $\nabla^2 u = 0$ on $\partial\Omega$, read

$$a(v, u) = \int_\Omega \nabla^2 v \nabla^2 u \, dx - \int_{\Gamma^0} [\![\nabla v]\!] \cdot \langle \nabla^2 u \rangle \, ds$$

$$- \int_{\Gamma^0} \langle \nabla^2 v \rangle \cdot [\![\nabla u]\!] \, ds + \int_{\Gamma^0} \frac{\alpha}{h} [\![\nabla v]\!] \cdot [\![\nabla u]\!] \, ds, \tag{9}$$

$$L(v) = \int_\Omega vf \, dx. \tag{10}$$

where $v \in V$ and $u \in V$, Γ^0 denotes interior facets, α is a penalty parameter and h is a measure of cell size defined as $h = (h^+ + h^-)/2$ for the two cells K^+ and K^- incident with the given interior facet. The size of a cell is defined here as twice the circumradius. The jump $[\![\cdot]\!]$ and average $\langle \cdot \rangle$ operators are defined as

```
element = FiniteElement("Lagrange", "tetrahedron", 4)

v = TestFunction(element)
u = TrialFunction(element)
f = Function(element)

n = FacetNormal("tetrahedron")
h = MeshSize("tetrahedron")

alpha = 16.0

a =   dot(div(grad(v)), div(grad(u)))*dx \
    - dot(jump(grad(v), n), avg(div(grad(u))))*dS \
    - dot(avg(div(grad(v))), jump(grad(u), n))*dS \
    + alpha/h('+')*dot(jump(grad(v), n), jump(grad(u), n))*dS

L = v*f*dx
```

Fig. 3 FFC input for the biharmonic equation using $k = 4$ in three dimensions.

$[\![\nabla v]\!] = \nabla v^+ \cdot \boldsymbol{n}^+ + \nabla v^- \cdot \boldsymbol{n}^-$ and $\langle \nabla^2 v \rangle = (\nabla^2 v^+ + \nabla^2 v^-)/2$ on Γ^0. The FFC input for this problem with $k = 4$ in three dimensions is shown in Figure 3. Again, the terms in the bilinear form resemble closely the mathematical notation of (9). This particular problem can be difficult to implement in a conventional framework, whereas through automation the problem is trivial.

3.4 Spinodal Decomposition

As a final problem we consider the Cahn–Hilliard equation [9] for modelling spinodal decomposition which is a common process in the evolution of microstructure in alloys. The Cahn–Hilliard equation is a phase field model which involves a concentration field c which represents the fraction of a given phase at a point. It is a nonlinear fourth-order parabolic equation, and we have solved it using a mixed finite element method (see for example [10]) for the case of constant mobility and a quartic free-energy function. Considering the function space

$$U = \{ v \in H^1(\Omega) : v \in P^1(K) \forall / K \in \mathcal{T} \}, \tag{11}$$

for the common case of homogeneous Neumann boundary conditions, the variational problem using a Crank–Nicolson scheme to advance in time reads: given c^n, find $c^{n+1} \in U$ and $k^{n+1} \in U$ such that

$$\int_\Omega q \frac{c^{n+1} - c^n}{\Delta t} \, dx + \int_\Omega \nabla q \cdot \nabla k^{n+1/2} \, dx = 0 \quad \forall q \in U, \tag{12}$$

$$\int_\Omega v k^{n+1} \, dx - \int_\Omega v \mu^{n+1} \, dx - \lambda \int_\Omega \nabla v \cdot \nabla c^{n+1} \, dx = 0 \quad \forall v \in U, \tag{13}$$

```
P1 = FiniteElement("Lagrange", "tetrahedron", 1)
P2 = FiniteElement("Lagrange", "tetrahedron", 1)
ME = P1 + P2

(v,  q)  = TestFunctions(ME)
(dk, dc) = TrialFunctions(ME)       # potential, concentration

(k1, c1) = Functions(ME)            # current solution
(k0, c0) = Functions(ME)            # solution from previous converged step

lmbda    = Constant("tetrahedron")  # surface parameter
muFactor = Constant("tetrahedron")  # chemical free energy multiplier

dt       = Constant("tetrahedron")  # time step
theta    = Constant("tetrahedron")  # time stepping parameter

# Chemical free-energy \phi = c^2*(1-c^2)
# mu = \phi,c
def mu(c, muFactor):
    return muFactor*(2*c*(1-c)*(1-c) - 2*c*c*(1-c))

# DmuDc = \phi,cc
def DmuDc(c, muFactor):
    return muFactor*(2.0*(1.0-c)*(1.0-c) \
           - 8.0*c*(1.0-c) + 2.0*c*c)

# k^(n+theta)
k_mid = (1-theta)*k0 + theta*k1

# \delta k^(n+theta)
dk_mid = theta*dk

a1 = q*dc*dx + dt*dot(grad(q), grad(dk_mid))*dx
a2 = v*dk*dx - v*DmuDc(c1, muFactor)*dc*dx \
             - lmbda*dot(grad(v), grad(dc))*dx

L1 = q*c1*dx - q*c0*dx + dt*dot(grad(q), grad(k_mid))*dx
L2 = v*k1*dx - v*mu(c1, muFactor)*dx \
             - lmbda*dot(grad(v), grad(c1))*dx

a =  a1 + a2
L = -L1 - L2
```

Fig. 4 FFC input for the Cahn–Hilliard equation in three dimensions using a mixed formulation and linear elements.

where $\lambda > 0$ is constant. For a simple case in which the chemical potential μ derives from a quartic chemical free-energy function, μ reads

$$\mu = 2c(1-c)^2 - 2c^2(1-c). \tag{14}$$

The interesting response of the Cahn–Hilliard equation is due to the non-convexity of the free-energy function from which μ is derived.

The bilinear and linear forms resulting from the linearisation of the weak form of the Cahn–Hilliard equation provide the input for FFC, which is shown in Figure 4. To illustrate a typical result from solving the Cahn–Hilliard equation, concentration contours ($c = 0.5$) are shown in Figure 5 for a three dimensional problem. The con-

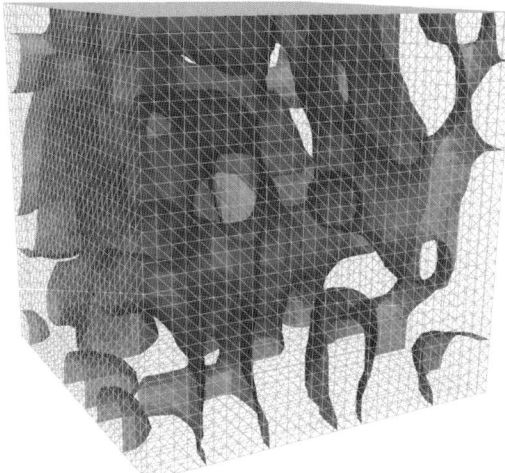

Fig. 5 Concentration contours $c = 0.5$ for the solution to the Cahn–Hilliard equation in three dimensions after evolution from a randomly perturbed constant initial condition.

tours represent the boundary between phases in the mixture. Using the discontinuous Galerkin operators, a primal formulation using recently developed techniques [11] could also be generated automatically.

4 Conclusions

An automated modelling approach with an emphasis on solid mechanics applications has been presented. A key element is the use of a form a compiler that translates variational forms expressed in a high-level language into low-level code. The compiler was demonstrated for various solid mechanics problems, although the approach generalises equally well to other variational forms. In particular, it was shown how close the high-level language and structure of the form compiler resembles the mathematical syntax which is used to define variational forms.

Acknowledgements

KBØ acknowledges the support of the Netherlands Technology Foundation STW, the Netherlands Organisation for Scientific Research and the Ministry of Public Works and Water Management. AL is supported by an Outstanding Young Investigator grant from the Research Council of Norway, NFR 180450.

References

1. The FEniCS Project. URL: http://www.fenics.org
2. Kirby RC, Logg A (2006) A compiler for variational forms. *ACM Transactions on Mathematical Software*, 32(3):417–444.
3. Ølgaard KB, Logg A, Wells GN (2008) Automated code generation for discontinuous Galerkin methods. Submitted.
4. Alnæs M, Logg A, Mardal KA, Skavhaug O, Langtangen HP (2007). Ufc specification and user manual. URL: http://www.fenics.org/ufc/
5. Logg A, Wells GN, Hoffman J, Jansson J, et al (2008) DOLFIN. URL: http://www.fenics.org/dolfin/
6. Kirby RC (2004) Algorithm 839: FIAT, a new paradigm for computing finite element basis functions. *ACM Transactions on Mathematical Software*, 30(4):502–516.
7. Engel G, Garikipati K, Hughes TJR, Larson MG, Mazzei L, Taylor RL (2002) Continuous/discontinuous finite element approximations of fourth-order elliptic problems in structural and continuum mechanics with applications to thin beams and plates, and strain gradient elasticity. *Computer Methods in Applied Mechanics and Engineering*, 191(34):3669–3750
8. Wells GN, Dung NT (2007) A C^0 discontinuous Galerkin formulation for Kirchhoff plates. *Computer Methods in Applied Mechanics and Engineering*, 196(35–36):3370–3380.
9. Cahn JW, Hilliard JE (1958) Free energy of a nonuniform system – I: Interfacial free energy. *The Journal of Chemical Physics*, 28(2):258–267.
10. Elliott CM, French DA, Milner FA (1989) A 2nd-order splitting method for the Cahn–Hilliard equation. *Numerische Mathematiek*, 54(5):575–590.
11. Wells GN, Kuhl E, Garikipati K (2006) A discontinuous Galerkin formulation for the Cahn–Hilliard equation. *Journal of Computational Physics*, 218(2):860–877.

Generalised Functions for Modelling Singularities: Direct and Inverse Problems

Salvatore Caddemi and Ivo Caliò

Dipartimento di Ingegneria Civile ed Ambientale, University of Catania,
Viale Andrea Doria 6, 95125 Catania, Italy
E-mail: scaddemi@dica.unict.it

Abstract. The possibility of utilising the generalised functions (distributions) and the related theory to model singularities showing a strongly local effect is explored in this work. In particular, singularities due to concentrated damages, to the presence of internal or external constraints or to abrupt changes of physical or geometrical properties, will be considered along the span of straight beams and modelled by means of Dirac's deltas or Heaviside functions. New developments of the distribution theory can be efficaciously adopted to provide closed form solutions, both in static and dynamic regime, in the case of damaged beams regardless of the number, position and intensity of the concentrated damages without the request of any additional computational work. Furthermore, the explicit expressions obtained in terms of eigenmodes are efficaciously adopted to address the identification problem of multiple damages by dynamic tests.

Key words: distributions, generalised functions, singularities, concentrated damages, damage identification.

1 Introduction

The possibility of utilising the generalised functions (distributions) and the related theory to model singularities, showing a strong local effect, is explored in this work. In particular, singularities due to concentrated damages, to the presence of internal or external constraints or to abrupt changes of physical or geometrical properties, will be considered along the span of straight beams and modelled by means of generalised functions such as the unit step (Heaviside) function, the Dirac's delta and the doublet function. There are examples in the literature of the adoption of generalised functions to treat such singularities, however enforcement of continuity conditions along the beam span is not completely avoided [10].

New developments of the distribution theory can be efficaciously adopted to provide closed form solutions of the direct analysis problem regardless of the number, position and intensity of the singularities. In this work, some models for different singularities will be presented, however, in order to show the potentiality of the generalised functions only the case of beams with concentrated cracks will be treated. The analysis is conducted for the case of open cracks both in static and dynamic regime. It is shown that the generalised function approach leads to closed form solutions of the direct analysis problem. In this case, the inverse problem concerning the identification of concentrated damages by non destructive dynamic tests can be solved by means of a novel procedure able to simplify the multiple damage identification problem.

2 Models of Singularities in the Beam Theory

In this section singularities affecting the response parameters of straight beams are presented by means of the introduction of suitable generalised functions.

Discontinuities in the curvature function, caused by an abrupt change of the geometrical or physical parameters of the beam, can be modelled by means of the well known unit step (Heaviside) function $U(x - x_\alpha)$ superimposed onto a uniform flexural stiffness $E_0 I_0$, as follows:

$$EI(x) = E_0 I_0 [1 - \alpha\, U(x - x_\alpha)]. \tag{1}$$

The case of a discontinuity of the rotation function at the abscissa x_γ can be reproduced by means of the following flexural stiffness model:

$$EI(x) = E_0 I_0 [1 - \gamma\, \delta(x - x_\gamma)], \tag{2}$$

where $\delta(x - x_\gamma)$ is the Dirac's delta.

On the other hand, if the shear stiffness is retained responsible of a transversal displacement discontinuity at x_ε, the following shear stiffness model can be adopted:

$$GA(x) = G_0 A_0 [1 - \varepsilon\, \delta(x - x_\varepsilon)]. \tag{3}$$

The presence of an external transversal spring along the beam span at x_u induces a concentrated load, dependent on the transversal displacement $u(x)$, which can be modelled by means of a Dirac's delta as follows:

$$R_{x_u} = -k_u\, u(x_u)\, \delta(x - x_u). \tag{4}$$

Finally, an external rotational spring at x_φ induces an external moment, dependent on the rotation function $\varphi(x)$, which can be thought as two adjacent opposite concentrated loads, and can be modelled as:

$$M_{x_\varphi} = -k_\varphi\, \varphi(x_\varphi)\, \delta'(x - x_\varphi), \tag{5}$$

where $\delta'(x - x_\varphi)$ (the apex means differentiation with respect to x) is the doublet distribution.

The models presented in Equations (1)–(5) can be adopted for the case of $n_\alpha, n_\gamma, n_\varepsilon, n_u, n_\varphi$ singularities of different types contemporarily present along the beam span. In the latter case the governing Equations of the Timoshenko beam with singularities can be given the following form:

$$\left[E_0 I_0 \left[1 - \sum_{j=1}^{n_\alpha} \alpha_j U(x - x_{\alpha j}) - \sum_{i=1}^{n_\gamma} \gamma_i \delta(x - x_{\gamma i}) \right] \varphi'(x) \right]''$$

$$= -q(x) + \sum_{r=1}^{n_u} k_{ur} u(x_{ur}) \delta(x - x_{ur}) + \sum_{s=1}^{n_\varphi} k_{\varphi s} \varphi(x_{\varphi s}) \delta'(x - x_{\varphi s}), \quad (6)$$

$$\left[G_0 A_0 \left[1 - \sum_{k=1}^{n_\varepsilon} \varepsilon_k \delta(x - x_{\varepsilon k}) \right] [u'(x) + \varphi(x)] \right]'$$

$$= -q(x) + \sum_{r=1}^{n_u} k_{ur} u(x_{ur}) \delta(x - x_{ur}) + \sum_{s=1}^{n_\varphi} k_{\varphi s} \varphi(x_{\varphi s}) \delta'(x - x_{\varphi s}). \quad (7)$$

Integration of Equations (6), (7), by means of the generalised function rules, provides explicit expressions for the response functions of beams with multiple singularities.

3 Beams with Multiple Damages

The potentiality of the generalised function approach is here investigated by treating the case of beams with concentrated cracks. The governing Equation of the Euler–Bernoulli beam in presence of multiple rotation discontinuities can be written, in view of the model proposed in Equation (2), as:

$$\left[E_0 I_0 \left[1 - \sum_{i=1}^{n_\gamma} \gamma_i \delta_i(x) \right] u''(x) \right]'' = q(x). \quad (8)$$

where $\delta_i(x) = \delta(x - x_{\gamma i})$ indicates the Dirac's delta function centred at $x_i = x_{\gamma i}$. The abbreviation $U_i(x) = U(x - x_{\gamma i})$ will be adopted for the unit step (Heaviside) function in what follows. Equation (8) has been solved in [3], by making use of the rules of the generalised functions, in terms of the following explicit expression for the rotation function $\varphi(x)$:

$$\varphi(x) = -u'(x) = -c_2 - c_3 \left[\frac{x}{E_0 I_0} + \sum_{i=1}^{n_\gamma} \frac{1}{K_i^{eq}} U_i(x) \right] - \frac{q^{[3]}(x)}{E_0 I_0}$$

$$- c_4 \left[\frac{x^2}{2 E_0 I_0} + \sum_{i=1}^{n_\gamma} \frac{1}{K_i^{eq}} x_i U_i(x) \right] - \sum_{i=1}^{n_\gamma} \frac{1}{K_i^{eq}} q^{[2]}(x_i) U_i(x). \quad (9)$$

The rotation function, given by Equation (9), shows discontinuities $\Delta\varphi_i$, where the unit step functions exhibit the jumps at abscissae $x_i = x_{\gamma i}$, which can be expressed as follows:

$$\Delta\varphi_i(x_i) = -\frac{1}{K_i^{eq}}[c_3 + c_4 x_i + q^{[2]}(x_i)] = \frac{M(x_i)}{K_i^{eq}}, \qquad (10)$$

where

$$K_i^{eq} = \frac{E_0 I_0}{\lambda_i} \quad \text{with} \quad \lambda_i = \frac{\gamma_i}{1 - \gamma_i B}. \qquad (11)$$

The B appearing in Equation (11) is a constant introduced by [2] to define the product of two Dirac's deltas; however its value is not influent in the analysis since the parameters λ_i are chosen as representative of the concentrated damages [4].

Equation (10) represents the relationship between the rotation discontinuities and the bending moments $M(x_i)$ at abscissae x_i by means of the stiffness K_i^{eq} of a rotational spring equivalent to the singularity introduced in the model of Equation (2). The equivalent stiffness provided by Equation (11), once equated to one of the macroscopic models of concentrated cracks available in the literature (e.g. [9]), is able to provide a relationship between the crack depth and the adopted singularity.

The integration of Equation (9) leads to the transversal displacement function $u(x)$ as follows:

$$u(x) = c_1 + c_2 x + \frac{c_3}{2}\left[\frac{x^2}{E_0 I_0} + 2\sum_{i=1}^{n_\gamma}\frac{1}{K_i^{eq}}(x - x_i)U_i(x)\right] + \frac{q^{[4]}(x)}{E_0 I_0} +$$

$$+ \frac{c_4}{6}\left[\frac{x^3}{E_0 I_0} + 6\sum_{i=1}^{n_\gamma}\frac{1}{K_i^{eq}}x_i(x - x_i)U_i(x)\right] +$$

$$+ \sum_{i=1}^{n_\gamma}\frac{1}{K_i^{eq}}q^{[2]}(x_i)(x - x_i)U_i(x). \qquad (12)$$

Equation (12) is the solution of a beam with multiple concentrated open cracks dependent on four integration constants c_1, c_2, c_3, c_4 to be determined by means of the boundary conditions without enforcement of any continuity conditions at the cracked cross-sections.

4 Free Vibration of Beams with Multiple Damages

The model for beams with multiple damages, adopted in Section 3, to solve explicitly the static governing Equations can be also conveniently adopted for vibrating beams with multiple damages as shown in this section.

The governing equation of free vibration of a beam with multiple cracks, with distributed mass m, can be written as follows:

$$\left[E_0 I_0 \left(1 - \sum_{i=1}^{n_\gamma} \gamma_i \delta_i(x) \right) u''(x,t) \right]'' + m\ddot{u}(x,t) = 0, \tag{13}$$

where the superimposed dot means differentiation with respect to time t. Separation of the spatial variable x and the time variable t in Equation (13), obtained by means of the well known position $u(x,t) = \phi(x) \cdot q(t)$, leads to the following spatial differential equation:

$$\phi^{IV}(x) - \beta^4 \phi(x) = \sum_{i=1}^{n_\gamma} \left[[\gamma_i \delta_i(x)] \phi^{IV}(x) + 2\gamma_i \phi'''(x) \delta_i'(x) + \gamma_i \phi''(x) \delta_i''(x) \right], \tag{14}$$

where $\beta = [m\omega^2/(E_0 I_0)]^{1/4}$ is the frequency parameter, and ω the natural frequency associated to the harmonic time dependence of $q(t)$.

The solution of Equation (14) is sought as the sum of the homegeneous solution $\phi_h(x)$ and a particular integral $\phi_p(x)$, defined as follows:

$$\phi_h(x) = \hat{c}_1 \sin \beta x + \hat{c}_2 \cos \beta x + \hat{c}_3 \sinh \beta x + \hat{c}_4 \cosh \beta x,$$

$$\phi_p(x) = d_1(x) \sin \beta x + d_2(x) \cos \beta x + d_3(x) \sinh \beta x + d_4(x) \cosh \beta x, \tag{15}$$

where $d_1(x), d_2(x), d_3(x), d_4(x)$, in the particular integral $\phi_p(x)$, are generalised functions responsible of the distribution terms in Equation (14), and, once evaluated, lead to the following expression for the eigen-modes:

$$\phi(x) = c_1 f_1(x) + c_2 f_2(x) + c_3 f_3(x) + c_4 f_4(x), \tag{16}$$

with:

$$f_1(x) = \sin(\beta x) + \sum_{i=1}^{n_\gamma} \mu_i H_i(x) U_i(x); \quad f_2(x) = \cos(\beta x) + \sum_{i=1}^{n_\gamma} \nu_i H_i(x) U_i(x)$$

$$f_3(x) = \sinh(\beta x) + \sum_{i=1}^{n_\gamma} \xi_i H_i(x) U_i(x); \quad f_4(x) = \cosh(\beta x) + \sum_{i=1}^{n_\gamma} \eta_i H_i(x) U_i(x)$$

$$\tag{17}$$

in which the terms $\mu_i, \nu_i, \xi_i, \eta_i$ are defined as follows:

$$\mu_i = \lambda_i \frac{\beta}{2} \left[-\sin(\beta x_i) + \sum_{j=1}^{i-1} \mu_j \widehat{H}(x_i - x_j) \right],$$

$$\nu_i = \lambda_i \frac{\beta}{2} \left[-\cos(\beta x_i) + \sum_{j=1}^{i-1} \nu_j \widehat{H}(x_i - x_j) \right],$$

$$\xi_i = \lambda_i \frac{\beta}{2} \left[\sinh(\beta x_i) + \sum_{j=1}^{i-1} \xi_j \widehat{H}(x_i - x_j) \right],$$

$$\eta_i = \lambda_i \frac{\beta}{2} \left[\cosh(\beta x_i) + \sum_{j=1}^{i-1} \eta_j \widehat{H}(x_i - x_j) \right], \qquad (18)$$

where

$$H_i(x) = \sin[\beta(x - x_i)] + \sinh[\beta(x - x_i)],$$

$$\widehat{H}(x_i - x_j) = -\sin[\beta(x_i - x_j)] + \sinh[\beta(x_i - x_j)]. \qquad (19)$$

Equation (16), solution of Equation (14), represents the explicit expression of the eigen-mode of a beam with multiple damages and is dependent on four integration constants only, that are obtained without enforcing any continuity condition at the cracked cross-sections. It is worth noticing that Equations (18), replaced in Equations (17), show that the functions $f_i(x)$, $i = 1, \ldots, 4$, at the generic abscissa x depend on the damages located at $x_i < x$ only.

The four integration constants c_1, c_2, c_3, c_4, once evaluated for specific boundary conditions, are such that the dependency character, on damages located at $x_i < x$, of the functions $f_i(x)$, $i = 1, \ldots, 4$, is lost. In fact, as an example, for the case of free boundary conditions the following integration constants are obtained: $c_1 = c_3 = C$, $c_2 = c_4 = \vartheta C$ with

$$\vartheta = -\frac{[\sinh(\beta L) - \sin(\beta L)] + \sum_{i=1}^{n_y}(\mu_i + \xi_i)[-\sin(\beta(L - x_i)) + \sinh(\beta(L - x_i))]}{[\cosh(\beta L) - \cos(\beta L)] + \sum_{i=1}^{n_y}(\nu_i + \eta_i)[-\sin(\beta(L - x_i)) + \sinh(\beta(L - x_i))]}. \qquad (20)$$

Since the eigen-modes, given by Equations (16), depend on the constants c_1, c_2, c_3, c_4, Equation (20) shows clearly that they are dependent on all the damages present along the beam span.

5 A Procedure for Damage Identification by Dynamic Tests

The problem of damage identification is of extreme importance and have been widely addressed in the literature. There are several diagnostic techniques based on dynamic data [1, 5–8]). They are mainly concerned with the identification of a single damage and, when the case of multiple damage identification is addressed, no explicit expressions are provided, on the contrary, they are based on complex numerical procedures.

Here a procedure for the identification of multiple concentrated damages on the basis of the eigen-mode explicit expressions provided in Section 4 is presented. More precisely, if a non destructive dynamic test is conducted on a multidamaged

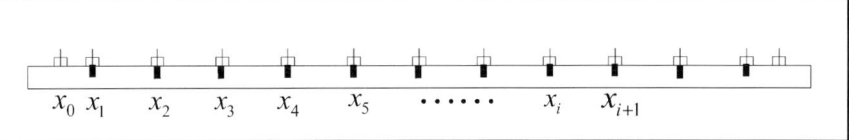

Fig. 1 Measurement positions coincident with damage locations.

beam, the first natural frequency ω_1^{ex} and the first eigen-mode $\phi_1^{ex}(x_i)$ at the cracked cross-sections x_i can be obtained by measured experimental data, as represented in Figure 1. In this case the latter values can be equated to the theoretical expressions of the first eigen-mode $\phi_1^{th}(x_i)$, given by Equation (16), as follows:

$$\phi_1^{ex}(x_0) = \phi_1^{th}(x_0), \quad \phi_1^{ex}(x_i) = \phi_1^{th}(x_i). \tag{21}$$

Equations (21) constitute the basis of the identification procedure.

It is important to note that, if we employ the first two measurements of the first eigen-mode $\phi_1^{ex}(x_0)$, $\phi_1^{ex}(x_1)$ under the assumption that $x_0 < x_1$, the dependence of $\phi_1^{th}(x_0)$, $\phi_1^{th}(x_1)$ on the unknown damages occurs only through the integration constants c_1, c_2, c_3, c_4 and not on the functions $f_1(x), f_2(x), f_3(x), f_4(x)$.

As a matter of example, Equations (21) at x_0, x_1, for the case of free boundary conditions, in view of Equations (16)-(18), lead to expressions which depend on the unknown damages through the variables ϑ, C only, as follows:

$$\phi_1^{ex}(x_0) = C\,\sin(x_0) + C\,\vartheta\,\cos(x_0) + C\,\sinh(x_0) + C\,\vartheta\,\cosh(x_0),$$
$$\phi_1^{ex}(x_1) = C\,\sin(x_1) + C\,\vartheta\,\cos(x_1) + C\,\sinh(x_1) + C\,\vartheta\,\cosh(x_1). \tag{22}$$

The dependence character of the eigen-mode on all the damages along the beam span is overcome by considering unknown, besides the damages, the two additional variables ϑ, C to be identified. The system of Equations (22) can be employed to evaluate the integration constants ϑ, C as functions of the first two measured data, as follows:

$$\vartheta = \frac{\phi_1^{ex}(x_0)[\sin(\beta x_1) + \sinh(\beta x_1)] - \phi_1^{ex}(x_1)[\sin(\beta x_0) + \sinh(\beta x_0)]}{\phi_1^{ex}(x_1)[\cos(\beta x_0) + \cosh(\beta x_0)] - \phi_1^{ex}(x_0)[\cos(\beta x_1) + \cosh(\beta x_1)]},$$

$$C = \frac{\phi_1^{ex}(x_0)}{[\sin(\beta x_0) + \sinh(\beta x_0)] + \vartheta\,[\cos(\beta x_0) + \cosh(\beta x_0)]}. \tag{23}$$

By means of the outlined procedure, the dependence of the eigen-function on all the damages through the integration constants is hidden in the identification of the integration constants, and the remaining character of the functions $f_1(x), f_2(x), f_3(x), f_4(x)$, dependent on the damages to one side of the argument, can now be exploited.

The evaluation of the integration constants ϑ, C by means of Equations (23), as part of the inverse identification problem, allows to write Equation (21) at the abscissa x_2 as function of the first damage only, present at x_1, hence providing the explicit expression of the first damage intensity λ_1, as:

$$\lambda_1 = \frac{[\phi_1^{ex}(x_2)/C] - [(\sin(\beta_1 x_2) + \sinh(\beta_1 x_2)) + \vartheta(\cos(\beta_1 x_2) + \cosh(\beta_1 x_2))]}{[(k_{\mu 1} + k_{\xi 1}) + \vartheta(k_{\nu 1} + k_{\eta 1})][\sin(\beta_1(x_2 - x_1)) + \sinh(\beta_1(x_2 - x_1))]}, \quad (24)$$

where

$$k_{\mu 1} = \frac{\beta_1}{2}[-\sin(\beta_1 x_1)]; \quad k_{\nu 1} = \frac{\beta_1}{2}[-\cos(\beta_1 x_1)];$$

$$k_{\xi 1} = \frac{\beta_1}{2}[\sinh(\beta_1 x_1)]; \quad k_{\eta 1} = \frac{\beta_1}{2}[\cosh(\beta_1 x_1)]. \quad (25)$$

Once the first damage has been identified by means of Equation (24), the measurement $\phi_1^{ex}(x_3)$ at x_3 can be employed to evaluate the second damage intensity λ_2, and so on. The intensity λ_i of the generic damage can be written explicitly as follows:

$$\lambda_i = \frac{[\phi_1^{ex}(x_{i+1})/C] - [(\sin(\beta_1 x_{i+1}) + \sinh(\beta_1 x_{i+1})) + \vartheta(\cos(\beta_1 x_{i+1}) + \cosh(\beta_1 x_{i+1}))]}{[(k_{\mu i} + k_{\xi i}) + \vartheta(k_{\nu i} + k_{\eta i})][\sin(\beta_1(x_{i+1} - x_i)) + \sinh(\beta_1(x_{i+1} - x_i))]} +$$

$$- \frac{\sum_{j=1}^{i-1} \gamma_j [(k_{\mu j} + k_{\xi j}) + \vartheta(k_{\nu j} + k_{\eta j})][\sin(\beta_1(x_{i+1} - x_j)) + \sinh(\beta_1(x_{i+1} - x_j))]}{[(k_{\mu i} + k_{\xi i}) + \vartheta(k_{\nu i} + k_{\eta i})][\sin(\beta_1(x_{i+1} - x_i)) + \sinh(\beta_1(x_{i+1} - x_i))]}$$

(26)

where

$$k_{\mu i} = \frac{\beta_1}{2}\left[-\sin(\beta_1 x_i) + \sum_{j=1}^{i-1} \mu_j \widehat{H}(x_i - x_j)\right];$$

$$k_{\nu i} = \frac{\beta_1}{2}\left[-\cos(\beta_1 x_i) + \sum_{j=1}^{i-1} \nu_j \widehat{H}(x_i - x_j)\right];$$

$$k_{\eta i} = \frac{\beta_1}{2}\left[\cosh(\beta_1 x_i) + \sum_{j=1}^{i-1} \eta_j \widehat{H}(x_i - x_j)\right];$$

$$k_{\xi i} = \frac{\beta_1}{2}\left[\sinh(\beta_1 x_i) + \sum_{j=1}^{i-1} \nu_j \widehat{H}(x_i - x_j)\right]. \quad (27)$$

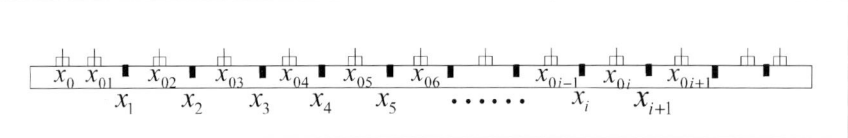

Fig. 2 Measurement positions between two damage locations.

Equation (26) provides the intensity of the generic damage based on the assumption that the experimental measurements of the first eigen-mode are given at the damage locations x_i. If there is no crack at the cross-section x_i, the identified damage parameter λ_i indicates the absence of damage.

For the case in Figure 2, in which the measurement positions, denoted as x_{0i}, are not coincident with the damage locations x_i, the measurements of the second eigen-mode are also required. In fact, in this case the generic damage position is a further unknown and the second eigen-mode measurements provide the sufficient additional data. In the generic interval between two measurements the following two equation system is set:

$$\phi_1^{ex}(x_{0i+1}) = \phi_1^{th}(x_i, \lambda_i), \quad \phi_2^{ex}(x_{0i+1}) = \phi_2^{th}(x_i, \lambda_i), \tag{28}$$

where $\phi_1^{th}(x_i, \lambda_i)$, $\phi_2^{th}(x_i, \lambda_i)$ are given by Equation (16). A numerical procedure can be applied to solve Equations (28) with respect to λ_i, x_i providing the solution to the damage identification problem between to measurement positions, once the damages at $x_j < x_{0i}$ have been identified with the same technique.

6 Conclusions

In this work the generalised functions (distributions) have been shown to be suitable for modelling singularities of different types introduced into the beam theory. Closed form solutions can be obtained for the direct analysis problems of beams with multiple damages. The latter solutions can be efficaciously employed to set an identification procedures for beams with multiple damages. In particular the advantage of the proposed procedure consists in a sequential identification of single damages either coincident with the position of the experimental measurements or lying in the beam intervals between two experimental measurements.

Acknowledgements

This work is part of the National Research Project (2006–2008), supported by MIUR, Grant n. 2006083134 whose title of the Unit of Palermo is "Identificazione del danneggiamento di strutture soggette a carichi statici e dinamici (deterministici e stocastici)".

References

1. Adams RD, Cawley P, Pye CJ, Stone BJ (1978) A vibration technique for non-destructively assessing the integrity of structures. *J Mech Eng Sci* **20**: 93–100.
2. Bagarello F (2002) Multiplication of distribution in one dimension and a first application to quantum field theory. *J Math Anal Appl* **266**: 298–320.
3. Biondi B, Caddemi S (2007) Euler–Bernoulli beams with multiple singularities in the flexural stiffness. *Eur J Mech A/Sol* **26**: 789–809.
4. Caddemi S, Caliò I (2008) Exact solution of the multi-cracked Euler–Bernoulli column. *Int J Sol Struct* **45**: 1332–1351.
5. Gladwell GML (2004) *Inverse Problems in Vibration*. Second ed. Kluwer Academic Publishers, Dordrecht, the Netherlands.
6. Hearn G, Testa RB (1991) Modal analysis for damage detection in structures. *J Struct Eng* **117**: 3042–3063.
7. Morassi A (1993) Crack-induced changes in eigenparameters of beam structures. *J Eng Mech* **119**: 1798–1803.
8. Narkis Y (1994) Identification of crack location in vibrating simply supported beams. *J Sound Vib* **172**: 549–558.
9. Rizos PF, Aspragatos N, Dimarogonas AD (1990) Identification of crack location and magnitude in a cantilever beam from vibration modes. *J Sound Vib* **138**: 381–388.
10. Yavari A, Sarkani S, Moyer ET (2000) On applications of generalised functions to beam bending problems. *Int J Sol Struct* **37**: 5675–5705.

Discontinuous Galerkin Methods

A Discontinuous Galerkin Method for an Incompatibility-Based Strain Gradient Plasticity Theory

J. Ostien[1] and K. Garikipati[2]

[1]Department of Mechanical Engineering, 2350 Hayward Street, University of Michigan, Ann Arbor, MI 48109, USA
E-mail: tostien@umich.edu
[2]Department of Mechanical Engineering & Michigan Center for Theoretical Physics, 2350 Hayward Street, University of Michigan, Ann Arbor, MI 48109, USA
E-mail: krishna@umich.edu

Abstract. We consider a recent strain gradient plasticity theory based on incompatibility of plastic strain due to the nature of lattice distortion around a dislocation (*J. Mech. Phys. Solids*, **52**, 2545–2568, 2004). The key features of this theory are an explicit treatment of the Burgers vector, a microforce balance that leads to a classical yield condition, and the inclusion of dissipation from plastic spin. The flow rule involves gradients of the plastic strain, and is therefore a partial differential equation. We apply recently-developed ideas on discontinuous Galerkin finite element methods to treat this higher-order nature of the yield condition, while retaining considerable flexibility in the mathematical space from which the plastic strain is drawn. In particular, despite the higher-order continuity apparent in the yield condition, it is possible to use plastic strain interpolations that are discontinuous across element edges. Two distinct approaches are outlined: the Interior Penalty Method and the Lifting Operator Method. The numerical implementation of the Interior Penalty Method is discussed, and a numerical example is presented.

Key words: interior penalty, lifting operator, microforces.

1 Introduction

Discontinuous Galerkin methods have become attractive in view of the difficulties associated with higher order partial differential or differential-algebraic equations, including the need for $\mathcal{C}^1$-continuous elements. Discontinuous Galerkin-based $\mathcal{C}^0$ finite element basis functions were developed for fourth-order elliptic problems related to thin beam and plate theory and gradient elasticity in [1]. Wells et al. [2] and Molari et al. [3] discuss strain gradient damage. The fourth order Cahn–Hilliard equation for phase segregation gets a treatment in [4]. DG formulations for Kirchoff–Love shells are presented in [5, 6]. Djoko et al. [7, 8] present a formulation for a different theory of strain gradient plasticity than that presented in this paper. Specifically, in the present work, a model of strain gradient plasticity based on incompatibilities in the plastic part of the deformation gradient [9] is implemented into a discontinuous Galerkin finite element method.

B.D. Reddy (ed.), *IUTAM Symposium on Theoretical, Modelling and Computational Aspects of Inelastic Media*, 217–226.
© Springer Science+Business Media B.V. 2008

Discontinuous Galerkin methods allow for a relaxation of the continuity requirement for basis functions by weakly imposing the desired order of continuity in the variational setting. This property facilitates the use of $\mathcal{C}^0$ interpolations of the primal field and avoids the difficulties associated with $\mathcal{C}^1$-continuous and mixed methods. In particular, strain gradient plasticity involves the gradient of the plastic part of the deformation, $\nabla \boldsymbol{H}^\mathrm{p}$, where $\boldsymbol{H}^\mathrm{p}$ lies in the space of deviatoric second-order tensors. DG implementations admit the use of piecewise linear, $\mathcal{C}^0$, or piecewise constant, $\mathcal{C}^{-1}$, interpolations of $\boldsymbol{H}^\mathrm{p}$, where the latter clearly offers relative numerical efficiency, and may provide added robustness to the formulation.

2 Strain Gradient Plasticity Model

As in the theory presented by Gurtin [9], restricting ourselves to infinitesimal strains, we assume a decomposition of the displacement gradient into elastic and plastic parts.

$$\nabla \boldsymbol{u} = \boldsymbol{H}^\mathrm{e} + \boldsymbol{H}^\mathrm{p} \tag{1}$$

A pertinent feature of the theory is that it does not assume, *a priori*, that the plastic spin, $\boldsymbol{W}^\mathrm{p} = \text{skew}\,\boldsymbol{H}^\mathrm{p}$, provides no contribution to the free energy of the plastically deformed body. The inclusion of the plastic spin occurs as a natural consequence of the characterization of the Burgers tensor in a continuum body. A measure of the incompatibility in the plastic distortion, $\boldsymbol{H}^\mathrm{p}$, can be related to the Burgers vector via Stokes' theorem and the definition of the Burgers tensor, $\boldsymbol{G} \equiv \text{curl}\,\boldsymbol{H}^\mathrm{p}$. For the boundary of a curve, S, with unit normal, $\boldsymbol{e}$, we have

$$\oint_{\partial S} \boldsymbol{H}^\mathrm{p} d X = \int_S \boldsymbol{G}^\mathsf{T} \boldsymbol{e}\, dA, \tag{2}$$

where the Burgers vector associated with S is $\boldsymbol{G}^\mathsf{T} \boldsymbol{e}$. For a theory to accurately capture the effect of the Burgers tensor, the evolution of plastic spin must be accounted for, since $\boldsymbol{G} = \text{curl}\,\boldsymbol{H}^\mathrm{p}$, and $\boldsymbol{H}^\mathrm{p} = \boldsymbol{E}^\mathrm{p} + \boldsymbol{W}^\mathrm{p}$.

Continuing to follow Gurtin [9], the macroscopic balance of momenta places two familiar restrictions on the Cauchy stress, $\boldsymbol{T}$, in the body, namely equilibrium in the absence of applied tractions and for a body force $\boldsymbol{b}$, and symmetry

$$\text{div}\,\boldsymbol{T} + \boldsymbol{b} = 0 \tag{3}$$
$$\boldsymbol{T} = \boldsymbol{T}^\mathsf{T} \tag{4}$$

Furthermore, upon definition of two microscopic stresses, $\boldsymbol{T}^\mathrm{p}$ and $\mathbb{S}$, a microforce balance is posed that admits the interpretation as a plastic flow rule.

$$\text{dev}\,\boldsymbol{T} = \boldsymbol{T}^\mathrm{p} + \text{dev}\,\text{curl}\,\mathbb{S}^\mathsf{T} \tag{5}$$

In formulating the constitutive theory, a free energy is chosen of the form $\Psi(E^e, G)$, and the macro and micro stresses are defined to be thermodynamically conjugate to the kinematic tensors E^e and G, respectively:

$$\Psi(E^e, G) = \Psi^e(E^e) + \frac{1}{2}k|G|^2, \tag{6}$$

$$T = \frac{\partial \Psi}{\partial E^e}, \tag{7}$$

$$\mathbb{S} = \frac{\partial \Psi}{\partial G}. \tag{8}$$

Next, a constitutive relation is assumed for the micro-stress,

$$T^p = Y(d^p)\left(\dot{E}^p + \chi \dot{W}^p\right), \tag{9}$$

where d^p is an effective distortion rate.

$$d^p = \sqrt{\dot{E}^p : \dot{E}^p + \chi \dot{W}^p : \dot{W}^p} \tag{10}$$

With (7), (8), and (9) in hand, we can revisit the flow rule (5).

$$\operatorname{dev} T - \operatorname{dev}\operatorname{curl}\left(k \operatorname{curl} H^p\right)^{\mathsf{T}} = Y(d^p)\left(\dot{E}^p + \chi \dot{W}^p\right)^{\mathsf{T}} \tag{11}$$

Attention is drawn to the form of (11), which is that of a flow rule with kinematic hardening for $k > 0$. In this interpretation $\operatorname{dev}\operatorname{curl}\left(k \operatorname{curl} H^p\right)^{\mathsf{T}}$ plays the role of a deviatoric back stress. Rate independent behavior is obtained when the function $Y(d^p)$ is specified to be Y/d^p. J2 flow theory is recovered by setting the parameter χ to zero.

For the partial differential equation describing the microforce balance, additional boundary conditionals are necessary. Consider the homogeneous essential boundary condition denoted the microhard boundary condition, which can be posed as follows.

$$\dot{H}^p(n\times) = 0 \text{ on } \Gamma_H \tag{12}$$

The microhard condition corresponds to a vanishing flux of the Burgers vector, $G^{\mathsf{T}} e$, for all planes with normal e intersecting Γ_H, where Γ_H is regarded as the microhard boundary. The complementary natural boundary condition corresponds to a microstress free boundary, Γ_S, and is referred to as the microfree boundary condition.

$$\operatorname{dev}(\mathbb{S}(n\times)) = 0 \text{ on } \Gamma_S \tag{13}$$

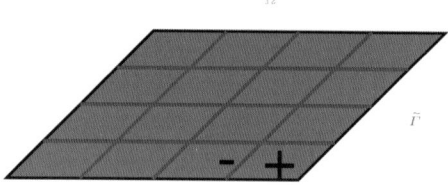

Fig. 1 Discontinuous domain.

3 Classical Formulation

The statement of the classical Galerkin weak form of the problem is the following: Find $\{u, H^p\} \in \mathcal{S} \times \mathcal{P} \subset H^1(\Omega) \times \operatorname{dev} H^1(\Omega)$ s.t. $\forall \{w, V\} \in \mathcal{V} \times \mathcal{Q} \subset H^1(\Omega) \times \operatorname{dev} H^1(\Omega)$

$$(\nabla w, T)_\Omega = (w, b)_\Omega + (w, t)_{\Gamma_t} \quad (14)$$

$$(V, T - T^p)_\Omega + (\operatorname{curl} V, k \operatorname{curl} H^p)_\Omega = (V, S(n))_{\Gamma_s} \quad (15)$$

The solution spaces, $\mathcal{P}, \mathcal{Q} \subset \operatorname{dev} H^1(\Omega)$ imply a minimum of 32 degrees of freedom for $H_h^p \in \mathcal{P}^1(\Omega_e)$. If the regularity assumptions can be relaxed, i.e. if we can choose $\mathcal{P}, \mathcal{Q} \subset \operatorname{dev} L^2(\Omega)$, then the minimum number of degrees of freedom can be reduced to 8, and, furthermore, larger functional spaces become available.

4 Discontinuous Galerkin Formulation

We will now cast the problem into the discontinuous Galerkin setting in order to alleviate the problems mentioned above with the Galerkin formulation. In preparation for this exercise we will first introduce some concepts required in the discontinuous Galerkin framework.

4.1 Preliminaries

Given a domain in $\mathcal{R}^n$ discretized into elements, we define the interior domains as in (16) and the inter-element boundaries as in (17).

$$\tilde{\Omega} = \bigcup_{e=1}^{n_{el}} \text{int}(\Omega_e) \tag{16}$$

$$\tilde{\Gamma} = \bigcup_{e_1,e_2=1}^{n_{el}} \left(\partial\Omega_{e_1} \bigcap \partial\Omega_{e_2}\right) \tag{17}$$

Next, we define the jump operator (19) and average operator (20).

$$\text{Discontinuous field}: \left. f^{\pm}\right|_{\tilde{\Gamma}} = \lim_{\varepsilon \to 0} f(x \mp \varepsilon n) \tag{18}$$

$$\text{Jump operator}: [\![f\,(n\times)^{\mathsf{T}}]\!] = f^+(n^+\times)^{\mathsf{T}} + f^-(n^-\times)^{\mathsf{T}} \tag{19}$$

$$\text{Average operator}: \langle f \rangle = \frac{1}{2}\left(f^+ + f^-\right) \tag{20}$$

$$\text{Element diameter}: \ h \tag{21}$$

4.2 Interior Penalty Method

Using the concepts and notation introduced above, we first pose the discontinuous Galerkin weak form using a penalty parameter to stabilize the discontinuous solution. The interior penalty discontinuous Galerkin formulation is the following: Find $\{u_h, H_h^p\} \in \mathcal{S}_h \times \mathcal{P}_h \subset H^1(\Omega) \times \text{dev}\,L^2(\Omega)$ s.t. $\forall \{w_h, V_h\} \in \mathcal{V}_h \times \mathcal{Q}_h \subset H^1(\Omega) \times \text{dev}\,L^2(\Omega)$

$$(\nabla w_h, T_h)_{\Omega} = (w_h, b)_{\Omega} + (w_h, t)_{\Gamma_t} \tag{22}$$

$$(V_h, T_h - T_h^{p^{\mathsf{T}}})_{\Omega} + (\text{curl}\,V_h, k\,\text{curl}\,H_h^p)_{\tilde{\Omega}}$$
$$-([\![V_h\,(n\times)^{\mathsf{T}}]\!], \langle k\,\text{curl}\,H_h^{p^{\mathsf{T}}}\rangle)_{\tilde{\Gamma}}$$
$$-(\langle k\,\text{curl}\,V_h^{\mathsf{T}}\rangle, [\![H_h^p\,(n\times)^{\mathsf{T}}]\!])_{\tilde{\Gamma}}$$
$$+\underbrace{\frac{\alpha\,k}{h}([\![V_h\,(n\times)^{\mathsf{T}}]\!], [\![H_h^p\,(n\times)^{\mathsf{T}}]\!])_{\tilde{\Gamma}}}_{\text{Int. penalty}} = (V_h, S(n))_{\Gamma_S} \tag{23}$$

Variational consistency of the formulation is demonstrated by applying integration by parts to arrive at (24)–(27). From these four equations it is clear that the exact solution also satisfies the interior penalty discontinuous Galerkin weak from (22) and (23), which is the classical requirement for consistency of a finite element formulation.

$$\left(V_h, T_h - T_h^{\text{p}} - \underbrace{(\text{curl}(k \text{ curl } H_h^{\text{p}})^{\text{T}})^{\text{T}}}_{\mathbb{S}_h}\right)_{\widetilde{\Omega}} = 0 \qquad (24)$$

$$(\langle k \text{ curl } V_h^{\text{T}} \rangle, [\![H_h^{\text{p}} (n \times)^{\text{T}}]\!])_{\widetilde{\Gamma}} = 0 \qquad (25)$$

$$\left(\langle V_h \rangle, [\![\underbrace{(k \text{ curl } H_h^{\text{p}})^{\text{T}}}_{\mathbb{S}_h} n \times]\!]\right)_{\widetilde{\Gamma}} = 0 \qquad (26)$$

$$\left(V_h, (\mathbb{S}_h^{\text{T}} n \times - S(n))\right)_{\Gamma_S} = 0 \qquad (27)$$

As mentioned above, For $\mathcal{C}^0$-continuous H_h^{p}, i.e. $H_h^{\text{p}} \in \mathcal{P}^1(\Omega_e)$, at least 32 degrees of freedom are needed. However for $\mathcal{C}^{-1}$-continuous H_h^{p}, i.e. $H_h^{\text{p}} \in \mathcal{P}^0(\Omega_e)$, only 8 degrees of freedom are needed. The reduction in degrees of freedom is attractive, but the cost is that for $H_h^{\text{p}} \in \mathcal{P}^0(\Omega_e)$, the gradient/backstress component of the model lies entirely within the interior penalty term, which depends on the penalty parameter α.

4.3 Lifting Operator Formulation

Recall that as in Figure 1 the inter-element boundaries are denoted by $\widetilde{\Gamma}$. We define the lifting operator $r \in \mathcal{R}$, s.t. $\forall A \in \mathcal{P}$

$$(A, r(B))_{\Omega} = - \left(\langle A \rangle, [\![B(n \times)^{\text{T}}]\!]\right)_{\widetilde{\Gamma}}$$
$$- \left(A, (B(n \times)^{\text{T}} - \mathbb{B}(n))\right)_{\Gamma_H}.$$

The lifting operator, $r(\bullet)$, so-defined, is an approximation to the curl operator in a distributional sense.

We now state the lifting operator formulation. Find $\{u_h, H_h^{\text{p}}\} \in \mathcal{S}_h \times \mathcal{P}_h \subset H^1(\Omega) \times \text{dev } L^2(\Omega)$ s.t. $\forall \{w_h, V_h\} \in \mathcal{V}_h \times \mathcal{Q}_h \subset H^1(\Omega) \times \text{dev } L^2(\Omega)$

$$(\nabla w_h, T_h)_{\Omega} = (w_h, b)_{\Omega}$$
$$+ (w_h, t)_{\Gamma_t} \qquad (28)$$

$$(V_h, T_h - T^{\text{p}_h})_{\Omega}$$
$$+ ((\text{curl } V_h + r(V_h)), k \text{ (curl } H_h^{\text{p}} + r(H_h^{\text{p}})))_{\Omega} = (V_h, S(n))_{\Gamma_S} \qquad (29)$$

While not included in (29), some numerical stabilization is also required in the lifting operator formulation. It usually takes the form of a term that is bilinear in the lifting operator and includes a stabilization parameter.

5 Implementation

The model was implemented, using the interior penalty formulation, into a non-linear iterative solver using a Newton algorithm. A yield surface was added as

$$\|2\mu(\boldsymbol{\varepsilon}_{n+1,k} - \boldsymbol{E}_n^p) - (\text{dev curl}\, \mathbb{S}^T)_n\| - \sqrt{\frac{2}{3}} Y_0 \leq 0.$$

Note that for $\boldsymbol{H}_h^p \in \mathcal{P}^0(\Omega_e)$, the terms $\mathbb{S} = k\,\text{curl}\,\boldsymbol{H}^p = \boldsymbol{0}$ and that $\text{dev curl}\,\mathbb{S}^T = \boldsymbol{0}$ in Ω_e. Lifting operators $\boldsymbol{r}_e(\mathbb{S})$ and $\text{dev}\,\boldsymbol{r}_e((\boldsymbol{r}_e(\mathbb{S}))^T)$ were used to evaluate the backstress in the yield surface evaluation.

5.1 Evolving elastic-plastic boundary

An interesting and difficult challenge associated with gradient plasticity models is the determination of the evolving boundary between the elastic and plastic domains. The boundary conditions discussed at the end of Section 2 are not to be applied to the whole domain, but just to that part of the domain undergoing plastic deformation. Here a strong argument can be made for the use of constants to represent the plastic distortion, as it simplifies the definition of the elastic-plastic boundary and provides a well defined outward normal vector to the plastic domain. If a linear space was used, it would be possible for the elastic-plastic boundary to exist within elements and more consideration would be necessary to properly define the normal vector and hence apply boundary conditions. Further complicating matters, the iterative solution procedure would then have to resolve both the plastic distortion and plastic domain. The current model, implemented with constants, simply tracks the elements undergoing plastic deformation and defines the elastic-plastic boundary using the outward normal of the facets of those elements in the plastic domain.

6 Numerical Results

The interior penalty discontinuous Galerkin implementation was used to solve a torsional boundary value problem on a cubic domain discretized into 6 tetrahedral elements. The parameters used in the model can be seen in Table 1.

Two views of the deformed mesh and contours of the magnitude of the plastic distortion can be seen in Figures 2 and 3.

Gurtin and Anand [10] show that for the microfree boundary condition (13), the model does not preclude softening, regardless of the body geometry or applied loading. In the present work where the microfree boundary conditions have been implemented, softening is indeed observed (Figure 4). The physical interpretation is that the microfree boundary does not cause dislocation pile up, and as a result

Table 1 Model parameters.

parameter	value
Young's Modulus E	200 GPa
Poisson's ratio ν	0.3
IP parameter α	1.0
Spin parameter χ	1.0
Yield stress σ_y	975 MPa
Gradient modulus k	1000 MPa-m^2

Fig. 2 Torsion of a cube, contours of $\|\boldsymbol{H}^{\mathrm{p}}\|$.

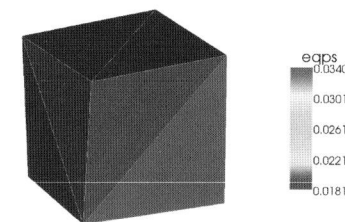

Fig. 3 Torsion of a cube, contours of $\|\boldsymbol{H}^{\mathrm{p}}\|$.

hardening is not obtained. Mathematically it is worth noting that while, for $k > 0$, (11) takes on the appearance of a flow rule with kinematic *hardening*, the "back stress" dev curl $\left(k \operatorname{curl} \boldsymbol{H}^{\mathrm{p}}\right)^{\mathrm{T}}$ does not evolve along the normal to the yield surface in stress space. Therefore hardening and softening cannot be explained solely in local terms, but only in the context of the full boundary value problem. It is worthwhile to note that despite the softening response, the boundary value problem remains well posed, as the necessary length scale has been introduced to the gradient term. For this reason convergent solutions are expected. The model also shows in Figure 4 that the amount of softening depends on the relative size of the body. This is the widely-known the size effect observed in plasticity experiments at micron scales.

Another characteristic of the model is the effect that k has on the softening behavior. Since the gradient plasticity term for this boundary value problem promotes softening, increasing k causes the model to become progressively unstable, finally failing to converge for $k = 1.2 \times 10^4$ and $k = 2 \times 10^4$ MPa-m^2 as seen in Figure 5.

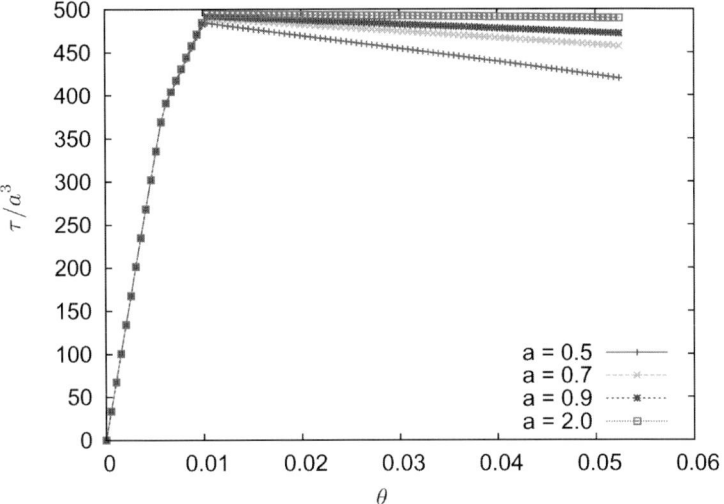

Fig. 4 Weakening under micro-free boundary conditions.

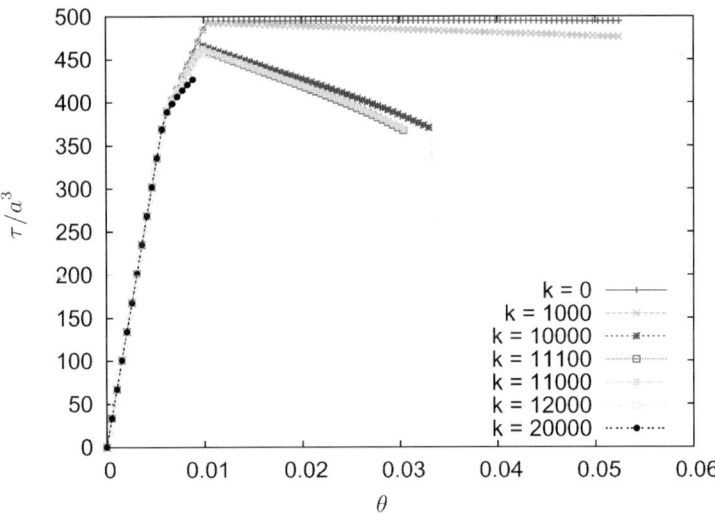

Fig. 5 Destabilizing effect of k due to softening.

7 Closing Remarks

In this paper we have proposed a formulation for incompatibility-based strain gradient plasticity that incorporates a discontinuous basis for interpolation of the solution. The formulation is variationally consistent and a version that included interior penalty stabilization was implemented into a non-linear finite element code using

$H_h^p \in \mathcal{P}^0(\Omega_e)$. The numerical results upend the intuition developed for hardening and softening effects from local models of plasticity. However, these results are not in violation of previous theoretical results on the softening behavior of the gradient plasticity model. They also admit a physical interpretation in terms of dislocation interactions. The model also displayed the size effect seen with plastic deformation at small scales, absent from classical treatments of plasticity.

In future work the response of a number of strain gradient plasticity theories will be explored using a similar framework. Implementation of the lifting operator formulation will be an important next step in the development of the discontinuous Galerkin treatment. This is primarily due to the undesirable consequence of the IP formulation where the penalty parameter affects the hardening response. Furthermore, an analysis to prove the optimal convergence rate is warranted, where we expect $\|\nabla \boldsymbol{u} - \nabla \boldsymbol{u}_h\| + \|\boldsymbol{H}^p - \boldsymbol{H}^{p_h}\| \leq Ch^k$, if $\boldsymbol{u}_h \in \mathcal{P}^k$, $\boldsymbol{H}^{p_h} \in \mathcal{P}^{k-1}$.

References

1. G. Engel, K. Garikipati, T.J.R. Hughes, M.G. Larson, L. Mazzei, R.L. Taylor. Continuous/discontinuous finite element approximations of fourth-order elliptic problems in structural and continuum mechanics with application to thin beams and plates, and strain gradient elasticity. *Computer Methods in Applied Mechanics and Engineering* **191**, 3669–3750, 2002.
2. G.N. Wells, K. Garikipati, L. Molari. A discontinuous Galerkin formulation for a strain gradient-dependent damage model. *Computer Methods in Applied Mechanics and Engineering* **193**, 3633–3645, 2004.
3. L. Molari, G.N. Wells, K. Garikipati, F. Ubertini. A discontinuous Galerkin method for strain gradient-dependent damage: Study of interpolations and convergence. *Computer Methods in Applied Mechanics and Engineering* **195**, 1480–1498, 2006.
4. G. Wells, E. Kuhl, K. Garikipati. A discontinuous Galerkin method for the Cahn–Hilliard equation. *Journal of Computational Physics* **218**, 860–877, 2006.
5. G.N. Wells, N. Tien Dung. A C^0 discontinuous Galerkin formulation for Kirchhoff plates. *Computer Methods in Applied Mechanics and Engineering* **196**, 3370–3380, 2007.
6. L. Noels, R. Radovitzky. An explicit discontinuous Galerkin method for non-linear solid dynamics: Formulation, parallel implementation and scalability properties. *International Journal for Numerical Methods in Engineering*, DOI: 10.1002/nme.2213.
7. J.K. Djoko, F. Ebobisse, A.T. McBride, B.D. Reddy. A discontinuous Galerkin formulation for classical and gradient plasticity – Part 1: Formulation and analysis. *Computer Methods in Applied Mechanics and Engineering* **196**, 3881–3897, 2007.
8. J.K. Djoko, F. Ebobisse, A.T. McBride, B.D. Reddy. A discontinuous Galerkin formulation for classical and gradient plasticity – Part 2: Algorithms and numerical analysis. *Computer Methods in Applied Mechanics and Engineering* **197**, 1–21, 2007.
9. M.E. Gurtin. A gradient theory of small-deformation isotropic plasticity that accounts for the Burgers vector and for dissipation due to plastic spin. *Journal of the Mechanics and Physics of Solids* **52**, 2545–2568, 2004.
10. M.E. Gurtin and L. Anand. A theory of strain-gradient plasticity for isotropic, plastically irrotational materials. Part I: Small deformations. *Jornal of the Mechanics and Physics of Solids* **53**, 1624–1649, 2005.

Some Applications of Discontinuous Galerkin Methods in Solid Mechanics

Adrian Lew, Alex Ten Eyck and Ramsharan Rangarajan

Mechanical Engineering, Stanford University, Stanford, CA 94305-4040, USA
E-mail: {lewa, alabute, rram}@stanford.edu

Abstract. We provide a brief overview of our recent work on applications of discontinuous Galerkin methods in solid mechanics. The discussion is light in technical details, and rather emphasizes key ideas, advantages and disadvantages of the approach, illustrating these with several numerical examples.

Key words: discontinuous Galerkin, nonlinear elasticity, immersed boundary methods, locking.

1 Introduction

Discontinuous Galerkin methods are finite element methods distinguished by the use of piecewise continuous functions that may be discontinuous across element boundaries to approximate the solution of a boundary value problem. Discontinuous Galerkin methods have been remarkably successful as high-order versions, and natural generalizations to unstructured meshes, of finite volume methods for hyperbolic conservation laws (see [5]). In this case, discontinuities across element boundaries permit the crafting of conservative and monotone schemes by adopting piecewise constant approximations within each element, as well as the simple use of slope limiters to achieve stability for high-order schemes. Hence, the introduction of discontinuous fields across element boundaries provides a clear set of algorithmic advantages.

Similar algorithmic advantages are encountered in the context of elliptic problems with constraints, for example, incompressible elasticity. In this case in addition to the mechanical equilibrium equation, the solution must satisfy the incompressibility constraint. It is well know that if piecewise affine conforming (i.e., continuous) finite elements are adopted, then the subset of the discrete space that satisfies the constraint almost everywhere may become so poor that its approximation properties are severely deteriorated. This is commonly known as locking. One way around this problem is to trade the importance of constraints, namely, it is possible to relax the continuity constraint across element boundaries to strongly impose the incompress-

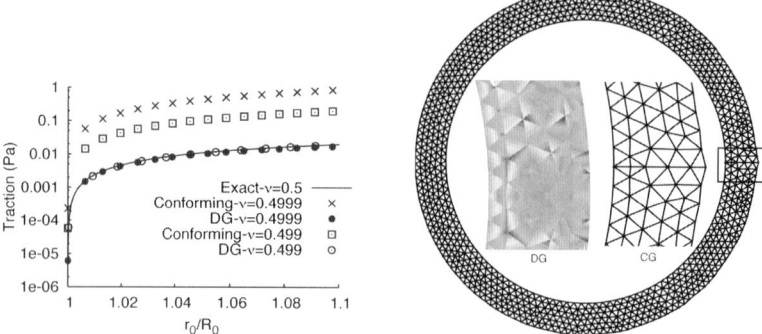

Fig. 1 On the left, evolution of the traction at the inner wall of the cylinder as a function of the relative expansion of the inner radius. Results computed with conforming (CG) and discontinuous (DG) Galerkin methods are shown, for two near incompressible cases, $\nu = 0.499, 0.4999$, where ν is the Poisson ratio. The exact solution is also shown for comparison. It is apparent from these results that the method is locking–free in the finite strain case. The use of discontinuities to enforce the incompressibility constraint is clearly revealed in the figure on the right, which depicts the deformed mesh in the CG case, with enlarged views of the area enclosed by a square for both CG and DG approximations. The small protruding cusp in the CG mesh testifies to the large stresses created within to satisfy the incompressibility constraint. In contrast, these are easily relaxed through discontinuities in the DG case, as the contour plots therein reveal

ibility one without any loss in accuracy (order of convergence), see [9, 11, 20]. This idea of *trading* the enforcement of the inter-element continuity constraint to strongly impose others, without any loss of accuracy, is a pattern that repeatedly appears in many applications for which DG methods display a clear algorithmic advantage.

A second such example is found in the solution of some structural models of shells, plates and beams [2,4,8,10,14,18]. The constraint in this case is the compatibility relation between the rotational and displacement degrees of freedom, which is nearly exactly enforced as the thickness of the structure becomes very small. Known as shear locking, it can be overcome, again, by relaxing the continuity constraint along element boundaries and switching to a DG formulation. A third such example arises in the context of immersed boundary methods, as we shall demonstrate later in the manuscript. In this case, the constraints are the Dirichlet boundary conditions on the immersed boundary, which may lead to a deteriorated convergence rate. Optimal convergence rates are recovered once a discontinuous Galerkin discretization is introduced. Finally, in the context of high-order equations, such as for gradient plasticity theories [6,7] and some phase transition models [19], the introduction of discontinuities in the approximation of the derivative helps overcome the difficulties in constructing conforming spaces with continuous derivatives on unstructured meshes.

The goal of this article is to highlight some of the main ideas and showcase some examples of our own work on the subject. With this in mind, Section 2 introduces a few highlights of the method, in the form of numerical examples. In

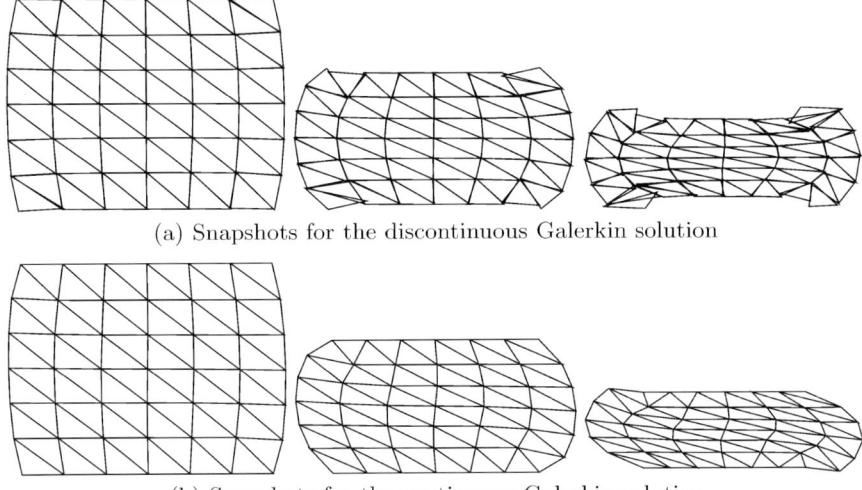

(a) Snapshots for the discontinuous Galerkin solution

(b) Snapshots for the continuous Galerkin solution

Fig. 2 Compression of an elastic block. Comparison between the continuous and discontinuous Galerkin solutions obtained with affine shape functions within each element.

Section 3, we formally present the class of discontinuous Galerkin methods which has been adopted in this article. Their application to nonlinear elasticity problems is described in Section 4, where we also briefly discuss the crucial subject of stabilization. Finally, Section 5 briefly describes a discontinuous-Galerkin-based immersed boundary method.

2 Examples

Incompressibility constraint at finite strains. The following example from [9] illustrates how discontinuities across element boundaries are actually used to accommodate the nonlinear incompressibility constraint at large deformations. We consider a cylinder under plane strain, made of a nearly incompressible neo-Hookean material, whose external wall is traction free. The internal boundary, in contrast, is deformed to acquire a new radius r_0 from its traction-free value of R_0. Both continuous and discontinuous Galerkin methods with an affine interpolation within each element were adopted. The results are depicted and explained in Figure 1.

Mesh-based kinematic constraints. The following example from [9] displays an unusual constraint, introduced by the choice of the mesh. An elastic block made of a compressible neo-Hookean material is squeezed by imposing displacements on its top and bottom faces, as shown in Figure 2. Because of the lack of symmetry of the mesh, the symmetry of the exact problem is lost at some point during the load-

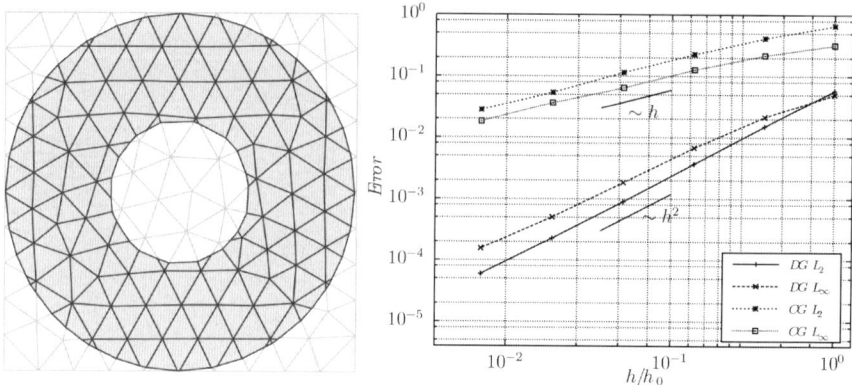

Fig. 3 A simulation with a DG-based immersed boundary method. The domain of the problem, a circular ring, is immersed in an arbitrary mesh, as shown on the left. Made of a linear elastic material, the ring is stretched by imposing Dirichlet boundary conditions on its entire boundary. The L_2 and L_∞ errors in the solutions as a function of the mesh size are shown on the right, for both CG and DG approximation spaces constrained to satisfy the Dirichlet boundary conditions on the approximate boundary of the domain. The CG method displays a sub-optimal convergence rate, again recovered by simply relaxing the continuity constraint across element boundaries. In this case, a DG approximation is only adopted in those elements intersected by the boundary.

ing path when continuity is enforced, while it is mostly restored when the latter is relaxed.

Boundary conditions as constraints. The term immersed boundary methods broadly describes methods in which the boundary of the domain may cut through elements in the mesh; see Figure 3. The problem of generating a mesh in a complicated geometry is, in this way, circumvented. It is replaced, however, by the need to devise strategies to impose boundary conditions on the immersed boundary. The natural idea of simply constraining the continuous finite element space over the mesh to satisfy prescribed Dirichlet boundary conditions generally leads to suboptimal approximation properties, known as boundary locking. This is not the case when a DG discretization is adopted, as clearly showcased in Figure 3 and described in [12].

Competitive performance. Not every problem presents a set of competing constraints for which it is convenient to relax the continuity across element boundaries. A known drawback of DG methods in these cases is that they often have a significantly larger number of degrees of freedom than CG methods *on the same mesh*. These additional degrees of freedom are generally used to obtain a better approximation of the solution. When CG and DG solutions with the same number of degrees of freedom are compared, the former is generally more accurate, but the latter is often competitive. This is illustrated in Figure 4 from [17], for a two-dimensional case. The contrast in the performance of both methods is often larger for three-dimensional computations.

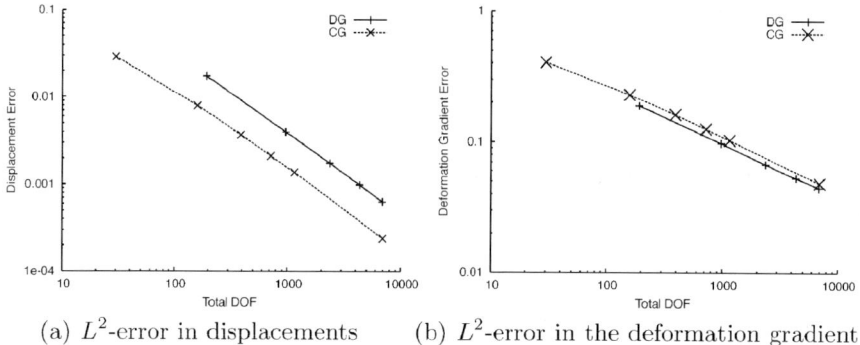

(a) L^2-error in displacements (b) L^2-error in the deformation gradient

Fig. 4 Convergence plot for the displacements (left) and deformation gradient (right) as a function of the number of degrees of freedom for a two-dimensional nonlinear elasticity example. For this case, the CG approximation is more efficient than the DG one for the computation of displacements, while the situation is reversed in the case of the deformation gradient

Large-scale simulations. We show next an application that benefits from the lack of locking for incompressible elastic materials of DG discretizations, and simultaneously demonstrates the possibility of utilizing them for the solution of large solid mechanics problems. It consists of the simulation of the mechanical response of a 50^3 μm^3 sample of blood-vessel microstructure (the media), obtained with novel scanning electron microscopy techniques [15]. Regions that contain elastin have been segmented, as shown in black in Figure 5. This rather heterogeneous microstructure was meshed with 175,616 trilinear hexahedra, totaling 4,214,784 degrees of freedom when the DG discretization was adopted. As a first step towards a more comprehensive study, elastin was assumed to be an isotropic linear elastic material immersed in incompressible water. With proper preconditioning of the linear system [17], the deformation of the sample could be solved using 512 processors in approximately one hour, allowing us to perform multiple simulations in one day.

3 Formulation of Discontinuous Galerkin Methods

The essential component in the construction of a DG method is the specification of how derivatives of functions are approximated. Since functions in the DG space may be discontinuous across element boundaries, their distributional derivatives may contain a singular part in the form of delta functions. Consequently, instead of approximating the derivative of a smooth function u with the exact distributional derivative of its discrete approximation u_h, we do so with another possibly piecewise discontinuous function, which we denote $D_{\text{DG}} u_h$.

We describe next the construction of $D_{\text{DG}} u_h$ in the simplest case, when $u_h = 0$ on $\partial \Omega$, where Ω is the domain, and the same discrete space V_h contains u_h and each component of $D_{\text{DG}} u_h$. The more general case can be found in [9]. The starting point

(a) Regions of elastin, in black.   (b) Shear strain under torsion

Fig. 5 Large-scale simulation to study the mechanical response of a of 50^3 μm^3 sample of blood-vessel microstructure. The material distribution in the mesh is shown on the left, with black for elastin and gray for everything else. Contour plots of the shear strain are depicted on the right. Each solution took approximately one hour in 512 processors, enabling the study of multiple loading conditions in a single day

is the following integration by parts identity:

$$\sum_{E \in \mathcal{T}_h} \int_E \nabla u_h \cdot w \, dV = \int_{\Gamma_I} (\llbracket u_h \rrbracket \cdot \{w\} + \{u_h\} \cdot \llbracket w \rrbracket) \, dS - \sum_{E \in \mathcal{T}_h} \int_E u_h \nabla \cdot w \, dV, \tag{1}$$

valid for any $w \in V_h^d$, where d is the number of components of ∇u_h. Here Γ_I indicates the set of all element faces in the mesh that do not lie on $\partial \Omega$. The jump $\llbracket \cdot \rrbracket$ and average $\{\cdot\}$ operators are defined as

$$\llbracket v \rrbracket = v^+ n^+ + v^- n^- \qquad \{v\} = \frac{1}{2}(v^+ + v^-), \tag{2}$$

where $v^\pm$ denotes the trace of v on either side of a face, and $n^\pm$ the corresponding external unit normal. An eminently intuitive idea for the definition of $D_{\mathrm{DG}} u_h$ is to assume that u is well approximated by u_h in the interior of every element, and by $\{u_h\}$ at element boundaries. An approximation of ∇u can then be constructed by requesting $D_{\mathrm{DG}} u_h$ to satisfy an equation similar to (1), namely

$$\sum_{E \in \mathcal{T}_h} \int_E D_{\mathrm{DG}} u_h \cdot w \, dV = \int_{\Gamma_I} \{u_h\} \cdot \llbracket w \rrbracket \, dS - \sum_{E \in \mathcal{T}_h} \int_E u_h \nabla \cdot w \, dV, \tag{3}$$

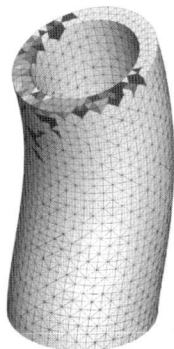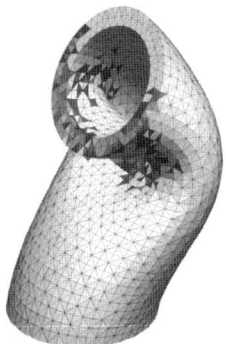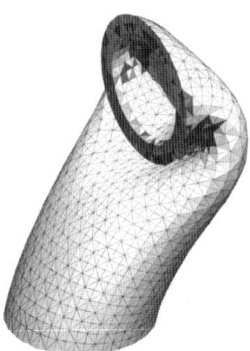

Fig. 6 Snapshots along the loading path of a nonlinear elastic cylinder. An adaptive stabilization strategy was adopted here, to automatically adjust the energetic cost of discontinuities as the deformation evolves. A nonzero stabilization term has been added only in the set of colored elements: the darker the color, the stiffer the term.

for any $w \in V_h^d$, where we have taken advantage of the fact that $[\![\{u_h\}]\!] = 0$. Alternative DG methods are obtained by assuming other approximations of u at element boundaries. Fortunately, it is possible to explicitly solve (1) and (3) for $D_{\mathrm{DG}} u_h$ as a function of u_h to obtain

$$D_{\mathrm{DG}} u_h = \nabla u_h + R([\![u_h]\!]) \tag{4}$$

in the interior of every element E. Here R is a linear operator on $[\![u]\!]$ that returns a function in V_h^d, and that can be precomputed for any given mesh, see [1, 9]. Notice that when u_h is continuous, (4) simply returns the standard definition of a derivative. Finally, when non-homogeneous Dirichlet boundary conditions are present, they can either be incorporated into $D_{\mathrm{DG}} u_h$ for weak enforcement, or simply constrain V_h to satisfy them.

4 Application: Nonlinear Elasticity

The nonlinear elasticity problem consists in finding local minimizers of the potential energy functional within some suitable functional space V^d. Its DG approximation is obtained by finding a local minimizer φ_h in V_h^d of the discrete potential energy functional

$$I_h[\varphi_h] = \sum_{E \in \mathcal{T}_h} \int_E \left[W(D_{\mathrm{DG}} \varphi_h) - f \cdot \varphi_h \right] dV. \tag{5}$$

Here W is the strain energy density and f the body force per unit volume. The examples in Figures 1, 2, 4 and 5 have all been obtained in this way.

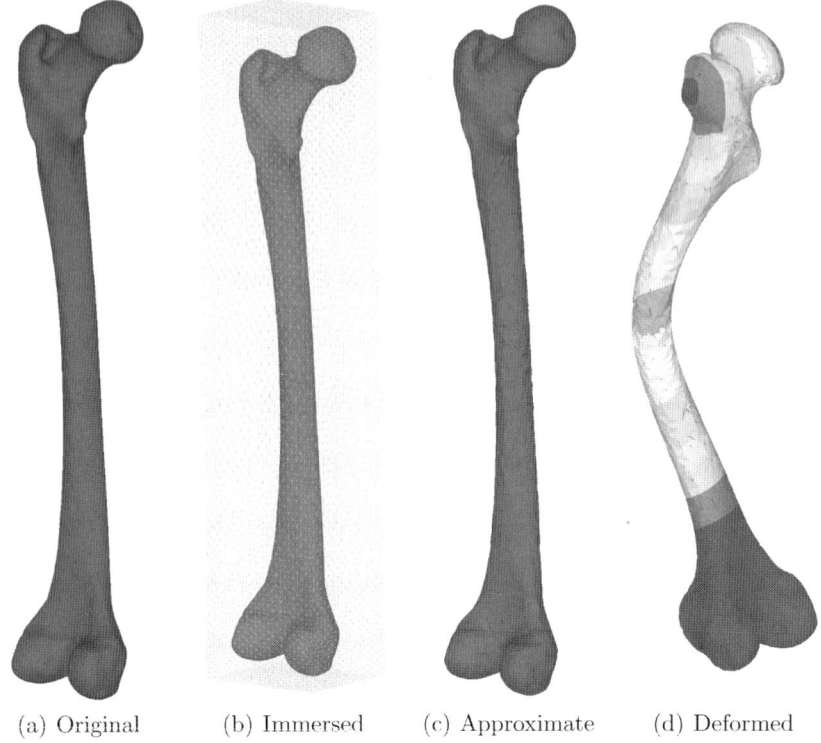

(a) Original (b) Immersed (c) Approximate (d) Deformed

Fig. 7 Three-dimensional example of a femur simulated with a DG-based immersed boundary method. The original geometry in (a) is immersed in a mesh of tetrahedra (b). An approximate geometry is extracted, shown in (c), which is then subjected to compressive loads on its two ends. Modeled as a linear elastic material, the amplitude of deformations in (d) have been amplified for clarity. The DG-IBM method sidesteps the creation of meshes in complicated geometry while retaining an optimal order of convergence

A crucial difficulty encountered when relaxing the continuity constraint across element boundaries is that often $V_h^d \not\subseteq V^d$. One of its consequences is that the potential energy I_h is not guaranteed to even have a finite lower bound in V_h^d, and often, it does not. The standard strategy in this case is to add a stabilization term to (5), in the form of a potential energy cost for each discontinuity in the solution, see [16, 17]. When the energetic cost of jumps is large enough, a *stable* scheme is recovered.

A delicate balance is required then. The additional term should be large enough to stabilize the problem, but not too large to essentially prevent any discontinuity from appearing. For classical linear elasticity the "right" size of the stabilization term is known [13], but in general, its fully automatic and efficient selection is still an open problem. We have recently made some progress by introducing the idea of adaptive stabilization [16, 17], but more comprehensive solutions may still be possible, see Figure 6.

Finally, the convergence of the method for the classical linear elasticity problem was proved in [13]. Since each approximate solution obtained as the mesh is refined is not even in H^1, the convergence of displacements was proved in the space of functions of bounded variations to any exact solution in H^2. Furthermore, displacements and stresses converge in L^2. In the nonlinear elastic case the convergence for the simpler case of a convex strain energy density was obtained in [3], by Gamma-convergence.

5 Application: A DG-Based Immersed Boundary Method

The two key ingredients in the construction of an immersed boundary method (IBM) are the approximation of the domain and how boundary conditions are imposed. Both need to work cooperatively to retain an optimal order of convergence. This is the reason for the profusion of first order methods, the crafting of complicated schemes to recover second-order, and the nearly absolute absence of high-order ones. These and other aspects in the case of homogeneous Dirichlet boundary conditions on $\partial \Omega$ are extensively discussed in [12], while the non-homogeneous case is the subject of an upcoming manuscript. A three-dimensional application of the DG-based IBM to elasticity is shown in Figure 7, obtained by introducing discontinuities across the boundaries of all elements intersected by the immersed boundary, and finding the stationary point of (5).

Acknowledgments

Support from the National Institutes of Health through the NIH Roadmap for Medical Research, Grant U54GM072970, and by the Department of the Army Research Grant W911NF-07-2-0027 is gratefully appreciated.

References

1. D. N. Arnold, F. Brezzi, B. Cockburn, and L. D. Marini. Unified analysis of discontinuous Galerkin methods for elliptic problems. *SIAM J. Numer. Anal.*, 39:1749–1779, 2002.
2. D.N. Arnold, F. Brezzi, and L.D. Marini. A family of discontinuous Galerkin Finite Elements for the Reissner–Mindlin plate. *Journal of Scientific Computing*, 22(1):25–45, 2005.
3. A. Buffa and C. Ortner. Variational convergence of IP–DGFEM. Technical Report 07/10, Oxford University Computing Laboratory, Numerical Analysis Group, Wolfson Building, Parks Road, Oxford, England OX1 3QD, April 2007.
4. F. Celiker, B. Cockburn, and H.K. Stolarski. Locking-free optimal discontinuous Galerkin methods for Timoshenko beams. *SIAM Journal on Numerical Analysis*, 44:2297, 2006.
5. B. Cockburn and C.W. Shu. Runge–Kutta discontinuous Galerkin methods for convection-dominated problems. *J. Sci. Comput*, 16(3):173–261, 2001.

6. J.K. Djoko, F. Ebobisse, A.T. McBride, and B.D. Reddy. A discontinuous Galerkin formulation for classical and gradient plasticity – Part 1: Formulation and analysis. *Computer Methods in Applied Mechanics and Engineering*, 196(37-40):3881–3897, 2007.
7. J.K. Djoko, F. Ebobisse, A.T. McBride, and B.D. Reddy. A discontinuous Galerkin formulation for classical and gradient plasticity. Part 2: Algorithms and numerical analysis. *Computer Methods in Applied Mechanics and Engineering*, 197(1-4):1–21, 2007.
8. G. Engel, K. Garikipati, T.J.R. Hughes, M.G. Larson, L. Mazzei, and R.L. Taylor. Continuous/discontinuous finite element approximations of fourth-order elliptic problems in structural and continuum mechanics with applications to thin beams and plates, strain gradient elasticity. *Computer Methods in Applied Mechanics and Engineering*, 191:3669–3750, 2002.
9. A. Ten Eyck and A. Lew. Discontinuous Galerkin methods for nonlinear elasticity. *International Journal for Numerical Methods in Engineering*, 67:1204–1243, 2006.
10. S. Guzey, B. Cockburn, and H.K. Stolarski. The embedded discontinuous Galerkin method: Application to linear shell problems. *International Journal for Numerical Methods in Engineering*, 70:757–790, 2007.
11. P. Hansbo and M.G. Larson. Discontinuous Galerkin methods for incompressible and nearly incompressible elasticity by Nitsche's method. *Computer Methods in Applied Mechanics and Engineering*, 191(17):1895–1908, 2002.
12. A. Lew and G. Buscaglia. A discontinuous-Galerkin-based immersed boundary method. *International Journal for Numerical Methods in Engineering*, 2008 (in press).
13. A. Lew, P. Neff, D. Sulsky, and M. Ortiz. Optimal BV estimates for a discontinuous Galerkin method in linear elasticity. *Applied Mathematics Research Express*, 3:73–106, 2004.
14. L. Noels and R. Radovitzky. A new discontinuous Galerkin method for Kirchhoff–Love shells. http://asap.mit.edu/publications/journal/cmame-2007.pdf, 2007.
15. M. O'Connel and C. Taylor. Personal communication, 2007. Stanford University.
16. A. Ten Eyck, F. Celiker, and A. Lew. Adaptive stabilization of discontinuous Galerkin methods for nonlinear elasticity: Analytical estimates. *Computer Methods in Applied Mechanics and Engineering*, 197(33–40):2989–3000, 2008.
17. A. Ten Eyck, F. Celiker, and A. Lew. Adaptive stabilization of discontinuous Galerkin methods for nonlinear elasticity: Motivation, formulation and numerical examples. *Computer Methods in Applied Mechanics and Engineering*, 2008 (in press).
18. G.N. Wells and N.T. Dung. A C0 discontinuous Galerkin formulation for Kirchhoff plates. *Computer Methods in Applied Mechanics and Engineering*, 196(35–36):3370–3380, 2007.
19. G.N. Wells, E. Kuhl, and K. Garikipati. A discontinuous Galerkin method for the Cahn–Hilliard equation. *Journal of Computational Physics*, 218(2):860–877, 2006.
20. T.P. Wihler. Locking–free DGFEM for elasticity problems in polygons. *IMA Journal of Numerical Analysis*, 24:45–75, 2004.

Some Aspects of a Discontinuous Galerkin Formulation for Gradient Plasticity at Finite Strains

Andrew McBride[1] and B. Daya Reddy[1,2]

[1] Centre for Research in Computational and Applied Mechanics, University of Cape Town, 7701 Rondebosch, South Africa
E-mail: Andrew.McBride@uct.ac.za
[2] Department of Mathematics and Applied Mathematics, University of Cape Town, 7701 Rondebosch, South Africa
E-mail: Daya.Reddy@uct.ac.za

Abstract. This work considers the extension of a model of gradient plasticity, previously analysed subject to the assumption of infinitesimal deformations, to the finite strain regime. The discontinuous Galerkin finite element method is used to solve the non-local expression of the plastic flow rule, thereby allowing the higher order terms that arise in the gradient formulation to be accommodated in an elegant manner.

Key words: gradient plasticity, finite deformations, discontinuous Galerkin.

1 Introduction

Motivated in large part by the inability of classical theories to model material behaviour at the meso-scale level, various plasticity theories that incorporate size-dependence via the inclusion of strain gradients have been developed. These theories include in a natural way a length scale, and permit phenomena such as shear banding to be captured. For example, in the early work of Aifantis [1], the yield function is augmented by a term involving the Laplacian of the equivalent plastic strain and possibly other higher-order terms. An overview and critical comparison of the features, both mathematical and physical, of various gradient plasticity formulations is given in [2, 3]. The non-standard higher-order contributions arising in gradient plasticity formulations render the conventional framework of classical finite elements inappropriate. The discontinuous Galerkin (DG) finite element method [4] allows the higher-order contributions to be treated in an elegant and effective manner as was demonstrated by the authors in [5, 6] for the aforementioned model of gradient plasticity [1] restricted to the infinitesimal strain regime.

In this work we consider the extension of the gradient plasticity model to the finite deformation regime. The extension uses a logarithmic hyperelastic-plastic model [7] that preserves the essential ingredients of the return mapping algorithms of the infinitesimal theory. The simplicity of this model of plasticity has been exploited by others as a basis for classical and non-local plasticity formulations (see e.g. [8] and references therein).

This work treats the formulation of the problem in the finite-strain context. Details of the thermodynamic framework, not previously discussed in detail, are presented. In addition, algorithmic and computational aspects are summarised, and two example problems discussed. Further results and details are presented in [9].

2 The Governing Equations for the Problem

Let Ω be a bounded domain in $\mathbb{R}^2$ with boundary $\partial\Omega$, and which is occupied by an elastoplastic body in its undeformed configuration. The current placement resulting from a motion $\varphi(X, t)$ is denoted $\mathcal{S}$. A material point in Ω is denoted X. Let $\mathcal{T}_h^0 = \{K^0\}$ be a shape-regular subdivision of the reference domain Ω where K^0 are quadrilateral elements. For the space of displacements we construct conforming basis functions via functions N_φ^A defined on the reference element $\square = [-1, -1] \times [1, 1]$ with coordinates $\boldsymbol{\xi} \in \square$. Then the interpolation of the reference domain and the displacement field over a typical element are given by

$$X \approx X_h = \sum_{A=1}^{n_{\text{node}}^e} N_\varphi^A(\boldsymbol{\xi}) X_A^e \quad \text{and} \quad u \approx u_h = \sum_{A=1}^{n_{\text{node}}^e} N_\varphi^A(\boldsymbol{\xi}) u_A^e , \qquad (1)$$

where n_{node}^e denotes the number of nodes per element, X_A^e the reference coordinates and u_A^e the displacement field associated with node A.

For the sake of convenience, the displacement field is approximated using conforming, bilinear elements. The classical deformation gradient is interpolated across an element from the nodal displacement $d_A^e := X_A^e + u_A^e$ by

$$\text{GRAD}_X[\varphi_h^e] = \sum_{A=1}^{n_{\text{node}}^e} d_A^e(t) \otimes \text{GRAD}_X[N_\varphi^A] , \qquad (2)$$

where $\text{GRAD}_X[(\cdot)] := \partial(\cdot)/\partial X$ is the gradient operator with respect to the reference configuration.

In order to circumvent volumetric locking associated with low-order elements we use an enhanced assumed strain formulation [10], in which the deformation gradient is additively composed of the Galerkin approximation (2) and an enhanced part as

$$F_h = \underbrace{\text{GRAD}_X[\varphi_h]}_{\text{Galerkin}} + \underbrace{\tilde{F}_h}_{\text{enhanced}} . \qquad (3)$$

The structure of $\tilde{F}_h$ chosen here is based on recommendations made in [11].

With P denoting the first Piola–Kirchhoff stress, we denote by $\boldsymbol{\tau} := PF_h^T$ the Kirchhoff stress which is related to the Cauchy stress σ by $\boldsymbol{\tau} = \det[F_h]\sigma$. Following

[10], the weak form of the governing equations are

$$\int_\Omega P_h : \text{GRAD}[v_h]\, dX = \int_\Omega B \cdot v_h\, dX + \int_{\Gamma_T} T \cdot v_h\, dS, \quad (4a)$$

$$-\int_\Omega P_h : g_h\, dX = 0, \quad (4b)$$

where v_h is an arbitrary test function, g_h an arbitrary enhanced assumed strain, and B is the body force per unit reference volume.

The right and left Cauchy–Green tensors are respectively defined by

$$C_h = F_h{}^T F_h \quad \text{and} \quad b_h = F_h F_h{}^T. \quad (5)$$

The deformation gradient is assumed to be multiplicatively decomposed into an elastic F_h^e and plastic part F_h^p as $F_h = F_h^e F_h^p$. The multiplicative decomposition of F_h motivates the definition of the elastic left, the plastic right, and the elastic right Cauchy–Green tensors as

$$b_h^e = F_h^e F_h^{eT}, \quad C_h^p = F_h^{pT} F_h^p \quad \text{and} \quad C_h^e = F_h^{eT} F_h^e. \quad (6)$$

The super- or subscript h will be dropped subsequently where convenient.

2.1 Constitutive Relations and the Flow Law

The system is characterised by a free energy function W additively composed of an elastic part $\bar{W}^e$ and plastic part W^p. The classical plastic free energy function, see e.g. [10], is extended to the gradient regime by setting

$$W = \bar{W}\left(F^e, \xi, \nabla\xi\right) = \bar{W}^e(C^e) + \underbrace{(1/2)k_1\xi^2 + (1/2)k_2|\nabla\xi|^2}_{\bar{W}^p(\xi, \nabla\xi)}, \quad (7)$$

where ξ is the isotropic hardening parameter, k_1 is the isotropic hardening constant and $k_2 > 0$ is the gradient hardening constant that effectively introduces a length scale $l = \sqrt{\text{abs}[k_2/k_1]}$ into the formulation. The classical plasticity formulation is recovered by setting $k_2 = 0$.

The problem of classical plasticity with hardening $k_1 > 0$ is well-posed. For the case of softening classical plasticity, i.e. when $k_1 < 0$, the problem is ill-posed. It is well documented that finite element approximations of softening materials are pathologically dependent upon the resolution of the discretisation as this provides the only length scale. Well-posedness of the gradient plasticity model analysed here for small strain softening problems was proven in [5]. A primary motivation for gradient plasticity formulations of this form is the ability to obtain mesh-independent solutions in the softening regime, see [6] and references therein.

Following classical thermodynamic arguments [12], the dissipation inequality $\tau : d - \dot{W} \geq 0$ yields

$$0 \leq \tau : d - \dot{W} \leq \tau : (d^e + d^p) - \frac{\partial \bar{W}^e(C^e)}{\partial C^e} : \dot{C}^e + \bar{g}\dot{\xi} + m \cdot \nabla\dot{\xi}$$

$$\leq \left(\tau - 2F^e \left[\frac{\partial \bar{W}^e(C^e)}{\partial C^e}\right] F^{eT}\right) : d^e + \tau : d^p + \bar{g}\dot{\xi} + m \cdot \nabla\dot{\xi}, \quad (8)$$

where $\bar{g} := -\partial W/\partial \xi$ and $m := -\partial W/\partial \nabla\xi$, and d is the symmetric rate of deformation tensor, which is additively decomposed into elastic and plastic parts d^e and d^p. The notation $\dot{(\cdot)}$ denotes the material time derivative of an arbitrary function $(\cdot)$. From the dissipation inequality (8) we obtain the elastic relation

$$\tau = 2F^e \frac{\partial \bar{W}^e(C^e)}{\partial C^e} F^{eT}, \quad (9)$$

and the reduced dissipation inequality

$$\tau : d^p + \bar{g}\dot{\xi} + m \cdot \nabla\dot{\xi} \geq 0. \quad (10)$$

We assume an isotropic elastic response, in which case the dependence of $\bar{W}^e$ on C^e or equivalently b^e can be expressed via the elastic principal stretches λ^e_A where $(\lambda^e_A)^2$ ($A = 1, 2, 3$) are the principal stretches of b^e. We assume a Hencky model for the elastic stored energy function, that is

$$\bar{W}^e(b^e) = \bar{w}^e(\varepsilon^e_A) := (\Lambda/2)(\varepsilon^e_1 + \varepsilon^e_2 + \varepsilon^e_3)^2 + \mu\left((\varepsilon^e_1)^2 + (\varepsilon^e_2)^2 + (\varepsilon^e_3)^2\right), \quad (11)$$

where $\Lambda > 0$ and $\mu > 0$ are the Lamé constants and $\varepsilon^e_A := \ln[\lambda^e_A]$ are the logarithmic elastic stretches. The constitutive relation for the components of τ in the principal directions of b^e, denoted $\bar{\tau}_A$, are obtained from (9) as

$$\bar{\tau}_A = \Lambda \left(\varepsilon^e_1 + \varepsilon^e_2 + \varepsilon^e_3\right) + 2\mu\varepsilon^e_A, \quad A = 1, 2, 3. \quad (12)$$

We note that although $\bar{w}$ is not a polyconvex function of F_h, the Hencky model has been shown to be acceptable for all but extreme elastic strains [7].

The gradient plasticity model considered here assumes, pointwise, an elastic domain $\mathcal{E}$ with boundary $\partial \mathcal{E}$, the yield surface, and a generalised normality law. For definiteness $\mathcal{E}$ is assumed here to be defined by the von Mises condition restricted to isotropic hardening and extended to the gradient regime. Plastic flow is considered incompressible, i.e. $\det[F^p] = 1$. The region of admissible generalised stresses is defined as the set (τ, g) that satisfies

$$f(\tau, g) := |\text{dev}[\tau]| - \sqrt{2/3}(\kappa - \underbrace{(\bar{g} - \text{div}[m])}_{g}) \leq 0, \quad (13)$$

where κ is a constant related to the initial tensile yield stress. The generalised plastic strains are thus defined via the flow law and Kuhn–Tucker conditions

$$\boldsymbol{d}^p = \lambda \frac{\partial f}{\partial \boldsymbol{\tau}}, \tag{14}$$

$$\dot{\xi} = \lambda \frac{\partial f}{\partial g} = \sqrt{2/3}\,\lambda, \tag{15}$$

$$\lambda \geq 0, \quad f(\boldsymbol{\tau}, g) \leq 0, \quad \text{and} \quad \lambda f(\boldsymbol{\tau}, g) = 0, \tag{16}$$

where λ is the plastic consistency parameter. The equivalent plastic strain $\gamma(t)$ is related to the plastic consistency parameter by $\dot{\gamma} = \lambda$.

The generalised normality law (14) can be restated in an equivalent dual form [13] by using the dissipation function D, which is given here by

$$D(\boldsymbol{d}^p, \dot{\xi}) = \begin{cases} \sqrt{2/3}\,\kappa\,|\boldsymbol{d}^p| & \text{if } |\boldsymbol{d}^p| \leq \dot{\xi}, \\ +\infty & \text{otherwise.} \end{cases} \tag{17}$$

Then, for arbitrary plastic deformations $\boldsymbol{q}$ and isotropic hardening parameter rates η, the flow rule becomes

$$D(\boldsymbol{q}, \eta) \geq D(\boldsymbol{d}^p, \dot{\xi}) + \boldsymbol{\tau} : (\boldsymbol{q} - \boldsymbol{d}^p) + g(\eta - \dot{\xi}). \tag{18}$$

The weak form of (18) leads, after substitution for g, integration by parts and application of the boundary conditions

$$\xi = 0 \text{ on } \partial \mathcal{S}_1, \quad \text{and} \quad -\boldsymbol{m} \cdot \boldsymbol{n} = k_2 \frac{\partial \xi}{\partial \boldsymbol{n}} = 0 \text{ on } \partial \mathcal{S}_2 \tag{19}$$

where $\partial \mathcal{S}_1$ and $\partial \mathcal{S}_2$ are complimentary subsets of $\partial \mathcal{S}$, to the expression

$$\int_{\mathcal{S}} D(\boldsymbol{q}, \eta)\, dx \geq \int_{\mathcal{S}} D\left(\boldsymbol{d}^p, \dot{\xi}\right) dx + \int_{\mathcal{S}} \boldsymbol{\tau} : \left(\boldsymbol{q} - \boldsymbol{d}^p\right) dx \\ + \int_{\mathcal{S}} \bar{g}\left(\eta - \dot{\xi}\right) dx + \int_{\mathcal{S}} \boldsymbol{m} \cdot \nabla[\eta - \dot{\xi}]\, dx, \tag{20}$$

which forms the basis for the analysis of the small strain problem in [5].

2.2 Time-Discrete Approximation of the Flow Law

We discretise the plastic flow law in time using a backward-Euler scheme. We denote an arbitrary function ψ evaluated at time t_n as ψ^n. Consider a partition of the time interval $[0, T]$ into N subintervals with node points $t_n = nk$, $0 \leq n \leq N$, where $\Delta t = t_{n+1} - t_n = T/N$ is the step-size. Following [10], we define

$$\boldsymbol{A}^p := \left[\boldsymbol{C}^p\right]^{-1} = \boldsymbol{F}_h^{-1} \boldsymbol{b}^e \boldsymbol{F}_h^{-T}. \tag{21}$$

Then for any $\tau \in [t_n, t_{n+1}]$, $\dot{A}^p = \underbrace{\left(-2\dot{\xi} F_h^{-1} \dfrac{\partial f}{\partial \tau} F_h\right)}_{\mathcal{Q}} A^p$, which motivates the approximation $A_\tau^p \approx (\exp[\mathcal{Q}])\, A_n^p$, and this in turn implies that

$$b^e \approx \exp\left[-2\Delta\xi \dfrac{\partial f}{\partial \tau}\right] \underbrace{\left(F_h A_n^p F_h^T\right)}_{b^{e\star}}, \qquad (22)$$

where $b^{e\star}$ is the trial elastic left Cauchy–Green tensor obtained by assuming that plastic flow is "frozen" at t_n. The principal trial stress $\bar{\tau}^\star$ is determined from the constitutive relationship (12) using the elastic principal stretches. Furthermore, we define the trial logarithmic strain components $\varepsilon_A^{e\star} := \log\left[\lambda_A^{e\star}\right]$ where $(\lambda_A^{e\star})^2$ are the eigenvalues of $\bar{\tau}^\star$. One then obtains the time-discrete approximation to the flow law as [7]

$$\varepsilon_{n+1}^e = \varepsilon^{e\star} - \Delta\gamma \dfrac{\partial \bar{f}(\bar{\tau}, g)}{\partial \operatorname{dev}[\bar{\tau}]} = \varepsilon^{e\star} - \Delta\gamma\, \nu\,, \qquad (23a)$$

$$\xi_{n+1} = \xi_n + \Delta\gamma \dfrac{\partial \bar{f}(\bar{\tau}, g)}{\partial g} = \xi_n + \sqrt{2/3}\,\Delta\gamma\,, \qquad (23b)$$

$$\Delta\gamma \geq 0\,, \quad \bar{f}(\bar{\tau}_{n+1}, g_{n+1}) \leq 0\,, \quad \Delta\gamma\, \bar{f}(\bar{\tau}_{n+1}, g_{n+1}) = 0\,, \qquad (23c)$$

where the normal to the yield surface is denoted ν and the von Mises yield function in the principal directions is denoted $\bar{f}$.

3 A Discontinuous Galerkin Formulation

We denote by $\mathcal{P}_k(K)$ the space of polynomials of degree at most $k \geq 0$ on K. Let $\mathcal{T}_h = \{K\}$ be a shape-regular subdivision of the current domain $\mathcal{S}$ where, here, K are quadrilaterals. We consider here the subdivision of the current configuration as this is the placement in which the non-local expression of the flow rule is defined, see (20). Let $\mathcal{E}_h = \{e\}$ denote the unique set of the edges of $\mathcal{T}_h$, and $\mathcal{E}_h^{\mathrm{int}} = \mathcal{E}_h \setminus \partial\mathcal{S}$ the unique set of interior edges. We associate with each edge e of an element the outward unit normal vector $n^{(e)}$. For an edge that lies on the boundary $\partial\mathcal{S}$ of the domain, $n^{(e)}$ is defined to be the outward normal to $\partial\mathcal{S}$. The jumps and averages, denoted $[\![\cdot]\!]$ and $\{\cdot\}$ respectively, of a scalar η and a vector v across an interior edge e_{12} common to elements K_1 and K_2 are defined as

$$\begin{aligned}
[\![\eta]\!] &= \eta_1 n_1 + \eta_2 n_2\,, & \{\eta\} &= 1/2(\eta_1 + \eta_2)\,, \\
[\![v]\!] &= v_1 \otimes n_1 + v_2 \otimes n_2\,, & \{v\} &= 1/2(v_1 + v_2)\,.
\end{aligned} \qquad (24)$$

Following [6] for the small strain problem, we obtain the non-local expression of the discrete consistency condition $\bar{f}_{n+1} = 0$ for $\Delta\gamma$ from the linearised, fully-discrete symmetric interior penalty DG formulation [4] of (20) as

$$\int_{\mathcal{T}} \left(\left(\left(2\mu + \frac{2}{3}k_1 \right) \Delta\gamma \right) \varrho + \frac{2}{3}k_2 \nabla[\Delta\gamma] \cdot \nabla\varrho \right) dx + \int_{\mathcal{E}_h} \frac{k_2 \beta}{h_e} [\![\Delta\gamma]\!] \cdot [\![\varrho]\!] \, ds$$
$$- \int_{\mathcal{E}_h} \frac{2}{3} k_2 \left([\![\varrho]\!] \cdot \{\nabla[\Delta\gamma]\} + \{\nabla\varrho\} \cdot [\![\Delta\gamma]\!] \right) ds = \int_{\mathcal{T}} \bar{f}(\bar{\tau}^\star, g_n) \varrho \, dx \,, \qquad (25)$$

where the parameter β is a positive penalty term and h_e the distance between the centroids of two elements that share a common edge.

Homogeneous Neumann type boundary conditions are assumed here on the external boundary of the current domain, i.e. $-\boldsymbol{m} \cdot \boldsymbol{n} = 0$ on $\partial \mathcal{S}_2$. While the choice of appropriate boundary conditions for the internal hardening parameter field is motivated by physical considerations, the emphasis in this work is to demonstrate the salient features of the gradient plasticity formulation. We remark however that any choice of boundary condition could be implemented. Also, motivated by the structure of the variational problem, we solve the non-local statement of the consistency condition over the complete current domain rather than over some plastic subset thereof.

4 Solution Procedure and Numerical Examples

A predictor-corrector type solution algorithm is used to solve the gradient plasticity problem for the increment in displacement and plastic deformation during a time-step of duration $\Delta t = t_{n+1} - t_n$. The complete system state is assumed known at t_n. In the predictor step, the increment in the nodal displacements is obtained by making an assumption as to the evolution of plastic flow and solving the governing equations (4a)–(4b). The purpose of the corrector step is to then determine the stress state based on the strain field arising from the predictor step taking into account the possible evolution of plastic deformation. For classical plasticity the corrector step involves checking the trial yield condition $f^\star := f(\tau^\star, g_n)$ at each quadrature point. If the quadrature point is active, i.e. $f^\star > 0$, we determine the increment in plastic flow by solving the algebraic consistency condition locally. The non-local expression for the increment in plastic deformation arising in the gradient formulation can not be solved at the level of the quadrature point. Instead, a search is performed to determine the active quadrature points and (25) solved to determine the increment in the internal hardening parameter.

The performance of the predictor–corrector solution scheme is dependent on the form of approximation as to the evolution of plastic flow in the predictor step. Following the approach pioneered in [7] for the classical problem at finite strains and drawing on the approach adopted in [6] for the gradient problem under the assumption of infinitesimal strain, we utilise a consistent tangent formulation rederived for the gradient plasticity problem under consideration. For details see [9].

The gradient plasticity formulation is applied to two example problems. The first, a rectangular plate with a small initial imperfection subjected to compressive loading wherein the material undergoes softening [14], is chosen to assess the ability of the gradient plasticity formulation to overcome the pathological mesh-dependence associated with classical formulations involving softening, i.e. $k_1 < 0$. The second example problem demonstrates features of the gradient formulation for a hardening problem involving more significant deformation. Plane strain conditions are assumed applicable.

Example 1: Softening response of a rectangular plate.
The formation of the shear band is induced via the introduction of a 10 mm square region in the lower left hand corner of the plate, see Figure 1(a), wherein the yield strength is reduced by 10% relative to the rest of the domain. The well-documented pathological localisation of the shear band to the scale of the discretisation (i.e. the mesh size) for the classical problem is evident in Figure 1(d). The relationship between the resulting force and the applied displacement on the upper edge of the plate, shown in Figure 1(e), illustrates, once again, the mesh dependent response. The post-peak response of the force displacement curve is governed by the discretisation: the finer the mesh, the greater the rate at which the material looses residual strength.

The ability of the gradient plasticity formulation to produce solutions independent of the mesh for softening problems is demonstrated in Figs 1(b)–1(c). As the mesh resolution is increased so the width of the shear band converges to a constant value prescribed by the internal length scale. It is evident that the pathological localisation demonstrated by the classical formulation has been overcome.

Example 2: Indentation test.
The second example problem entails a rectangular domain subjected to loading via a frictionless rigid indenter. The domain is composed of a hardening elastoplastic material. The indenter moves downwards at a constant rate into the specimen resulting in significant deformation. The final indentation depth is 8% of the initial specimen height.

The results predicted using the classical and gradient plasticity formulations at the final stage of the deformation process are compared in Figure 2(a). The results obtained using the gradient plasticity formulation with an internal length scale of 2.83 mm are mirrored around the symmetry axis for the purpose of direct comparison. The positions of the quadrature points are shown and the subset of those quadrature points that are active indicated. The relationship between the resulting force on the indenter and the imposed displacement is shown in Figure 2(b). The results obtained using the gradient formulation are clearly influenced by the relative scale of the problem, that is the ratio of the internal length scale to a characteristic dimension of the domain. The material offers increased resistance to deformation with increasing internal length scale.

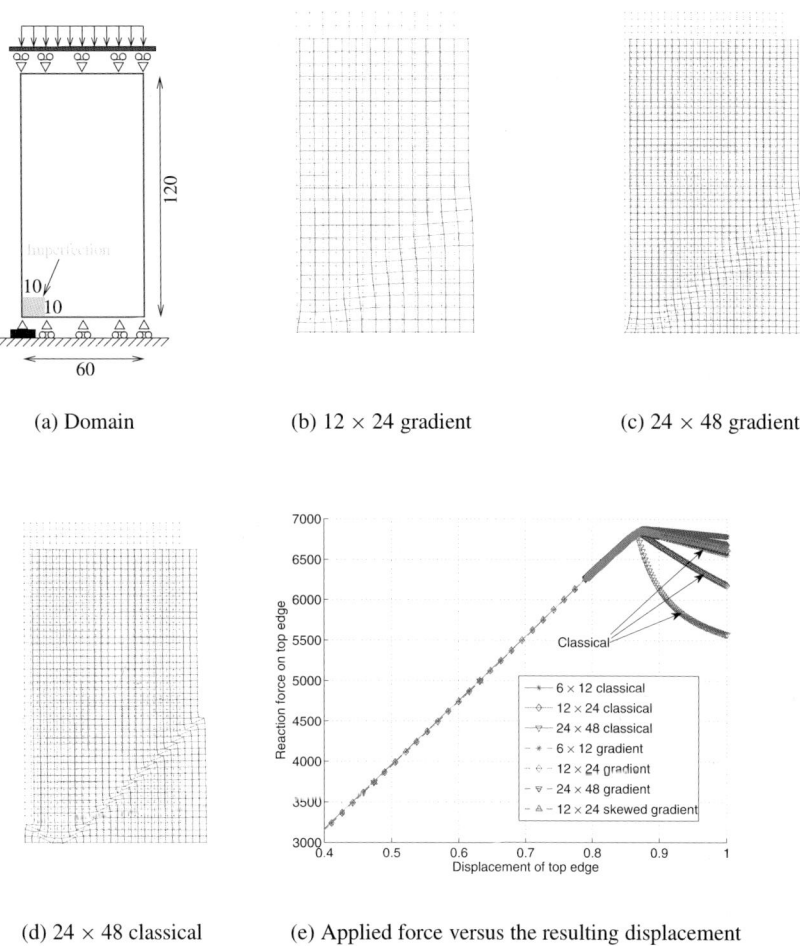

Fig. 1 Schematic of the domain, the deformed domain obtained using various discretisations, and the resulting force versus displacement relationship.

5 Conclusion

There remain a number of avenues which merit further exploration and study. First among these would be the extension of the present work to three space dimensions, the framework for which exists in much of the work reported here.

The major advantage of adopting a DG approach is that it offers the flexibility to accommodate more complex gradient plasticity models. The gradient model considered here, for example, could be readily extended to include a biharmonic term in

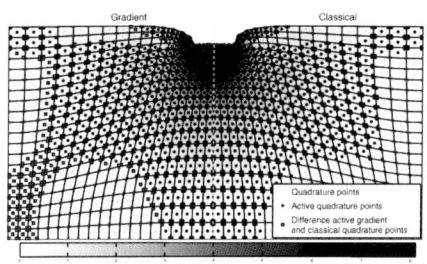 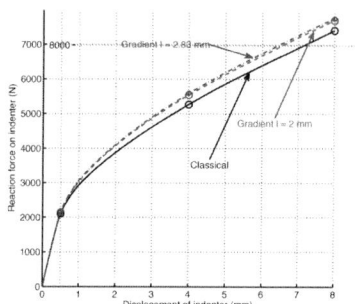

(a) Deformed domain using gradient (left) and classical (right) plasticity formulations

(b) Resulting force on the indenter and the imposed displacement

Fig. 2 Deformed domain and the force versus displacement relationship for the indentation test.

the yield condition by using a continuous/discontinuous Galerkin formulation rather than a cumbersome classical approach based on Hermite interpolation. Again, the groundwork for such a model has been presented here and in previous works.

Further questions worth addressing include the extension of the work reported here to other models of strain gradient plasticity. For example, the class of problems that involve the Burgers vector, or curl of plastic strain, represents a group that is worth studying computationally (see, e.g., [15]).

References

1. E. C. Aifantis. On the microstructural origin of certain inelastic models. *J. Eng. Mater. Tech.*, 106:326–330, 1984.
2. P. Gudmundson. A unified treatment of strain gradient plasticity. *J. Mech. Phys. Solids*, 52:1379–1406, 2004.
3. R. A. B. Engelen, N. A. Fleck, R. H. J. Peerlings, and M. G. D. Geers. An evaluation of higher-order plasticity theories for predicting size effects and localisation. *Int. J. Solids Struct.*, 43:1857–1877, 2006.
4. D. N. Arnold. An interior penalty finite element method with discontinuous elements. *SIAM J. Numer. Anal.*, 19:742–760, 1982.
5. J. K. Djoko, F. Ebobisse, A. T. McBride, and B. D. Reddy. A discontinuous Galerkin formulation for classical and gradient plasticity – Part 1: Formulation and analysis. *Comput. Methods Appl. Mech. Engrg.*, 196:3881–3897, 2007.
6. J. K. Djoko, F. Ebobisse, A. T. McBride, and B. D. Reddy. A discontinuous Galerkin formulation for classical and gradient plasticity. Part 2: Algorithms and numerical analysis. *Comput. Methods Appl. Mech. Engrg.*, 197:1–21, 2007.
7. J. C. Simo. Algorithms for static and dynamic multiplicative plasticity that preserve the classical return mapping schemes of the infinitesimal theory. *Comput. Methods Appl. Mech. Engrg.*, 99:61–112, 1992.

8. M. G. D. Geers. Finite strain logarithmic hyperelasto-plasticity with softening: a strongly non-local implicit gradient framework. *Comput. Methods Appl. Mech. Engrg.*, 193:3377–3401, 2004.
9. A. T. McBride and B. D. Reddy. A discontinuous Galerkin formulation for gradient plasticity at finite strains. In review.
10. J. C. Simo and F. Armero. Geometrically non-linear enhanced strain mixed methods and the method of incompatible modes. *Int. J. Numer. Methods Engrg.*, 33:1413–1449, 1992.
11. S. Glaser and F. Armero. On the formulation of enhanced strain finite elements in finite deformations. *Eng. Comput.*, 14(7):759–791, 1997.
12. B. D. Coleman and M. Gurtin. Thermodynamics with internal variables. *J. Chem. Phys.*, 47:597–613, 1967.
13. W. Han and B. D. Reddy. *Plasticity: Mathematical Theory and Numerical Analysis*, Vol. 9. Springer, 1999.
14. J. Pamin. *Gradient-Dependent Plasticity in Numerical Simulation of Localization Phenomena*. PhD thesis, Delft University of Technology, 1994.
15. M. E. Gurtin and L. Anand. A theory of strain gradient plasticity for isotropic, plastically irrotational materials. Part I: Small deformations. *Int. J. Plasticity*, 53:1624–1649, 2005.

Computational Dynamics

Energy-Momentum Algorithms for the Nonlinear Dynamics of Elastoplastic Solids

Francisco Armero

Department of Civil and Environmental Engineering, University of California, Berkeley, CA 94720-1710, USA
E-mail: armero@ce.berkeley.edu

Abstract. This paper presents the formulation of energy-dissipative momentum-conserving time-stepping algorithms for finite strain dynamic plasticity. These methods require special return mapping algorithms for the integration of the plastic evolution equations, as well as the proper assumed strain treatment to arrive at fully conserving, locking-free assumed strain B-bar finite element methods.

Key words: dynamic plasticity, energy-dissipative momentum-conserving time-stepping algorithms, assumed strain finite element methods.

1 Introduction

Classical time-stepping algorithms like Newmark, HHT and similar are known to develop numerical instabilities in the geometrically nonlinear range. The instabilities are characterized by an unlimited growth of energy [12], and occur even when the same scheme is shown to be unconditionally stable in the linear range. This situation has motivated the development of time-stepping algorithms that exactly conserve the energy for nonlinear elasticity, as presented in [5, 6, 12] among others. The conservation of linear and angular momenta, and the associated relative equilibria [7], is also of the main interest in this nonlinear range, as it is the incorporation of a controllable high-frequency energy dissipation to handle the high numerical stiffness of the mechanical systems of interest; see [3] for recent developments along these lines.

The same numerical instabilities have been observed in the elastoplastic range for existing classical schemes. Despite the dissipative character of the physical system, the energy evolution in the discrete problem is not monotonic leading also to its unstable growth. Algorithms that avoid this situation have been presented in [9, 10]. We present here a summary of the results presented in [1, 4] identifying a new return mapping algorithm for the integration of the plastic evolution equations that leads to the exact energy dissipation and to global energy-dissipative momentum-conserving (EDMC) schemes. Elastoplastic problems require the consideration of locking-free

B.D. Reddy (ed.), IUTAM Symposium on Theoretical, Modelling and Computational Aspects of Inelastic Media, 251–262.
© *Springer Science+Business Media B.V. 2008*

assumed strain B-bar finite elements. As shown in [2], standard treatments like the one presented in [13] for static problems destroy the conservation properties gained by the conserving temporal integration. This motivates the development of an alternative conserving assumed strain treatment as also presented in this paper.

2 The Continuum Problem

We summarize briefly in this section the equations governing the motion of an elastoplastic solid in the finite deformation range, and the characterization of the momentum conservation laws and energy dissipation along these motions.

2.1 The Equation of Motion and the Momentum Conservation Laws

The aim is to solve for the motion $\varphi(X, t) \in \mathbb{R}^3$ of a solid with reference placement $X \in \mathcal{B} \subset \mathbb{R}^3$ in time t, when subjected to an external body force $\rho_o b$ (with reference density ρ_o), external tractions $\bar{T}$ on the part of the boundary $\partial_T \mathcal{B}$ and an imposed deformation $\varphi = \bar{\varphi}$ on the complementary part $\partial_\varphi \mathcal{B}$. The equation describing this motion reads in weak form

$$\int_\mathcal{B} \rho_o \ddot{\varphi} \cdot \eta \, dV + \int_\mathcal{B} \underbrace{S : \left(F^T \mathrm{GRAD}\,[\eta]\right)^s}_{\frac{1}{2}\delta C(\varphi, \eta)} dV = \int_\mathcal{B} \rho_o b \cdot \eta \, dV + \int_{\partial_T \mathcal{B}} \bar{T} \cdot \eta \, dA \,, \tag{1}$$

for all admissible variations $\eta(X) \in \mathbb{R}^3$, that is, vanishing on $\partial_\varphi \mathcal{B}$. Here we have introduced the material acceleration $\ddot{\varphi} := \left(\partial^2 \varphi / \partial t^2\right)\big|_X$, the deformation gradient $F = \mathrm{GRAD}\,[\varphi] := \left(\partial \varphi / \partial X\right)\big|_t$, with the right Cauchy–Green tensor $C = F^T F$, and the symmetric second Piola–Kirchhoff stress tensor S given by the constitutive relations discussed below. We point out the notation $\delta C(\varphi, \eta)$ for the indicated combination of those two functions as arguments.

The structure of equation (1) leads directly to the conservation of the linear l and angular j momenta, defined in terms of the material velocity $V = \dot{\varphi} := \left(\partial \varphi / \partial t\right)\big|_X$ as

$$l := \int_\mathcal{B} \rho_o V \, dV = \text{constant}\,, \quad \text{and} \quad j := \int_\mathcal{B} \rho_o \varphi \times V \, dV = \text{constant}\,, \tag{2}$$

along the motions with, say, no external loading ($\rho b = 0$, $\bar{T} = 0$ and $\partial_\varphi \mathcal{B} = \emptyset$). These properties follow from the vanishing of the stress term in (1) for the variations corresponding to a translation and infinitesimal rotation, that is,

$$\delta C(\boldsymbol{\varphi}, \boldsymbol{\eta}) = 0 \quad \text{for} \quad \boldsymbol{\eta} = \boldsymbol{c} \quad \text{and} \quad \boldsymbol{\eta} = \boldsymbol{c} \times \boldsymbol{\varphi}, \tag{3}$$

for all (constant) vectors $\boldsymbol{c} \in \mathbb{R}^3$.

The symmetry of the governing equations (1) under the action of the group of rigid body deformations $\mathbb{R}^3 \times SO(3)$ (translations × rotations) lies beneath these considerations. This situation leads to the existence of the so-called relative equilibria corresponding to motions evolving entirely in the symmetry group as fixed points of the reduced dynamics (modulo the symmetry group); see e.g. [8] for details in general Hamiltonian systems. For the mechanical problem of interest here, these group motions are expressed as

$$\boldsymbol{\varphi}_{et}(\boldsymbol{X}, t) = \underbrace{\text{EXP}\left[t\, \text{SPIN}\left[\boldsymbol{\Omega}_e\right]\right]}_{\boldsymbol{\Lambda}(t)} \boldsymbol{\varphi}_e(\boldsymbol{X}) + \underbrace{\left(\int_0^t \text{EXP}\left[\eta\, \text{SPIN}\left[\boldsymbol{\Omega}_e\right]\right] d\eta\right) \boldsymbol{V}_e}_{\boldsymbol{u}(t)} \tag{4}$$

for the exponential map $\text{EXP} : so(3) \to SO(3)$, spin (axial vector to skew tensor) $\text{SPIN}[:]\mathbb{R}^3 \to so(3)$, and constant angular and translational velocities $\boldsymbol{\Omega}_e$ and $\boldsymbol{V}_e$. The motion (4) corresponds to a rigid body motion superposed to a fixed deformation $\boldsymbol{\varphi}_e(\boldsymbol{X})$, which is defined by the equilibrium equation

$$\int_{\mathcal{B}} \rho_o \boldsymbol{\Omega}_e \times [\boldsymbol{\Omega}_e \times \boldsymbol{\varphi}_e + \boldsymbol{V}_e] \cdot \boldsymbol{\eta}\, dV + \int_{\mathcal{B}} \boldsymbol{S}_e : \left(\boldsymbol{F}_e^T \text{GRAD}[\boldsymbol{\eta}]\right)^s dV = 0, \tag{5}$$

for all variations $\boldsymbol{\eta}(\boldsymbol{X})$, with $\boldsymbol{F}_e = \text{GRAD}[\boldsymbol{\varphi}_e]$ and the corresponding equilibrium stress $\boldsymbol{S}_e$. Equation (5) follows by inserting the group motion (4) in (1), and the elimination of the group rotation after using the relation

$$\delta C(\boldsymbol{\varphi}_{et}, \boldsymbol{\eta}) = \delta C(\boldsymbol{\varphi}_e, \boldsymbol{\Lambda}^T \boldsymbol{\eta}) \quad \text{along the group motion (4)}, \tag{6}$$

and for all variations $\boldsymbol{\eta}$; see [2]. In the elastoplastic case considered below, the plastic evolution equations must also exhibit a fixed point along the motion (4), that is, the stresses $\boldsymbol{S}_e$ involve a fixed set of plastic internal variables.

2.2 Multiplicative Finite Strain Plasticity and Energy Dissipation

We are interested in constitutive models of finite strain plasticity characterized by the multiplicative decomposition $\boldsymbol{F} = \boldsymbol{F}^e \boldsymbol{F}^p$. The elastic part $\boldsymbol{F}^e$ (not to be confused with the equilibrium $\boldsymbol{F}_e$ above) defines the stresses $\boldsymbol{S}$ as

$$\widetilde{\boldsymbol{S}} := \boldsymbol{F}^p \boldsymbol{S} \boldsymbol{F}^{p^T} = 2\frac{\partial W^e}{\partial \boldsymbol{C}^e}, \tag{7}$$

in the intermediate configuration defined by $\boldsymbol{F}^p$ through an elastic potential $W^e(\boldsymbol{C}^e)$ in terms of the elastic right Cauchy–Green tensor $\boldsymbol{C}^e := \boldsymbol{F}^{e^T} \boldsymbol{F}^e$.

Considering the stored energy function $W(C^e, \alpha) = W^e(C^e) + \mathcal{H}(\alpha)$ for a hardening potential depending on the internal variable α (isotropic hardening assumed without loss of generality), equation (1) leads to the energy evolution

$$\frac{d}{dt}\underbrace{\left(\int_{\mathcal{B}} \frac{1}{2}\rho_0 \|V\|^2 \, dV + \int_{\mathcal{B}} W \, dV\right)}_{H \,=\, \text{total energy}} = -\underbrace{\int_{\mathcal{B}} \left(S : \frac{1}{2}\dot{C} - \dot{W}\right) dV}_{\mathcal{D} \,=\, \text{dissipation} \,\geq\, 0} + \mathcal{P}_{\text{ext}}, \quad (8)$$

for the external power $\mathcal{P}_{\text{ext}}$. Critical to obtain the relation (8) is the result

$$\delta C(\varphi, \dot{\varphi}) = \dot{C}, \quad (9)$$

for the strain variation formula in terms of φ and $\dot{\varphi}$, and the strain rate $\dot{C}$.

For the case of multiplicative plasticity of interest here, the energy dissipation reads, after some algebraic manipulations,

$$S : \frac{1}{2}\dot{C} - \dot{W} = \widetilde{S} : D^p + q\,\dot{\alpha}, \quad \text{for } D^p := \left(C^e L^p\right)^s = \frac{1}{2}\left[F^{p^{-T}} \dot{C} F^{p^{-1}} - \dot{C}^e\right] \quad (10)$$

the plastic strain rate with $L^p := \dot{F}^p F^{p^{-1}}$, and $q := -d\mathcal{H}/d\alpha$. It is precisely this form of the dissipation that motivates the plastic evolution equations

$$\left.\begin{array}{c} D^p = \gamma\, N_\phi, \quad W^p = \gamma\, N_{W^p}, \quad \dot{\alpha} = \gamma\, n_\phi, \\ \phi \leq 0, \quad \gamma \geq 0, \quad \gamma \phi = 0 \quad \text{and} \quad \gamma \dot{\phi} = 0, \end{array}\right\} \quad (11)$$

for the yield surface $\phi(\widetilde{S}, q; G^e)$, and the general flow vectors N_ϕ, N_{W^p} and n_ϕ, depending also on $\{\widetilde{S}, q; G^e\}$. Note the presence of the elastic metric $G^e = C^{e^{-1}}$ in the intermediate configuration. These flow vectors are general but must lead to the fundamental physical requirement $\mathcal{D} \geq 0$ (second law) in (8). In the case of associated plasticity (i.e. N_ϕ and n_ϕ orthogonal to ϕ), this condition can be easily accomplished by the special form of the yield surface ϕ. The evolution of the plastic spin $W^p := \text{skew}\left[C^e L^p\right]$ is needed to fully define the rate $\dot{F}^p$. See e.g. [1, 11] for complete details on all these considerations.

3 The Discrete Problem

We study next the discrete equations resulting from a finite element interpolation in space and a general one-step scheme in time.

3.1 The Finite Element Equations and Discrete Conservation Laws

We consider the classical finite element interpolations

$$\varphi^h(X, t_i) = X + \sum_{A=1}^{n_{\text{node}}} N^A d_i^A \quad \text{and} \quad V^h(X, t_i) = \sum_{A=1}^{n_{\text{node}}} N^A v_i^A, \quad (12)$$

of the deformation and velocity, respectively, in terms of the nodal displacements and velocities (collected in the global arrays d and v, respectively) for a set of shape functions $N^A(X)$ and at discrete times $\{t_0, t_1, \ldots\}$. Considering a one-step approximation in time, the resulting discrete equations for a generic increment $[t_n, t_{n+1}]$ (with $\Delta t = t_{n+1} - t_n$, not necessarily constant) read

$$\left. \begin{aligned} \frac{1}{\Delta t} M \left(v_{n+1} - v_n \right) + \int_{\mathcal{B}^h} \overline{B}_*^T S_* \, dV &= f_{\text{ext}} \\ d_{n+1} - d_n &= \Delta t \, v_* \end{aligned} \right\} \quad (13)$$

for the mass matrix $M = \mathbf{A} \left(M^{AB} \mathbf{1} \right)$ (consistent or lumped with nodal contributions M^{AB}), the external nodal force vector f_{ext}, and a general linearized strain operator $\overline{B}_*$. The quantities marked by $(\cdot)_*$ need to be defined in terms of the solution values at t_n and t_{n+1}. In particular, the stress S_* is to be obtained by integrating the constitutive equations identified above.

The linear and angular momenta (2) are given in this discrete case by

$$l_{n+i}^h = \sum_{A,B=1}^{n_{\text{node}}} M^{AB} v_{n+i}^B \quad \text{and} \quad j_{n+i}^h = \sum_{A,B=1}^{n_{\text{node}}} M^{AB} x_{n+i}^A \times v_{n+i}^B, \quad (14)$$

for $i = 0, 1$ in the time increment $[t_n, t_{n+1}]$. The analog of conditions (3) on the strain variations for their conservation in the discrete case are respectively

$$\overline{B}_* \{c\} = 0 \quad \text{and} \quad \overline{B}_* \left\{ c \times x_{n+\frac{1}{2}} \right\} = 0, \quad (15)$$

for all vectors $c \in \mathbb{R}^3$, with $\{\cdot\}$ denoting the vector with the nodal values for the enclosed quantity. Angular momentum conservation requires in addition that the velocity approximation v_* is parallel to the mid-point value $v_{n+\frac{1}{2}}$, which we write as

$$v_* = v_{n+\frac{1}{2}} \left(1 + 2 \frac{\mathcal{D}_{\text{num}}^v}{\|v_{n+1}\|_M^2 - \|v_n\|_M^2} \right), \quad (16)$$

for a dissipation function $\mathcal{D}_{\text{num}}^v$ that allows the incorporation of a controllable high-frequency numerical dissipation if desired (e.g. $\mathcal{D}_{\text{num}}^v = 0$); see [3].

The discrete relative equilibria of (13), analog to (4), were found in [3] as

$$x^A_{e_n} = \Lambda_n \varphi^{hA}_e + u_n \quad \text{for} \quad \begin{cases} u_{n+1} = u_n + \Delta t\, \Lambda_{n+\frac{1}{2}}\, v_e \\ \Lambda_{n+1} = \Lambda_n\, \text{CAY}\left[\Delta t\, \text{SPIN}\left[\Omega_e\right]\right] \end{cases} \quad (17)$$

in terms of the Cayley transform $\text{CAY} : so(3) \to SO(3)$ defining a sequence of rotations, with $\Lambda_0 = 1$, $u_0 = 0$ and $\Lambda_{n+\frac{1}{2}} = (\Lambda_n + \Lambda_{n+1})/2$ (not a rotation in general). The equilibrium deformation $\varphi^{hA}_e = \varphi^h_e(X^A)\big|^{n_{\text{node}}}_{A=1}$ is given by

$$M\left\{\Omega_e \times \left(\Omega_e \times \varphi^h_e + v_e\right)\right\} + \int_{\mathcal{B}^h} \overline{B}^T_e S(\varphi^h_e) dV = 0, \quad (18)$$

the exact discrete counterpart of equation (5), as long as we have

$$\overline{B}_{*e} = \overline{B}_e \Lambda^T_{n+\frac{1}{2}} \quad \text{along the dicrete group motion (17)} \quad (19)$$

for $\overline{B}_e$ evaluated at the equilibrium configuration; see [2] for complete details. Condition (19) is the discrete analog of relation (6) in the continuum problem.

Finally, combining equations (13) leads to the discrete energy evolution

$$\frac{1}{\Delta t}(H_{n+1} - H_n) = -\underbrace{\int_{\mathcal{B}}\left(S_* : \frac{1}{2}\Delta\overline{C} - \Delta W\right) dV}_{\mathcal{D}^h\, =\, \text{discrete dissipation}} + \mathcal{D}^v_{\text{num}} + \mathcal{P}^h_{\text{ext}}, \quad (20)$$

for the assumed strain $\overline{C}$ approximating the continuum C, as long as we have

$$\overline{B}_*(d_{n+1} - d_n) = \frac{1}{2}\Delta\overline{C} = \frac{1}{2}\left(\overline{C}_{n+1} - \overline{C}_n\right), \quad (21)$$

the analog of condition (9) for the continuum problem.

3.2 Conserving Assumed Strain B-Bar Elements

Equations (15), (19) and (21) define the conditions that the linearized strain operator $\overline{B}_*$ needs to satisfy in order to arrive at a momentum conserving, relative equilibria preserving approximation that exhibits the correct energy evolution (20). Note that the latter is to be satisfied with the assumed strain $\overline{C}$ defining the stress. A simple choice that satisfies these requirements is the basic displacement model, with the assumed strain given directly by the interpolated deformation φ^h (that is, $C^h = F^{h^T} F^h$ for $F^h = \text{GRAD}[\varphi^h]$), and the linearized strain operator defined by its standard variations (that is, $B_{n+\frac{1}{2}} \delta d = (F^T_{n+\frac{1}{2}} \text{GRAD}[\delta\varphi^h])^s$) evaluated at the midpoint configuration.

It is well-known that this choice leads, however, to serious difficulties (volumetric locking) when used with J_2 plasticity models, as it is the interest in this paper.

Assumed strain formulations based on a mixed treatment of the Jacobian have been identified as a solution to this problem. Following [13], this can be accomplished with the use of the assumed deformation gradient

$$\overline{F} = \left(\frac{\Theta}{J}\right)^{\frac{1}{3}} \text{GRAD}\left[\varphi^h\right] \quad \text{for} \quad J := \det\left[\text{GRAD}\left[\varphi^h\right]\right], \quad (22)$$

and the mixed Jacobian Θ defined at the element level $\mathcal{B}_e^h$ by

$$\Theta(X) := \boldsymbol{\Gamma}^T(X) H^{-1} \int_{\mathcal{B}_e^h} \boldsymbol{\Gamma}(Y) J(Y) dV \quad \text{for} \quad H = \int_{\mathcal{B}_e^h} \boldsymbol{\Gamma}(Y) \boldsymbol{\Gamma}^T(Y) dV, \quad (23)$$

and element interpolation functions $\boldsymbol{\Gamma}$. Typical choices include bilinear quads or trilinear bricks with piece-wise constant volume ($Q1/A0$ with $\boldsymbol{\Gamma} = 1$), and quadratic elements with linear volume ($Q2/A1$ quads and $P2/A1$ triangles with $\boldsymbol{\Gamma} = [1\ \xi\ \eta]^T$ for the isoparametric coordinates (ξ, η) in the plane).

However, we showed in [2] that the standard strain variations based on the assumed deformation gradient (22) do not lead to an energy-conserving approximation nor a relative equilibria-preserving formulation (i.e. relations (21) and (19) are not satisfied) even though the momentum conservation conditions (15) can be satisfied by an evaluation of these variations at the mid-point configuration. This situation can be corrected by the consideration of the modified linearized strain operator

$$\overline{B}_*^A = \left[\Theta^{\frac{2}{3}}\right]_{n+\frac{1}{2}} \left[J^{-\frac{2}{3}}\right]_{n+\frac{1}{2}} B_{n+\frac{1}{2}}^A + \frac{1}{2}\left[\Theta^{\frac{2}{3}}\right]_{n+\frac{1}{2}} D_J^{(-\frac{2}{3})} [C]_{n+\frac{1}{2}} \otimes \widehat{g}_{n+\frac{1}{2}}^A$$

$$+ \frac{1}{2} D_\Theta^{(\frac{2}{3})} \left[J^{-\frac{2}{3}} C\right]_{n+\frac{1}{2}} \otimes \overline{g}_{n+\frac{1}{2}}^A, \quad (24)$$

for each node $A = 1, n_{\text{node}}$, with the notation

$$D_{(\cdot)}^{(a)} := \begin{cases} [(\cdot)]_{n+\frac{1}{2}} \frac{((\cdot)_{n+1})^a - ((\cdot)_n)^a}{(\cdot)_{n+1} - (\cdot)_n} & \text{for } (\cdot)_{n+1} \neq (\cdot)_n, \\ a\left([(\cdot)]_{n+\frac{1}{2}}\right)^a & \text{for } (\cdot)_{n+1} = (\cdot)_n, \end{cases} \quad (25)$$

and $[(\cdot)]_{n+\frac{1}{2}} := \frac{1}{2}[(\cdot)_n + (\cdot)_{n+1}] \neq (\cdot)_{n+\frac{1}{2}}$ at the mid-point configuration. Crucial to the conserving properties of (24) are the modified spatial gradients

$$\widehat{g}_{n+\frac{1}{2}}^A := F_{n+\frac{1}{2}}^h [C]_{n+\frac{1}{2}}^{-1} G^A$$

$$+ 2 \frac{\frac{J_{n+1}\breve{U} - J_n}{[J]_{n+\frac{1}{2}}} - \frac{1}{2}[C]_{n+\frac{1}{2}}^{-1} : \Delta C}{[C]_{n+\frac{1}{2}}^{-1} \Delta C : \Delta C [C]_{n+\frac{1}{2}}^{-1}} F_{n+\frac{1}{2}}^h [C]_{n+\frac{1}{2}}^{-1} \Delta C [C]_{n+\frac{1}{2}}^{-1} G^A, \quad (26)$$

for the material gradients $G^A := \text{GRAD} N^A$, and their assumed values

$$\left(\bar{\widehat{g}}^A_{n+\frac{1}{2}}\right)_k = \frac{1}{[\Theta]_{n+\frac{1}{2}}} \boldsymbol{\Gamma}^T \boldsymbol{H}^{-1} \int_{\mathcal{B}^h_e} \boldsymbol{\Gamma} [J]_{n+\frac{1}{2}} \left(\widehat{g}^A_{n+\frac{1}{2}}\right)_k dV , \quad (27)$$

for $k = 1, 3$. A key aspect of the modified spatial gradients (26) is that $\sum_A \widehat{g}^A_{n+\frac{1}{2}} \cdot (d^A_{n+1} - d^A_n) = (J_{n+1} - J_n) / [J]_{n+\frac{1}{2}}$, in contrast with the spatial gradients $g^A_{n+\frac{1}{2}} = F^{h^{-T}}_{n+\frac{1}{2}} G^A \neq \widehat{g}^A_{n+\frac{1}{2}}$. The final formulas result in a second order approximation in time, as desired. See [2] for a complete discussion.

4 The EDMC Return-Mapping Algorithm

It remains to define the stress approximation S_*. The goal is to satisfy the analogs of relations (10) for the continuum problem, identifying in the process the discrete counterpart of plastic strain rate D^p and, hence, the discrete form of the plastic evolution equations (11), that is, the return mapping algorithm.

We start with the approximation of the hyperelastic relation (7) given by

$$\widetilde{S}_* = \widetilde{S}_\sharp + 2 \frac{W^e_{n+1} - W^e_n - \widetilde{S}_\sharp : \Delta C^e + \mathcal{D}^S_{num}}{[C^e]^{-1}_{n+\frac{1}{2}} \Delta C^e : \Delta C^e [C^e]^{-1}_{n+\frac{1}{2}}} [C^e]^{-1}_{n+\frac{1}{2}} \Delta C^e [C^e]^{-1}_{n+\frac{1}{2}} , \quad (28)$$

with $\Delta C^e := C^e_{n+1} - C^e_n$ and

$$\widetilde{S}_\sharp := 2 \frac{\partial W^e}{\partial C^e} \left([C^e]_{n+\frac{1}{2}}\right) .$$

Equation (28) is an extension of the stress formula originally proposed in [6] with two important modifications. First, it includes a numerical dissipation $\mathcal{D}^S_{num}$, a non-negative function of the solution values at t_n and t_{n+1} and controllable (i.e. $\mathcal{D}^S_{num} = 0$ as a particular case; see [3] for details. Second, it makes use of the convected metric $[C^e]^{-1}_{n+\frac{1}{2}}$ in the intermediate configuration to arrive to a fully invariant (two-contravariant) stress approximation.

The stress (28) satisfies $\widetilde{S}_* : \frac{1}{2} \Delta C^e = \Delta W^e + \mathcal{D}^S_{num}$ which with (20) results

$$S_* : \frac{1}{2} \Delta \overline{C} - \Delta W = \widetilde{S}_* : \frac{1}{2} \left(F^{p^{-T}}_* \Delta \overline{C} F^{p^{-1}}_* - \Delta C^e\right) + q_* \Delta \alpha + \mathcal{D}^S_{num} , \quad (29)$$

for $q_* := -\left(\mathcal{H}(\alpha_{n+1}) - \mathcal{H}(\alpha_n)\right)/\Delta\alpha$ and with $\widetilde{S}_* = F^{p^T}_* S_* F^p_*$. Equation (29) identifies the discrete approximation of D^p, which motivates the consideration of the discrete plastic evolution equations (return mapping algorithm)

$$\frac{1}{2}\left(\boldsymbol{F}_*^{p^{-T}}\Delta\overline{\boldsymbol{C}}\boldsymbol{F}_*^{p^{-1}} - \Delta\boldsymbol{C}^e\right) = \Delta\gamma \; \boldsymbol{N}_\phi(\widetilde{\boldsymbol{S}}_*, q_*; \boldsymbol{C}_*^e) \;, \tag{30}$$

$$\text{skew}\left[\boldsymbol{C}_*^e\left(\boldsymbol{F}_{n+1}^p - \boldsymbol{F}_n^p\right)\boldsymbol{F}_*^{p^{-1}}\right] = \Delta\gamma \; \boldsymbol{M}_{WP}(\widetilde{\boldsymbol{S}}_*, q_*; \boldsymbol{C}_*^e) \;, \tag{31}$$

$$\phi_* := \phi(\widetilde{\boldsymbol{S}}_*, q_*; \boldsymbol{C}_*^e) \leq 0, \quad \Delta\gamma \geq 0, \quad \Delta\gamma\phi_* = 0, \tag{32}$$

to be solved for $\{\boldsymbol{F}_{n+1}^p, \alpha_{n+1}\}$ from their converged values at t_n, obtaining $\boldsymbol{C}_{n+1}^e$ from $\boldsymbol{F}_{n+1}^e = \overline{\boldsymbol{F}}_{n+1}\boldsymbol{F}_{n+1}^{p^{-1}}$ and the stresses $\widetilde{\boldsymbol{S}}_*$ with (28).

We can readily observe that inserting the relations (30)-(32) in (29) shows that the exact plastic dissipation is obtained in terms of the plastic flow vectors, recovering in particular exact energy conservation for an elastic step for $\mathcal{D}_{num} = 0$, with the added option of a controllable high-frequency energy dissipation. We see that this property is independent of the choice $\boldsymbol{F}_*^p$ defining the intermediate configuration and the associated elastic metric $\boldsymbol{G}_*^e = \boldsymbol{C}_*^{e^{-1}}$. As discussed in [1], a simple choice for these quantities is given by the mid-point values $\boldsymbol{F}_*^p = (\boldsymbol{F}_n^p + \boldsymbol{F}_{n+1}^p)/2$ and $\boldsymbol{C}_*^e = (\boldsymbol{C}_n^e + \boldsymbol{C}_{n+1}^e)/2$.

Alternatively, we can consider the elastic metric ($\boldsymbol{G}_*^e = \boldsymbol{C}_*^{e^{-1}}$)

$$\boldsymbol{G}_*^e = [\boldsymbol{C}^e]_{n+\frac{1}{2}} + 2\frac{\log\left(J_{n+1}^e/J_n^e\right) - [\boldsymbol{C}^e]_{n+\frac{1}{2}} : \frac{1}{2}\Delta\boldsymbol{C}^e}{[\boldsymbol{C}^e]_{n+\frac{1}{2}}\Delta\boldsymbol{C}^e : \Delta\boldsymbol{C}^e[\boldsymbol{C}^e]_{n+\frac{1}{2}}}[\boldsymbol{C}^e]_{n+\frac{1}{2}}\Delta\boldsymbol{C}^e[\boldsymbol{C}^e]_{n+\frac{1}{2}} \;, \tag{33}$$

and similarly for the convected metric $\boldsymbol{G}_*$ in terms of $\overline{\boldsymbol{C}}$, and

$$\boldsymbol{F}_*^p = \boldsymbol{G}_*^{e^{\frac{1}{2}}}\left(\boldsymbol{\Lambda}_{n+1}\boldsymbol{\Lambda}_n^T\right)^{\frac{1}{2}}\boldsymbol{\Lambda}_n\boldsymbol{G}_*^{-\frac{1}{2}} \;, \text{ with } \boldsymbol{\Lambda}_{n+i} = \boldsymbol{R}_{n+i}^{e^T}\boldsymbol{R}_{n+i} \quad (i=0,1) \tag{34}$$

for the rotation tensors of the polar decompositions of $\overline{\boldsymbol{F}}_{n+i}$ and $\boldsymbol{F}_{n+i}^e$. This choice leads to a volume-preserving approximation, in the sense that no plastic change of volume occurs for deviatoric plastic flow vectors, as it is the case for J_2 plasticity of metals. Note that the metric (33) is such that $\boldsymbol{G}_*^e : \Delta\boldsymbol{C}^e/2 = \Delta\varepsilon_v^e$ for the natural volumetric strain $\varepsilon_v^e := \log J^e$. See [4] for details.

5 Representative Numerical Simulation

Figure 1 shows the solution obtained with the energy-momentum algorithm after giving the depicted solid an initial angular velocity of $0.2 \; rad/ms$ to the rigid ring about its axis. The distribution of the equivalent plastic strain α is also shown. Associated J_2-flow theory (i.e. Mises yield function) with Hencky hyperelastic law is considered for the arms, with Young modulus $10 \; GPa$, Poisson ratio 0.2, yield limit $975 \; MPa$, linear hardening modulus $200 \; MPa$ and reference density $2000 \; kg/m^3$ (10 times less than the ring's density); see [2]. We use Q1/A0 bricks with the conserving B-bar treatment of Section 3.2.

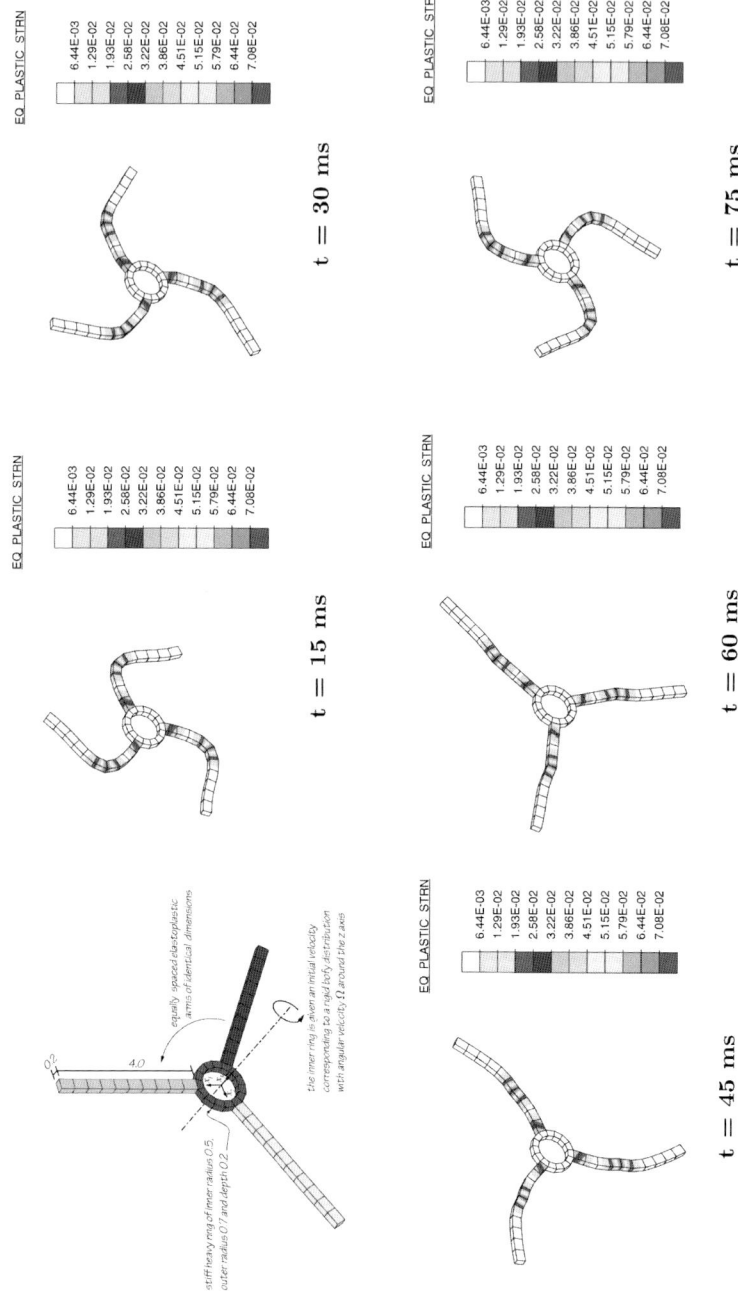

Fig. 1. Three-dimensional solid in free flight [2]. Geometry and loading definition, and the early deformations of the motion, showing the distribution of the equivalent plastic strain α.

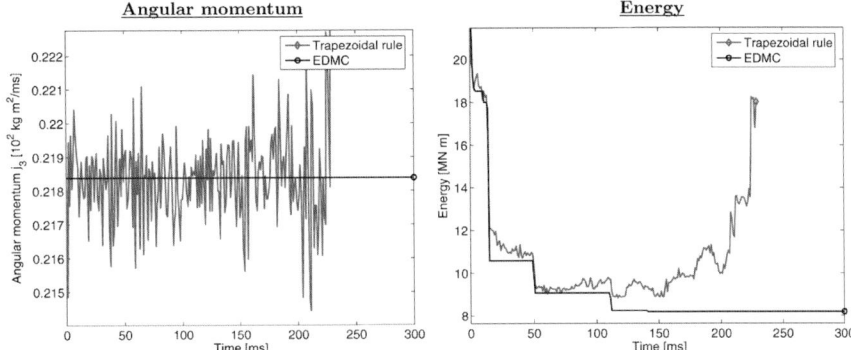

Fig. 2 Three-dimensional solid in free flight [2]. Evolution in time of the angular momentum j_3^h (left) and total energy H (right) for the new conserving EDMC scheme and the standard Newmark trapezoidal rule. The instability of the latter with a non-physical energy growth is to be noted.

Figure 2 shows the evolution of the angular momentum j^h and the total energy H. The exact conservation of the former is confirmed and so is the non-negative character of the energy dissipation, with exact conservation for elastic steps. This situation is to be contrasted with the solution obtained with the Newmark trapezoidal rule in combination with an exponential return mapping [11], with the lack of a strictly non-negative energy dissipation leading eventually to a numerical instability. This situation clearly illustrates the improvement gained by the energy-momentum methods discussed in this paper. They can be seen to be computationally demanding but, as shown by the comparative cost study presented in [4], the improved stability properties together with their generality (e.g. applicable to general anisotropic models) lead to efficient methods for the nonlinear dynamics of elastoplastic solids.

Acknowledgement

The financial support of the AFOSR under grant no. FA9550-05-1-0117 with UC Berkeley is gratefully acknowledged.

References

1. Armero, F. (2006) Energy-dissipative momentum-conserving time-stepping algorithms for finite strain multiplicative plasticity. *Comp. Meth. Appl. Mech. Engrg.* **195**: 4862–4889.
2. Armero, F. (2007) Assumed strain finite element methods for conserving temporal integrations in nonlinear solid dynamics. *Int. J. Num. Mech. Engr.*, in press.
3. Armero, F., Romero, I. (2001) On the formulation of high-frequency dissipative time-stepping algorithms for nonlinear dynamics. Part I: Low order methods for two model problems and

nonlinear elastodynamics. Part II: High order methods. *Comp. Meth. Appl. Mech. Engrg.* **190**: 2603–2649 & 6783–6824.
4. Armero, F., Zambrana, C. (2005) Volume-preserving energy-dissipative momentum-conserving algorithms for isochoric multiplicative plasticity. *Comp. Meth. Appl. Mech. Engrg.* **196**: 4130–4159.
5. Crisfield, M., Shi, J. (1994) A co-rotational element/time-integration strategy for non-linear dynamics. *Int. J. Num. Meth. Engrg.* **37**: 1897–1913.
6. González, O. (2000) Exact energy-momentum conserving algorithms for general models in nonlinear elasticity. *Comp. Meth. Appl. Mech. Engrg.* **190**: 1763–1783.
7. González, O. (1996) Time integration and discrete Hamiltonian systems. *J. Nonlinear Sci.* **6**: 449–467.
8. Marsden, J.E. (1992) *Lectures on Mechanics*, London Math. Soc. Lect. Note Series, Vol. 174, Cambridge University Press.
9. Meng, X., Laursen, T.A. (2002) Energy consistent algorithms for dynamic finite deformation plasticity. *Comp. Meth. Appl. Mech. Engrg.* **191**: 1639–1675.
10. Noels, L., Stainier, L., Ponthot, J.P. (2006) An energy momentum conserving algorithm using the variational formulation of visco-plastic updates. *Int. J. Num. Meth. Engrg.* **65**: 904–942.
11. Simo, J.C. (1998) Numerical analysis and simulation of plasticity. In: Ciarlet, P.G., Lions, J.J. (Eds.), *Handbook on Numerical Analysis VI*, Elsevier, Amsterdam.
12. Simo, J.C., Tarnow, N. (1992) The discrete energy–momentum method: Conserving algorithms for nonlinear elastodynamics. *ZAMP* **43**: 757–793.
13. Simo, J.C., Taylor, R.L., Pister, K. (1988) Variational and projection methods for the volume constraint in finite deformation elastoplasticity. *Comp. Meth. Appl. Mech. Engrg.* **57**: 177–208.

Internal Variable Formulations of Problems in Elastoplastic Dynamics

Modify A.E. Kaunda

School of Mechanical Engineering & Applied Mathematics, Central University of Technology, Free State, Private Bag X20539, Bloemfontein, 9300, South Africa
E-mail: kaundam@cut.ac.za

Abstract. The fundamental problem of an elastic-plastic body subjected to incremental loading is reviewed using a compact internal variable approach which is expressed as a convex nonlinear mathematical programming problem. The approach is based on work carried out by Martin and co-workers [1–6] at the University of Cape Town. Additional contributions are introduced in the area of solution algorithms. Algorithms are developed aimed at approximating kinematic variables. Static and dynamic differential equations of motion and algorithms in form of a convex nonlinear mathematical programming problem are developed using the application of Liapunov functions. Stability analysis via energy methods with the help of Liapunov functions is performed to determine integration parameters.

Key words: dynamics, plasticity, internal variables, integration of algorithms, Liapunov function.

1 Introduction

Recent work, based on the original formulations by Kestin and Rice [7] and Lemaitre and Chaboche [8], on the incremental problem for statically loaded elastic-plastic bodies using an internal variable framework has provided a coherent, compact and physically based formulation which permits several useful insights into the structure of the problem [9–14]. The formulation [1] initially makes use of a backward difference approximation in time of the constitutive equations, and leads to a non-linear programming problem which is convex for stable materials. The simplest two step iterative scheme for the solution of the non-linear programming problem can be recognized as the Newton–Raphson algorithm, and the issue of convergence can be investigated.

Subsequently, Kaunda [3] and Kaunda and Martin [4–6] have considered a broad class of "forward-backward" difference algorithms for the approximation of both the velocities and accelerations and the internal variable rates. The basic framework of the internal variable formulation permits a discussion of convergence and stability for a wide range of choices of the integration parameters. This work is supported by calculations using a modified version of PCFEAP, and shows some promise of more efficient integration procedures.

It was shown [1] that the formulation can be extended to include dynamic effects in a very simple way, and hence the integration of the dynamic problem has the same structure as the static problem previously investigated. This permits, again, some insights into the nature of the problem, and also leads to the practical result that any static finite element program for elastic-plastic bodies can be easily modified to analyze dynamic problems.

In this paper, we further develop differential equations of motion and algorithms using the applications of Liapunov functions for both static and dynamic problems.

2 Internal Variable Formulation of Static Elastic-plastic Problems

We limit consideration to a fairly simple class of elastic-plastic materials which includes perfect plasticity, linear kinematic hardening and isotropic hardening. We also conceive of a spatially discrete representation of the body (say by a finite element approximation) in which kinematic behaviour is represented by the variables u (a vector of the components of nodal point displacements) and λ (a vector of internal variable components at Gauss points).

The free energy of the body is assumed [1] to be a quadratic function $F(u, \lambda)$ of the kinematic variables. A small change in the kinematic variables leads to:

$$dF = \frac{\partial F}{\partial u} du + \frac{\partial F}{\partial \lambda} d\lambda = \mathbf{R} du - \chi d\lambda \tag{1}$$

The internal forces $\mathbf{R}(u, \lambda)$, acting on the nodes, are conjugate to the nodal displacements u, and the slip forces $\chi(u, \lambda)$ are conjugate to the internal variables λ. The minus sign is chosen so that $\chi(u, \lambda)$ are the forces applied by the structure to the slips. As a consequence of the assumption that $F(u, \lambda)$ is quadratic, both the internal forces $\mathbf{R}(u, \lambda)$ and slip forces $\chi(u, \lambda)$ are linear in u and λ. A small change in the kinematic variables leads to:

$$d\mathbf{R} = \frac{\partial \mathbf{R}}{\partial u} du + \frac{\partial \mathbf{R}}{\partial \lambda} d\lambda \tag{2}$$

$$-d\chi = \frac{\partial \chi}{\partial u} du + \frac{\partial \chi}{\partial \lambda} d\lambda \tag{3}$$

For linear functions we deduce from equations (2) and (3) the following expressions:

$$\mathbf{R} = \mathbf{K}u + \mathbf{L}\lambda, \quad \text{where} \quad \mathbf{K} = \frac{\partial \mathbf{R}}{\partial u}, \quad \text{and} \quad \mathbf{L} = \frac{\partial \mathbf{R}}{\partial \lambda} \tag{4}$$

$$-\chi = \mathbf{L}^T u + \mathbf{H}\lambda, \quad \text{where} \quad \mathbf{L}^T = \frac{\partial \chi}{\partial u}, \quad \text{and} \quad \mathbf{H} = \frac{\partial \chi}{\partial \lambda} \tag{5}$$

The superscript T indicates transpose. We substitute (4) and (5) into (1) and integrate to get

$$\int dF = \int (\mathbf{K}u + \mathbf{L}\lambda)du + \int (\mathbf{L}^T u + \mathbf{H}\lambda)d\lambda \tag{6}$$

It follows that the free energy $F(u, \lambda)$ can be written as

$$F(u, \lambda) = \frac{1}{2}u^T \mathbf{K}u + \frac{1}{2}u^T \mathbf{L}\lambda + \frac{1}{2}\lambda^T \mathbf{L}^T u + \frac{1}{2}\lambda^T \mathbf{H}\lambda \tag{7}$$

$$= \frac{1}{2}\begin{Bmatrix} u \\ \lambda \end{Bmatrix}^T \begin{bmatrix} \mathbf{K} & \mathbf{L} \\ \mathbf{L}^T & \mathbf{H} \end{bmatrix} \begin{Bmatrix} u \\ \lambda \end{Bmatrix} > 0$$

Assuming a quadratic functional, we deduce [23] from Taylor series expansion the functional of the form:

$$F(u, \lambda) = \frac{1}{2}\{u^T \cdot \frac{\partial^2 F}{\partial u \partial u} \cdot u + u^T \cdot \frac{\partial^2 F}{\partial u \partial \lambda} \cdot \lambda + \lambda^T \cdot \frac{\partial^2 F}{\partial \lambda \partial u} \cdot u + \lambda^T \cdot \frac{\partial^2 F}{\partial \lambda \partial \lambda} \cdot \lambda\} > 0 \tag{8}$$

The free energy is positive definite. From equations (4) to (8), we further define:

$$\mathbf{K} = \frac{\partial \mathbf{R}}{\partial u} = \frac{\partial^2 F}{\partial u \partial u}, \quad \mathbf{L} = \frac{\partial \mathbf{R}}{\partial \lambda} = \frac{\partial^2 F}{\partial \lambda \partial u},$$

$$\mathbf{L}^T = \frac{\partial \boldsymbol{\chi}}{\partial u} = \frac{\partial^2 F}{\partial u \partial \lambda}, \quad \text{and} \quad \mathbf{H} = \frac{\partial \boldsymbol{\chi}}{\partial \lambda} = \frac{\partial^2 F}{\partial \lambda \partial \lambda} \tag{9}$$

We introduce a global dissipation function, $D(\dot{\lambda})$ which is assumed to be homogeneous, convex and of degree one in the internal variable rates $\dot{\lambda}$. The slip forces $\boldsymbol{\chi}(u, \lambda)$ are then given by

$$\boldsymbol{\chi} = \partial D(\dot{\lambda}) \tag{10}$$

The operator $\partial(\ )$ indicates the sub-differential. Thus $\boldsymbol{\chi}(u, \lambda)$ is homogeneous and of degree zero in the components of the internal variable rates. To be consistent with thermodynamic requirements, $D(\dot{\lambda})$ must be positive definite; $D(\dot{\lambda}) = 0$ iff $\dot{\lambda} = 0$.

The dissipation function $D(\dot{\lambda})$ is in fact [1] the sum of dissipation functions $D_g(\dot{\lambda}_g)$ at individual Gauss or integration points, where $\dot{\lambda}_g$ are the internal variable components associated with that particular Gauss point. Each Gauss point dissipation function is homogeneous and of degree one, and is a generalized cone in the $\dot{\lambda}_g$ space. For simplicity, we shall assume that $D_g(\dot{\lambda}_g)$ is differentiable at all points where $\dot{\lambda}_g \neq 0$. Slip forces $\boldsymbol{\chi}_g(u, \lambda)$ are associated with each Gauss point, and collectively make up the vector $\boldsymbol{\chi}(u, \lambda)$.

The set of values of $\boldsymbol{\chi}(u, \lambda)$ which are admissible at the origin can be plotted in $\boldsymbol{\chi}(u, \lambda)$ space. This set is bounded by a yield surface which we define [1, 14] as

$$\phi(\chi) = 0 \tag{11}$$

The yield surface is convex. The slip forces are constrained to lie on or within this convex surface; if the forces lie within the surface, $\dot{\lambda} = 0$, while if the forces lie on the surface the internal variable rate may be different from zero, and will have the direction of the outward normal to $\phi(\chi)$ (or an admissible outward normal if the normal is not uniquely defined).

The external nodal forces may be functions of time, and are denoted by $\mathbf{P}(t)$. In static problems, the time simply measures the order of events rather than real time. For static analysis, equilibrium is attained when

$$\mathbf{R}(t) = \mathbf{P}(t) \tag{12}$$

For the functional defined in equations (7) and (8) to be a Liapunov function [3, 29], we need to show that:

$$\frac{dF}{dt} \leq 0 \quad \text{or} \quad \frac{\Delta F}{\Delta t} \leq 0 \tag{13}$$

Consider a Taylor series expansion of the quadratic functional defined in equations (7) and (8):

$$F(u + \Delta u, \lambda + \Delta \lambda)$$
$$= F(u, \lambda) + \Delta u^T \frac{\partial F}{\partial u} + \Delta \lambda^T \frac{\partial F}{\partial \lambda}$$
$$+ \frac{1}{2} \left\{ \Delta u^T \frac{\partial^2 F}{\partial u \partial u} \Delta u + \Delta u^T \frac{\partial^2 F}{\partial u \partial \lambda} \Delta \lambda + \Delta \lambda^T \frac{\partial^2 F}{\partial \lambda \partial u} \Delta u + \Delta \lambda^T \frac{\partial^2 F}{\partial \lambda \partial \lambda} \Delta \lambda \right\}$$
$$\tag{14}$$

Equations (13) and (14) with the definitions given in equations (4), (5) and (9) enforce the Liapunov stability criterion [3, 29] for an incremental static problem:

$$\frac{\Delta F}{\Delta t} = \frac{1}{\Delta t} \left\{ \Delta u^T \mathbf{R} - \Delta \lambda^T \chi \right.$$
$$\left. + \frac{1}{2} \{ \Delta u^T \cdot \mathbf{K} \cdot \Delta u + \Delta u^T \cdot \mathbf{L}^T \cdot \Delta \lambda + \Delta \lambda^T \cdot \mathbf{L} \cdot \Delta u + \Delta \lambda^T \cdot \mathbf{H} \cdot \Delta \lambda \} \right\}$$
$$\leq 0 \tag{15}$$

In equation (15) the coefficient of $1/\Delta t$ is negative semi-definite. Equation (15) then states that the positive definite functional represented by equation (7) attain a minimum value, which is equivalent to the result obtained in [1] using a simple backward difference scheme for an incremental problem. We also note in equation (15) that the dissipation function is positive definite:

$$\Delta \lambda^T \chi = D(\Delta \lambda) \quad \text{or} \quad \dot{\lambda}^T \chi = D(\dot{\lambda}) > 0 \tag{16}$$

3 Dynamic Behaviour

For dynamic behaviour, the total energy is the sum of the free energy, kinetic energy, work done by dissipative viscous damping forces and the plastic energy dissipation function, minus the work done by external forces:

$$E(u, \dot{u}, \lambda) = \frac{1}{2}u^T \mathbf{K} u + \frac{1}{2}u^T \mathbf{L}\lambda + \frac{1}{2}\lambda^T \mathbf{L}^T u + \frac{1}{2}\lambda^T \mathbf{H}\lambda + \frac{1}{2}\dot{u}^T \mathbf{M}\dot{u}$$
$$+ \int \dot{u}^T \mathbf{C}\dot{u}\, dt + \int D(\dot{\lambda})\, dt - \int \mathbf{P}(t) \cdot \dot{u}\, dt > 0 \quad (17)$$

For the functional defined in equation (17) to be a Liapunov function [3, 29], we need to show that:

$$\frac{dE}{dt} \leq 0 \quad \text{or} \quad \frac{\Delta E}{\Delta t} \leq 0 \quad (18)$$

Using the first expression in equation (18) we enforce the Liapunov stability criterion [3, 29] for a dynamic problem:

$$\frac{dE}{dt} = \dot{u}^T \{\mathbf{M}\ddot{u}(t) + \mathbf{C}\dot{u}(t) + \mathbf{K}u(t) + \mathbf{L}\lambda(t) - \mathbf{P}(t)\}$$
$$+ \dot{\lambda}^T \{\mathbf{L}^T u(t) + \mathbf{H}\lambda(t) + \chi(u, \lambda)\} \leq 0 \quad (19)$$

Applying Lagrange's equations or Hamilton's principle to equation (17), or indeed using the equality of equation (19) while realizing the arbitrariness of $\dot{u}^T$ and $\dot{\lambda}^T$, we can derive a set of differential equations of motion:

$$\mathbf{M}\ddot{u}(t) + \mathbf{C}\dot{u}(t) + \mathbf{K}u(t) + \mathbf{L}\lambda(t) = \mathbf{P}(t) \quad (20)$$

$$\mathbf{L}^T u(t) + \mathbf{H}\lambda(t) = -\partial D(\dot{\lambda}(t), \quad \text{where} \quad \chi = \partial D(\dot{\lambda}) \quad (21)$$

Using the second expression in equation (18) we enforce the Liapunov stability criterion [3, 29] for an incremental dynamic problem:

$$\frac{\Delta E}{\Delta t} = \frac{1}{\Delta t} \left\{ \frac{1}{2}(\dot{u}^T \mathbf{M}\dot{u})\Big|_{\dot{u}_{n-1}}^{\dot{u}_n} + \int_{t_{n-1}}^{t_n} \{\dot{u}^T \mathbf{C}\dot{u} + D(\dot{\lambda}) - \mathbf{P}(t)\dot{u}\} dt \right.$$
$$+ \Delta u^T \mathbf{R} - \Delta \lambda^T \chi + \frac{1}{2}\{\Delta u^T \cdot \mathbf{K} \cdot \Delta u + \Delta u^T \cdot \mathbf{L}^T \cdot \Delta \lambda$$
$$\left. + \Delta \lambda^T \cdot \mathbf{L} \cdot \Delta u + \Delta \lambda^T \cdot \mathbf{H} \cdot \Delta \lambda\} \right\} \leq 0 \quad (22)$$

For dynamic behaviour, equation (22) then states that the positive definite functional represented by equation (17) attain a minimum value for an incremental problem.

4 The Iterative Solution Procedure

Iterative solution procedures for static and dynamic problems are well established and are based on the standard forward, central or backward difference schemes. Modern procedures use integration parameters that can be adjusted to reproduce the standard difference schemes as special cases or to provide a family of new schemes that may enhance accuracy, convergence as well as stability characteristics.

We first rewrite equations (20) and (21) as follows:

$$\begin{bmatrix} \mathbf{M} & 0 \\ 0 & 0 \end{bmatrix} \begin{Bmatrix} \ddot{u}(t) \\ \ddot{\lambda}(t) \end{Bmatrix} + \begin{bmatrix} \mathbf{C} & 0 \\ 0 & 0 \end{bmatrix} \begin{Bmatrix} \dot{u}(t) \\ \dot{\lambda}(t) \end{Bmatrix} + \begin{bmatrix} \mathbf{K} & \mathbf{L} \\ \mathbf{L}^T & \mathbf{H} \end{bmatrix} \begin{Bmatrix} u(t) \\ \lambda(t) \end{Bmatrix} = \begin{Bmatrix} \mathbf{P}(t) \\ -\partial D(\dot{\lambda}) \end{Bmatrix} \quad (23)$$

To develop a Newton–Raphson scheme, we further simplify equation (23) as follows:

$$\bar{\mathbf{M}}\tilde{a}(t) + \bar{\mathbf{C}}\tilde{v}(t) + \bar{\mathbf{K}}\tilde{s}(t) = \tilde{f}(t) \quad (24)$$

The symbols in equation (24) represent matrices (with over-bar) and vectors (with tilde) of equation (23). The nonlinear time-dependent equation may be rewritten in the discrete form:

$$\Psi_n(t_n, \tilde{s}_n, \tilde{v}_n, \tilde{a}_n) = \bar{\mathbf{M}}\tilde{a}_n + \bar{\mathbf{C}}\tilde{v}_n + \bar{\mathbf{K}}\tilde{s}_n - \tilde{f}_n(t_n) = \tilde{0} \quad (25)$$

The residual Ψ can be expanded by using Taylor series in a forward-backward difference scheme [3].

$$\Psi_n = \Psi_{n-1}$$

$$+ \left\{ \sum_{k=1}^{k=\infty} \left\{ \frac{1}{k!} \left[(1-\mu_1)\Delta s \cdot \frac{\partial}{\partial s} + (1-\mu_2)\Delta v \frac{\partial}{\partial v} + (1-\mu_3)\Delta a \frac{\partial}{\partial a} \right]^k \right\} \cdot \Psi_{n-1} \right\}$$

$$- \left\{ \sum_{k=1}^{k=\infty} \left\{ \frac{(-1)^k}{k!} \left[\mu_1 \cdot \Delta s \cdot \frac{\partial}{\partial s} + \mu_2 \cdot \Delta v \cdot \frac{\partial}{\partial v} + \mu_3 \cdot \Delta a \cdot \frac{\partial}{\partial a} \right]^k \right\} \cdot \Psi_n \right\} \quad (26)$$

The second term on the right hand side represents the forward-difference expansion whereas the third term represents the backward-difference expansion. The integration parameters $\mu_1, \mu_2, \mu_3 \in [0, 1]$ are introduced to form tuneable integration schemes. With the upper limit of the summation interval set to $k = 1$ the resulting implicit Newton–Raphson scheme is given by the following expression:

$$0 = \Psi_{n-1} + (1-\mu_1) \cdot \Delta s \cdot \frac{\partial \Psi_{n-1}}{\partial s} + (1-\mu_2) \cdot \Delta v \cdot \frac{\partial \Psi_{n-1}}{\partial v}$$

$$+ (1-\mu_3) \cdot \Delta a \cdot \frac{\partial \Psi_{n-1}}{\partial a} + \mu_1 \cdot \Delta s \cdot \frac{\partial \Psi_n}{\partial s}$$

$$+ \mu_2 \cdot \Delta v \cdot \frac{\partial \Psi_n}{\partial v} + \mu_3 \cdot \Delta a \cdot \frac{\partial \Psi_n}{\partial a} \quad (27)$$

The partial derivatives are defined as follows:

$$\frac{\partial \Psi_{n-1}}{\partial s} = \bar{\mathbf{K}}_{n-1}, \quad \frac{\partial \Psi_n}{\partial s} = \bar{\mathbf{K}}_n, \quad \frac{\partial \Psi_n}{\partial v} = \bar{\mathbf{C}}_n, \quad \frac{\partial \Psi_n}{\partial a} = \bar{\mathbf{M}}_n \quad (28)$$

By setting $\mu_1 = \mu_2 = \mu_3 = 0$, the standard Newton–Raphson scheme is recovered whereas by only setting $\mu_2 = \mu_3 = 0$, a special case of the generalized Newton–Raphson scheme may be investigated where the mass and damping constants do no vary significantly within a time-step while changes in the stiffness may not be ignored, as implied in nonlinear spring behaviour [24].

5 Displacement, Velocity, Acceleration and Generalized Trapezoidal Scheme

From the definitions of velocity and acceleration, the following relations hold: $v = ds/dt$, $a = dv/dt$ and $v\,dv = a\,ds$. Based on these definitions, a generalized trapezoidal scheme [3] consists of the following recurrence equations:

$$s_n = s_{n-1} + \Delta t v_{n-1} + \frac{\Delta t^2}{2!}\{(1-\gamma_1)a_{n-1} + \gamma_1 a_n\} \quad (29)$$

$$v_n = v_{n-1} + \Delta t\{(1-\gamma_2)a_{n-1} + \gamma_2 a_n\} \quad (30)$$

$$v_n^2 = v_{n-1}^2 + 2\Delta s[(1-\gamma_3)a_{n-1} + \gamma_3 a_n] \quad (31)$$

The scalars $\gamma_1, \gamma_2, \gamma_3 \in [0, 1]$ form tuneable integration parameters [28] for the generalized trapezoidal scheme. The first two equations of the generalized trapezoidal scheme are solved simultaneously together with the generalized Newton-Raphson algorithm or they can be solved in a predictor-corrector iterative algorithm. The third equation is used to enhance the accuracy of the algorithm and also used in stability analysis.

6 Accuracy of Recurrence Equations

By comparing the recurrence relation (29) with the corresponding complete Taylor series expansion, we arrive at a local displacement truncation error.

$$\varepsilon_s = s^*_{n+1} - s_{n+1} = O\left(\frac{\Delta t^4}{24}\ddot{a}_n\right) \quad \text{for} \quad \gamma_1 = \frac{1}{3}$$

$$\varepsilon_s = O\left(\left(\frac{\gamma_1}{2} - \frac{1}{3!}\right)\Delta t^3 \dot{a}_n\right) \quad \text{for} \quad \gamma_1 \neq \frac{1}{3} \quad (32)$$

Similarly, the local velocity truncation error for (30) is estimated as follows:

$$\varepsilon_v = v_{n+1}^* - v_{n+1} = O\left(\frac{\Delta t^3}{12}\ddot{a}_n\right), \quad \text{for} \quad \gamma_2 = \frac{1}{2}, \quad \text{or}$$

$$\varepsilon_v = O\left(\left(\gamma_2 - \frac{1}{2!}\right)\Delta t^2 \dot{a}_n\right) \quad \text{for} \quad \gamma_2 \neq \frac{1}{2} \tag{33}$$

Stability analysis of the algorithm needs to be done to complement this set of choices.

7 Convergence

It was shown in [1] that a sufficient condition for monotonic convergence is that the functional U_n being minimized should decrease in each of the predictor and corrector steps. However, while some choices of predictor step satisfy the sufficient condition, the consistent tangent predictor does not necessarily do so. The consistent tangent predictor does lead to quadratic convergence, while the other choices converge monotonically but very slowly. As a result, the consistent tangent predictor is the best choice, but it should be combined with a line search algorithm if convergence is to be guaranteed [13, 17].

8 Stability Analysis

Algorithm stability analysis based on the norms, energy and other measures may be investigated by examining the following relationship:

$$v_{n+1}^2 = v_n^2 + 2\int_{S_n}^{S_{n+1}} a\,ds = v_n^2 + 2 * \text{Area under } a\text{-}s \text{ curve } (A_{a-s})$$

The area under the a-s curve represents the work done by external forces on unit mass through displacement ds. The energy-balance equation is then given by:

$$m \cdot \left(\frac{v_{n+1}^2 - v_n^2}{2}\right) = \int_{S_n}^{S_{n+1}} m \cdot a\,ds = \int_{S_n}^{S_{n+1}} \sum F \cdot ds \tag{34}$$

This equation states that the change in kinetic energy, ΔT_n, equals the work done by the resultant forces. Consider a spring hardening model [24] and the case where the external forces $f(t)$ and forces resulting from dissipative functions are deformation-independent [25] within a time step. Then:

$$m \cdot \left(\frac{v_{n+1}^2 - v_n^2}{2}\right) + k_1 \cdot \left(\frac{s_{n+1}^2 - s_n^2}{2}\right) + k_2 \cdot \left(\frac{s_{n+1}^4 - s_n^4}{4}\right)$$

Problems in Elastoplastic Dynamics

$$= \int_{t_n}^{t_{n+1}} (f(t) \cdot v - c \cdot v^2) \cdot dt \qquad (35)$$

The sum of the kinetic and elastic energies at the beginning and end of the iteration is positive-definite, evident from the fact that the left hand side of equation (36) is positive-definite.

$$m \cdot \left(\frac{v_{n+1}^2}{2}\right) + k_1 \cdot \left(\frac{s_{n+1}^2}{2}\right) + k_2 \cdot \left(\frac{s_{n+1}^4}{4}\right)$$

$$= m \cdot \left(\frac{v_n^2}{2}\right) + k_1 \cdot \left(\frac{s_n^2}{2}\right) + k_2 \cdot \left(\frac{s_n^4}{4}\right) + \int_{t_n}^{t_{n+1}} (f(t) \cdot v - c \cdot v^2) \cdot dt$$

$$> 0 \qquad (36)$$

A Liapunov function is positive definite and that its derivative with respect to time is negative semi-definite [3, 29]. This is called the Liapunov stability criterion. It then follows that:

$$\frac{d}{dt}\left\{m \cdot \left(\frac{v_n^2}{2}\right) + k_1 \cdot \left(\frac{s_n^2}{2}\right) + k_2 \cdot \left(\frac{s_n^4}{4}\right) + \int_{t_n}^{t_{n+1}} (f(t) \cdot v - c \cdot v^2) \cdot dt\right\} \leq 0$$

An energy method of stability analysis is given in [24]. Consider the *sum* of the changes in kinetic (ΔT_n) and potential (ΔV_n) energies.

$$\langle v_n \rangle \cdot m \cdot \Delta v + \langle s_n \rangle \cdot \left[k_1 + k_2 \cdot \left(\frac{s_{n+1}^2 + s_n^2}{2}\right)\right] \cdot \Delta s = \int_{t_n}^{t_{n+1}} (f(t) \cdot v - c \cdot v^2) \cdot dt \qquad (37)$$

where

$$\langle v_n \rangle = \frac{(v_{n+1} + v_n)}{2}, \quad \langle s_n \rangle = \frac{(s_{n+1} + s_n)}{2},$$

$$\Delta v = v_{n+1} - v_n \quad \text{and} \quad \Delta s = s_{n+1} - s_n \qquad (38)$$

Assuming that the deformation-independent external forces are prescribed within a time-step, and are considered as fixed and bounded in each time interval, we may rewrite the above equation as follows:

$$\langle v_n \rangle \cdot m \cdot \Delta v + \langle s_n \rangle \cdot \left[k_1 + k_2 \cdot \left(\frac{s_{n+1}^2 + s_n^2}{2}\right)\right] \cdot \Delta s = (f(t_n) \cdot v_n - c \cdot v_n^2) \cdot \Delta t \qquad (39)$$

In the absence of both the external and dissipative forces, the energy is positive definite and remains constant so that the *sum* of the changes in kinetic and potential energies is zero. In the absence of external forces alone, the energy is positive definite but decreases through the dissipative agents. Therefore the *sum* is non-increasing.

$$\langle v_n\rangle \cdot m \cdot \Delta v + \langle s_n\rangle \cdot \left[k_1 + k_2 \cdot \left(\frac{s_{n+1}^2 + s_n^2}{2}\right)\right] \cdot \Delta s \leq 0 \tag{40}$$

The recurrence equation of velocity may be transformed into the following:

$$\Delta v = \Delta t \cdot \langle a_n\rangle + \left(\gamma_2 - \frac{1}{2}\right) \cdot \Delta t \cdot \Delta a \quad \text{where} \quad \langle a_n\rangle = \frac{(a_{n+1} + a_n)}{2}$$

Multiplying this equation by $\Delta t/2$ and subtracting the recurrence equation of displacement we have:

$$\Delta t \langle v_n\rangle = \Delta s - (\gamma_1 - \gamma_2)\frac{\Delta t^2}{2}\Delta a \quad \text{where} \quad \langle v_n\rangle = \frac{(v_{n+1} + v_n)}{2}$$

Substituting the transformed recurrence equations into the equation for the *sum* of the changes in kinetic (ΔT_n) and potential (ΔV_n) energies, we have:

$$\Delta T_n + \Delta V_n = \frac{1}{\Delta t}\left(\Delta s - (\gamma_1 - \gamma_2)\frac{\Delta t^2}{2}\Delta a\right)$$

$$\cdot m \cdot \left(\Delta t \cdot \langle a_n\rangle + \left(\gamma_2 - \frac{1}{2}\right) \cdot \Delta t \cdot \Delta a\right)$$

$$+ \langle s_n\rangle \cdot \left[k_1 + k_2 \cdot \left(\frac{s_{n+1}^2 + s_n^2}{2}\right)\right] \cdot \Delta s \leq 0 \tag{41}$$

From this equation we may choose the integration parameters that result in a stable algorithm. For unconditional stability $\gamma_1 \geq \gamma_2$ and $\gamma_2 \geq 1/2$.

9 Conclusions

1. The fundamental problem of an elastic-plastic body subjected to incremental loading has been reviewed using a compact internal variable approach which may be expressed as a convex nonlinear mathematical programming problem.
2. Static and dynamic differential equations of motion and algorithms in form of a convex nonlinear mathematical programming problem have been developed using the application of Liapunov functions.
3. Stability analysis via energy methods with the help of Liapunov functions has been performed to determine integration parameters. For unconditional stability $\gamma_1 \geq \gamma_2$ and $\gamma_2 \geq 1/2$.

References

1. Martin, J.B., Kaunda, M.A.E. and Isted, R.D., Internal variable formulations of elastic-plastic dynamic problems. *Int. J. Impact Engng.* **18**(7–8), 849–858, 1996.
2. Isted, R.D., *A Method for Including Transient Dynamics in the finite Element Analysis of Elastic-Plastic Bodies*. M.Sc. (Eng) Thesis, University of Cape Town, 1988.
3. Kaunda, M.A.E., *Finite Element Algorithms for the Static and Dynamic Analysis of Time-Dependent and Time-Independent Plastic Bodies*. Ph.D. Thesis, University of Cape Town, 1994.
4. Kaunda, M.A.E. and Martin, J.B., *Finite Element solution Algorithms for Nonlinear Time-dependent/Independent Problems in Solid Mechanics*, Tech. Rep. No. 199, Centre for Research in Computational and Applied Mechanics, University of Cape Town, July 1993; Presented at 18th South African Symposium on Numerical Mathematics, Durban, July 1993.
5. Kaunda, M.A.E. and Martin, J.B., *Finite Element Solution Algorithms for Time-Independent and Time-dependent plasticity*. Tech. Rep. No. 232, Centre for Research in Computational and Applied Mechanics, University of Cape Town, January 1994; Presented at 5th Int. Conf. on Modern Group Analysis: Theory and Applications in Mathematical Modelling, University of the Witswatersrand, January 1994.
6. Kaunda, M.A.E. and Martin, J.B., Finite element solution algorithms for an internal variable framework in elastoplastic-dynamics. In *Dynamic Plasticity and Structural Behaviours*, S. Tanimura and A.S. Khan (Eds.), Gordon & Breach, 1995, pp. 883–886.
7. Kestin, J. and Rice, J.R., Paradoxes in the application of thermodynamics to strained solids. In *A Critical Review of Thermodynamics*, E.B. Stuart, B. Gal-Or and A.J. Brainard (Eds.), Mono Book, 1970, pp. 275–298.
8. Lemaitre, J. and Chaboche, J., Aspects phenomenologiques de la rupture par endommagement. *J. Mech. App.* **2**, 1978, 317–365.
9. Martin, J.B., An internal variable approach to finite element problems in plasticity. In *Physical Nonlinearities in Structural Analysis*, J. Hult and J. Lemaitre (Eds.), Springer, Berlin, 1981, pp. 165–176.
10. Martin, J.B., Reddy, B.D., Griffin, T.B. and Bird, W.W., Application of mathematical programming concepts to incremental elastic-plastic analysis. *Engrg. Struct.* **9**, 1987, 171–176.
11. Martin, J.B. and Reddy, B.D., Variational principles and solution algorithms for internal variable formulations of problems on plasticity. In *Omaggio a Giulio Ceradini*, U. Andreaus et al. (Eds.), University La Sapienza Rome, 1988, pp. 465–477.
12. Nappi, A. and Martin, J.B., An internal variable approach to perfectly plastic and linear hardening relations in plasticity. *Eur. J. Mech.* **A9**, 1991, 107–131.
13. Martin, J.B. and Caddemi, S., Sufficient conditions for the convergence of the Newton–Raphson iterative algorithm in incremental elastic-plastic analysis. *Eur. J. Mech.* **A13**, 1995, 114–133.
14. Reddy, B.D. and Martin, J.B., Internal variable formulations of problems in elastoplasticity: Constitutive and algorithmic aspects. *Appl. Mech. Rev.* **47**, 1994, 429–456.
15. Comi, C. and Maier, G., On the convergence of a backward difference iterative procedure in elastoplasticity with nonlinear kinematic and isotropic hardening. In *Computational Plasticity: Models, Software and Applications*, D.R.J. Owen, E. Hinton and E. Onate (Eds.), Pineridge Press, Swansea, 1989.
16. Simo, J. and Taylor, J.L., Consistent tangent operators for rate-independent-elastoplasticity. *Comp. Meth. App. Mech. Engng.* **48**, 1985, 101–118.
17. Crisfield, M.A., Accelerating and damping the modified Newton–Raphson method. *Comput. & Struct.* **18**, 1984, 395–407.
18. Belytschko, T. and Choeberle, D.F., On the unconditional stability of an implicit algorithm for nonlinear structural dynamics. *J. Appl. Mech.* **45**, 1975, 865–869.
19. Rajgelj, S. Amadio, C. and Nappi, A., An internal variable approach applied to the dynamic analysis of elastic-plastic structural systems. *Earthquake Engrg. Structural Dynam.* **22**, 1993, 885–903.

20. Rencontre, L., Martin, J.B. and Caddemi, S., The relationship between the generalised mid point and trapezoidal rules in incremental elastoplasticity. *Comp. Meth. Appl. Mech. Engng.* **96**, 1992, 201–212.
21. Simo, J.C. and Govindjee, S., Non-linear B-stability and symmetry preserving return mapping algorithms for plasticity and viscoplasticity. *Int. J. Num. Meth. Engng.* **31**, 1991, 151–176.
22. Corigliano, A. and Perego, U., Unconditionally stable mid-point time integration in elastic-plastic dynamics. *Rendi Acc Lincei, s. 9* **1**, 1990, 367–376.
23. Irons, B.M. and Shrive, N.G., *Numerical Methods in Engineering and Applied Science: Numbers are Fun*, Ellis Horwood, 1987.
24. Wood, W.L., *Practical Time-Stepping Schemes*, Clarendon Press, Oxford, 1990.
25. Bathe, K-J., *Finite Element Procedures*, Prentice-Hall, 1996.
26. Zienkiewicz, O.C. Taylor, R.L. and Zhu, J.Z., *The Finite Element Method: Its Basis & Fundamentals*, 6th edition, Elsevier/Butterworth-Heinemann, 2005.
27. Hughes, T.J.R., *The Finite Element Method: Linear Static and Dynamic Finite Element Analysis*, Dover Publication, 2000.
28. Smith, J.M., *Mathematical Modelling and Digital Simulation for Engineers and Scientists*, John Wiley & Sons, 1977.
29. Liapunov, A.M., *On the General Problem of Stability of Motion*. PhD Thesis, Kharkov, 1892. Princeton University Press, Princeton, NJ, 1949.

Time-FE Methods for the Nonlinear Dynamics of Constrained Inelastic Systems

Rouven Mohr[1], Stefan Uhlar[2], Andreas Menzel[3,4] and Paul Steinmann[5]

[1] *Chair of Applied Mechanics, University of Kaiserslautern,
D-67653 Kaiserslautern, Germany
E-mail: rmohr@rhrk.uni-kl.de*
[2] *Chair of Computational Mechanics, University of Siegen,
D-57068 Siegen, Germany
E-mail: uhlar@imr.mb.uni-siegen.de*
[3] *Group of Mechanics and Machine Dynamics, Dortmund University
of Technology, D-44227 Dortmund, Germany
E-mail: andreas.menzel@udo.edu*
[4] *Division of Solid Mechanics, Lund University, SE-221 00 Lund, Sweden
E-mail: andreas.menzel@solid.lth.se*
[5] *Chair of Applied Mechanics, University of Erlangen-Nuremberg,
D-91058 Erlangen, Germany
E-mail: paul.steinmann@ltm.uni-erlangen.de*

Abstract. In the following, a general framework for the completely consistent integration of constrained dissipative dynamics is proposed, that essentially relies on Finite Element methods in space and time.

In this context, hybrid systems consisting of rigid bodies and inelastic flexible parts are considered, where special emphasis is placed on the resulting algorithmic fulfilment of fundamental balance equations.

Finally, the excellent performance of the presented concepts will be demonstrated by means of a representative numerical example, involving finite elasto-plastic deformations.

Key words: constrained dynamics, inelastic deformations, Finite Elements in time, consistent integration.

1 Introduction

It is well known that the performance of classical time integration schemes for structural dynamics, as for instance developed by Newmark [13], is strongly restricted when dealing with highly nonlinear systems. In a nonlinear setting, advanced numerical techniques are required to satisfy classical balance laws of the underlying mechanical model, being important for the physical quality of the results and also for the robustness of the resulting integrators. In this area, most of the existing formulations have been restricted to the conservative case. Nevertheless, the incorporation of dissipation effects in dynamical systems is of cardinal importance, representing a demanding task from the computational point of view. In the last years, notable contributions dealing with elasto-plastic structural dynamics have been published by Meng and Laursen [10], Noels et al. [14, 15], and Armero [1–3]. Furthermore, the simulation of mechanical systems subject to constraints represents an additional challenge, whereby particularly the modelling of inelastic effects in flexible multibody systems has only been addressed by few authors, see e.g. [7, 8, 17].

In this contribution, we follow the concepts which have been proposed for conservative systems by Betsch and Steinmann [4, 5] and pick-up the powerful framework of time-FE methods, developing integrators for constrained dissipative systems that essentially rely on an internal variable approach. Hereby, we aim at a general formulation that enables a completely consistent time-integration. In this context, one can distinguish between 'mechanical consistency' including the conservation of both momentum maps, 'energy-consistency' related to a conservation of the sum of the total energy and the (strictly non-negative) dissipation, and 'kinematic consistency' corresponding to a fulfilment of the involved constraints.

2 Inelastic Flexible Parts

To set the stage, we start with some basic notation of geometrically nonlinear continuum mechanics, considering a deformable solid body $\mathcal{B}^d \subset \mathbb{R}^3$. First, the nonlinear deformation map $\boldsymbol{\varphi}(X, t) : \mathcal{B}_0^d \times [0, T] \to \mathcal{B}_t^d$ is introduced as a mapping from the reference to the spatial configuration, whereby $[0, T]$ denotes the time period of interest. Furthermore, the deformation gradient reads

$$F := \nabla_X \boldsymbol{\varphi}(X, t). \tag{1}$$

Since we intend to consider dissipative systems, additional internal variables enter into the Helmholtz energy density ψ to cover also inelastic effects. In the following, we assume that all relevant internal variables are represented by the general quantity κ. By incorporating invariance requirements, we obtain the representation $\psi(C, \kappa)$ based on the right Cauchy–Green tensor $C = F^t \cdot F$. Following standard arguments based on the Clausius–Planck inequality, the Piola–Kirchhoff stresses are given by $S := 2 \nabla_C \psi$. Referring to the second law of thermodynamics, it is indispensable that the dissipation is strictly non-negative, resulting in

$$\mathcal{D} = \langle \boldsymbol{\beta}, \dot{\boldsymbol{\kappa}} \rangle \geq 0. \tag{2}$$

Therein, we have introduced a generalised scalar product $\langle \bullet, \bullet \rangle$ and the conjugated thermodynamical forces $\boldsymbol{\beta} := -\nabla_\kappa \psi$. Finally, the rate of the internal variables $\dot{\boldsymbol{\kappa}}$ has to be defined via a specific evolution equation, depending on the considered dissipation effect and on the particular model of interest.

In a next step, a standard Finite Element discretisation in space with n_{el} elements is applied, rendering the partition of the continuum

$$\mathcal{B}_0^d \approx \bigcup_{el=1}^{n_{el}} {}^{el}\mathcal{B}_0^d. \tag{3}$$

Additionally, the global shape functions $N_I(X) : \mathcal{B}_0^d \to \mathbb{R}$ are introduced interpolating the values $\boldsymbol{q}_I(t) : [0, T] \to \mathbb{R}^3$ which are referred to the spatial nodes $I = 1, \ldots, n_{\text{node}}$. Including these definitions, the spatial approximations of the nonlinear deformation map $\boldsymbol{\varphi}$ and the right Cauchy–Green tensor $\boldsymbol{C}$ result in

$$\boldsymbol{\varphi}(X,t) \approx \sum_{I=1}^{n_{\text{node}}} \boldsymbol{q}_I N_I \quad \text{and} \quad \boldsymbol{C}(X,t) \approx \sum_{I,J=1}^{n_{\text{node}}} \boldsymbol{q}_I \cdot \boldsymbol{q}_J \nabla_X N_I \otimes \nabla_X N_J \tag{4}$$

respectively. Consequently, the semidiscrete system of interest can be characterised by a vector of nodal coordinates $\boldsymbol{q}^d(t) := [\boldsymbol{q}_1(t), \ldots, \boldsymbol{q}_{n_{\text{node}}}(t)]^t \in \mathbb{R}^{n_{\text{dof}}}$ and a vector of nodal velocities $\boldsymbol{v}^d(t) := [\dot{\boldsymbol{q}}_1(t), \ldots, \dot{\boldsymbol{q}}_{n_{\text{node}}}(t)]^t \in \mathbb{R}^{n_{\text{dof}}}$ with the number of degrees of freedom $n_{\text{dof}} = 3\,n_{\text{node}}$. Furthermore, the mass matrix $\mathbb{M}^d \in \mathbb{R}^{n_{\text{dof}}} \times \mathbb{R}^{n_{\text{dof}}}$ of the semidiscrete system is introduced which consists of the sub-matrices

$$\boldsymbol{M}_{IJ} := \int_{\mathcal{B}_0^d} \rho^d\, N_I N_J \,\mathrm{d}V\, \boldsymbol{I} \tag{5}$$

for $I, J = 1, \ldots, n_{\text{node}}$, using the mass density of the deformable body ρ^d and the identity matrix $\boldsymbol{I} \in \mathbb{R}^3 \times \mathbb{R}^3$. If we define additionally a vector consisting of nodal momenta $\boldsymbol{p}^d = [\boldsymbol{p}_1, \ldots, \boldsymbol{p}_{n_{\text{node}}}]^t := \mathbb{M}^d \cdot \boldsymbol{v}^d \in \mathbb{R}^{n_{\text{dof}}}$ and assume conservative external loads based on a potential V, a semidiscrete function

$$H^d(\boldsymbol{q}^d, \boldsymbol{p}^d; \boldsymbol{\kappa}) = \frac{1}{2}\boldsymbol{p}^d \cdot \mathbb{M}^{d-1} \cdot \boldsymbol{p}^d + \Psi(\boldsymbol{q}^d; \boldsymbol{\kappa}) + V(\boldsymbol{q}^d) \tag{6}$$

can be introduced analogously to the classical Hamiltonian for the hyperelastic case, wherein the free energy of the semidiscrete system $\Psi := \int_{\mathcal{B}_0^d} \psi \, \mathrm{d}V$ has been incorporated.

3 Rigid Bodies – Internal Constraints

In contrast to standard approaches in the multibody community, which usually rely on the classical Euler equations or on local coordinates for the special orthogonal group SO(3) respectively, in this work a rotationless formulation is preferred to describe the rigid body kinematics, as advocated for instance by Betsch and Steinmann [5] or Betsch and Uhlar [6]. For the following investigations, a representative rigid body $\mathcal{B}^r \subset \mathbb{R}^3$ with the mass density ρ^r is considered. Moreover, let $X = \sum_{\alpha=1}^{3} X^\alpha\, \boldsymbol{e}_\alpha \in \mathbb{R}^3$ denote the placement of a material point of the body in the corresponding reference configuration $\mathcal{B}_0^r$, whereby we assume for the sake of simplicity that the centre of mass of $\mathcal{B}_0^r$ coincides with the origin of the space-fixed orthonormal frame $\{e_\alpha\}$. Consequently, the current placement in the spatial configuration $\bar{\mathcal{D}}_t^r$ can be formulated by means of the mapping

$$q(X, t) = \varrho(t) + \sum_{\alpha=1}^{3} X^\alpha \, d_\alpha(t). \tag{7}$$

Herein, the vector $\varrho(t) : \mathbb{R}^+ \to \mathbb{R}^3$ refers to the position of the centre of mass and $X(t) := R(t) \cdot X = \sum_{\alpha=1}^{3} X^\alpha \, d_\alpha(t)$ with the rotation matrix $R \in SO(3)$ denotes the (relative) placement with respect to the director triad $\{d_\alpha(t)\}$. Hereby, the directors constitute a right-handed body-fixed frame which is assumed to be aligned with the principal axes of the rigid body. Hence, the current configuration $\mathcal{B}_t^r$ might be specified by the vector $q^r(t) = [\varrho(t) \quad d_1(t) \quad d_2(t) \quad d_3(t)]^t \in \mathbb{R}^{12}$, including a set of redundant coordinates. Additionally, six orthonormality conditions $d_\alpha \cdot d_\beta = \delta_{\alpha\beta}$, which reflect the requirement of rigidity, have to be taken into account, rendering the vector of so-called internal constraints

$$g^{\text{int}} := \begin{bmatrix} \frac{1}{2}[d_1 \cdot d_1 - 1] \\ \frac{1}{2}[d_2 \cdot d_2 - 1] \\ \frac{1}{2}[d_3 \cdot d_3 - 1] \\ d_1 \cdot d_2 \\ d_1 \cdot d_3 \\ d_2 \cdot d_3 \end{bmatrix}. \tag{8}$$

Next, the constant mass matrix of the rigid body $\mathbb{M}^r \in \mathbb{R}^{12} \times \mathbb{R}^{12}$ is introduced as

$$\mathbb{M}^r := \begin{bmatrix} M\boldsymbol{I} & \boldsymbol{0} & \boldsymbol{0} & \boldsymbol{0} \\ \boldsymbol{0} & E_1 \boldsymbol{I} & \boldsymbol{0} & \boldsymbol{0} \\ \boldsymbol{0} & \boldsymbol{0} & E_2 \boldsymbol{I} & \boldsymbol{0} \\ \boldsymbol{0} & \boldsymbol{0} & \boldsymbol{0} & E_3 \boldsymbol{I} \end{bmatrix}, \tag{9}$$

wherein we have incorporated the identity and zero matrices $\boldsymbol{I}, \boldsymbol{0} \in \mathbb{R}^3 \times \mathbb{R}^3$, the total mass M, as well as the principal values of the convected Euler tensor E_α which are given by

$$M = \int_{\mathcal{B}_0^r} \rho^r(X) \, \mathrm{d}V \quad \text{and} \quad E_\alpha = \int_{\mathcal{B}_0^r} X_\alpha^2 \, \rho^r(X) \, \mathrm{d}V \tag{10}$$

respectively. Once more, a vector of momenta $p^r(t) := \mathbb{M}^r \cdot \dot{q}^r(t) \in \mathbb{R}^{12}$ is defined, so that the classical Hamiltonian for one rigid body might be written as

$$H^r(q^r, p^r) := \frac{1}{2} p^r \cdot \mathbb{M}^{r-1} \cdot p^r + V(q^r). \tag{11}$$

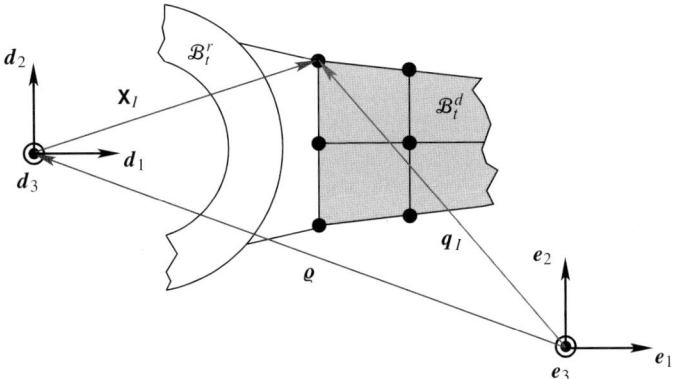

Fig. 1 Coupling between rigid and flexible parts.

4 Rigid vs. Flexible Parts – Coupling Constraints

So far, we have considered separately, on the one hand, the modelling of inelastic flexible structures based on a fully nonlinear continuum theory and, on the other hand, a rotationless formulation to model the dynamics of rigid bodies. Consequently, the next step is the investigation of dynamical systems which consist of flexible as well as of rigid parts, considering a partitioned solid body $\mathcal{B} = \mathcal{B}^r \cup \mathcal{B}^d$. As sketched in Figure 1, the actual coupling between the rigid body and a spatial node I of the deformable part can be formulated by incorporating the so-called (nodal) coupling constraints

$$g_I^{\text{cou}} := \varrho(t) + \sum_{\alpha=1}^{3} X_I^\alpha \, d_\alpha(t) - q_I(t). \tag{12}$$

For a system with n_c coupled FE nodes, we obtain straightforwardly the (global) vector of the coupling constraints $g^{\text{cou}} := [g_1^{\text{cou}}, \ldots, g_{n_c}^{\text{cou}}]^{\text{t}} \in \mathbb{R}^{3n_c}$. Regarding the system dynamics, we introduce the global (system) vectors $\bar{q}(t) := [q^r(t), q^d(t)]^{\text{t}}$ and $\bar{p}(t) := [p^r(t), p^d(t)]^{\text{t}}$. Abbreviating the notation by means of the vector $\bar{z}(t) := [\bar{q}(t), \bar{p}(t)]^{\text{t}} \in \mathbb{R}^{2n_{\text{sys}}}$ with $n_{\text{sys}} = 12 + n_{\text{dof}}$, the resulting function H for the coupled system reads

$$H(\bar{z}; \kappa) := H^d(q^d, p^d; \kappa) + H^r(q^r, p^r), \tag{13}$$

using the flexible and rigid components H^d and H^r given by (6), (11).

5 Constrained Dissipative Dynamics

Based on the foregoing investigations, we consider a finite-dimensional dynamical system consisting of flexible and rigid parts which is constrained by the general function

$$\bar{g}(\bar{q}(t)) \stackrel{!}{=} \mathbf{0} \tag{14}$$

with $\bar{g} := [g^{\text{int}}, g^{\text{cou}}]^{\text{t}}$. Taking the constraints (14) into account, we introduce an extended function

$$\mathsf{H}(\bar{z}, \bar{\mu}; \kappa) = H(\bar{z}; \kappa) + \bar{\mu} \cdot \bar{g}(\bar{q}), \tag{15}$$

wherein the vector of Lagrange multipliers $\bar{\mu} \in \mathbb{R}^m$ has been incorporated. In this context, m denotes the total number of constraints, depending on the number of coupled FE nodes as well as on the number of involved rigid bodies. To formulate the semidiscrete system of equations of motion in first-order format, a Hamiltonian-type notation is applied next that enables technically the same compact representation as in the purely elastic case due to the unchanged structure of the equations of motion.[1] In fact, the internal variables κ are not directly 'visible' on the global level, so that the resulting (semidiscrete) system of equations of motion can still be written as

$$\begin{aligned} \dot{\bar{z}}(t) &= \mathbb{J} \cdot \nabla_{\bar{z}} \mathsf{H}(\bar{z}(t), \bar{\mu}(t); \kappa(t)) \\ \mathbf{0} &= \bar{g}(\bar{q}(t)), \end{aligned} \tag{16}$$

using the skew-symmetric matrix

$$\mathbb{J} := \begin{bmatrix} \mathbf{0} & I \\ -I & \mathbf{0} \end{bmatrix}. \tag{17}$$

Hereby, the system (16) represents a set of differential algebraic equations – or short a set of 'DAEs' – which will be directly integrated in time instead of applying a reformulation.

Concerning the time discretisation, we extend the time-FE concepts for inelastic media by Mohr et al. [11, 12] to constrained dissipative systems, following the approach by Betsch and Steinmann [5]. Initially, we apply a decomposition of the considered time interval

---

[1] Once more, it is important to emphasise that the application of this Hamiltonian-type setting is indeed only used as a formalism of notation. Naturally, the conservative character of the system and, consequently, the classical Hamiltonian interpretation gets lost when dissipation effects are involved.

$$[0, T] = \bigcup_{n=0}^{N} [t_n, t_{n+1}] \tag{18}$$

and a map of each physical sub-interval $\mathcal{T} := [t_n, t_{n+1}]$ to the reference interval $\mathcal{I} := [0, 1]$ by means of the mapping $\alpha(t) := [t - t_n]/h_n$, involving the time-step size $h_n = t_{n+1} - t_n$. For the time approximation, a mixed Galerkin method – abbreviated by 'mG(k) method' – shall be applied. Therefore, we introduce next approximations in time for the trial and test functions of the phase-space variables

$$\bar{z}^h = \sum_{j=1}^{k+1} M_j(\alpha)\, \bar{z}_j, \qquad \delta \bar{z}^h = \sum_{i=1}^{k} \widetilde{M}_i(\alpha)\, \delta \bar{z}_i \tag{19}$$

as well as of the Lagrange multipliers

$$\bar{\mu}^h = \sum_{i=1}^{k} \widetilde{M}_i(\alpha)\, \bar{\mu}_i, \qquad \delta \bar{\mu}^h = \sum_{i=1}^{k} \widetilde{M}_i(\alpha)\, \delta \bar{\mu}_i . \tag{20}$$

Please note, that the time shape functions $M_j \in \mathcal{P}^k$ are polynomials of degree k, whereas the reduced shape functions $\widetilde{M}_i \in \mathcal{P}^{k-1}$ are only of degree $k - 1$. Obviously, the resulting approximations of the (unknown) trial functions of the Lagrange multipliers $\bar{\mu}^h \in \mathcal{P}^{k-1}$ belong to the same function space as the approximated test functions $\delta \bar{\mu}^h \in \mathcal{P}^{k-1}$. To the contrary, the corresponding approximations of the phase-space variables belong to different function spaces. In general, the resulting weak forms in time of the global governing equations (16) read

$$\int_0^1 \left[\mathbb{J} \cdot \delta \bar{z}^h \right] \cdot \left[D_\alpha \bar{z}^h - h_n\, \mathbb{J} \cdot \nabla_{\bar{z}} H(\bar{z}^h, \bar{\mu}^h; \kappa) \right] d\alpha = 0$$

$$\int_0^1 \delta \bar{\mu}^h \cdot \mathbb{G}(\bar{q}^h) \cdot D_\alpha \bar{z}^h\, d\alpha = 0, \tag{21}$$

wherein the constraint Jacobian $\mathbb{G}(\bar{z}^h) := \nabla_{\bar{z}}\, \bar{g}(\bar{q}^h)$ has been introduced and $D_\alpha[\bullet]$ denotes the derivative in time of $[\bullet]$ with respect to the reference-time parameter α. By inserting the time approximations (19), (20) in (21), we obtain for arbitrary polynomial degree k the system of equations

$$\sum_{j=1}^{k+1} \int_0^1 \widetilde{M}_i M_j'\, d\alpha\, \bar{z}_j - h_n\, \mathbb{J} \cdot \int_0^1 \widetilde{M}_i \left[\nabla_{\bar{z}} H(\bar{z}^h; \kappa) + \bar{\mu}^h \cdot \mathbb{G}(\bar{q}^h) \right] d\alpha = \mathbf{0}$$

$$\sum_{j=1}^{k+1} \int_0^1 \widetilde{M}_i M_j'\, \mathbb{G}(\bar{q}^h)\, d\alpha \cdot \bar{z}_j = \mathbf{0} \tag{22}$$

for $i = 1, \ldots, k$. Eventually, the remaining task is to evaluate the related time-integrals in (22). Hereby, the applied option directly influences the resulting conservation properties of the integrator, depending crucially on the specific format of the underlying function H and the involved constraints g.

With regard to *mechanical consistency*, in particular the conjunction of the mG(k) method with a k-stage Gauss Runge-Kutta method, being related to k Gaussian integration points in time, has turned out to be helpful, as discussed in detail in [5]. Moreover, *kinematic consistency* is inherent to the system of equations (22) when time-integrals which include the constraint Jacobian $\mathbb{G}$ are calculated exactly, recapturing the fundamental theorem of calculus for the rate $D_\alpha \bar{g}$. Actually, already a standard Gauss quadrature rule is sufficient in the case of the here considered constraints, being at most quadratic in $\bar{q}$. However, the claimed guarantee of *energy-consistency* requires in general the design of adequate non-standard quadrature rules, since the gradient $\nabla_{\bar{z}} H$ involves for the flexible parts so-called internal load vectors, resulting in highly nonlinear time-integrals. As proposed by Mohr et al. [12], consistent non-standard quadrature rules are obtained by introducing a modified algorithmic stress tensor

$$S^{alg} := S + 2 \frac{\psi_{\alpha=1} - \psi_{\alpha=0} + \int_0^1 \mathcal{D} \, d\alpha - \int_0^1 S : \frac{1}{2} D_\alpha C \, d\alpha}{\int_0^1 D_\alpha C : D_\alpha C \, d\alpha} D_\alpha C \quad (23)$$

which is given as the solution of a constrained optimisation problem, analogously to the 'eG(k) method' for elastodynamics by Gross et al. [9]. Summarising, the applied mG(k) method provides a general framework to design *completely consistent* time-stepping schemes for constrained inelastic dynamics.

6 Example – Elasto-Plastic Conrod

Finally, we present a representative numerical example, considering the spatial motion of a 'Flying Conrod'. Hereby, the conrod consists of two rigid rings connected via a flexible shaft, which has been discretised in space by means of 48 isoparametric 8-node elements. As a fundamental example, linear finite elements in time related to $k = 1$ – involving $M_1 = 1 - \alpha$, $M_2 = \alpha$, and $\widetilde{M}_1 = 1$ – have been applied for the global time-integration, using furthermore an appropriate non-standard quadrature rule based on (23).

Concerning the inelastic material behaviour, an elasto-plastic model has been chosen based on a multiplicative decomposition of the deformation gradient into elastic and plastic parts.[2] Moreover, the constitutive response relies on a Helmholtz energy density of 'Hencky type' combined with a 'v. Mises' yield function, incorporating linear isotropic hardening effects. The local (associated) evolution equations have been integrated exemplarily by means of a well-established exponential update. For further details, we refer to [16].

---

[2] This approach is indeed able to capture finite deformations including large strains, being not always the case in classical multibody formulations.

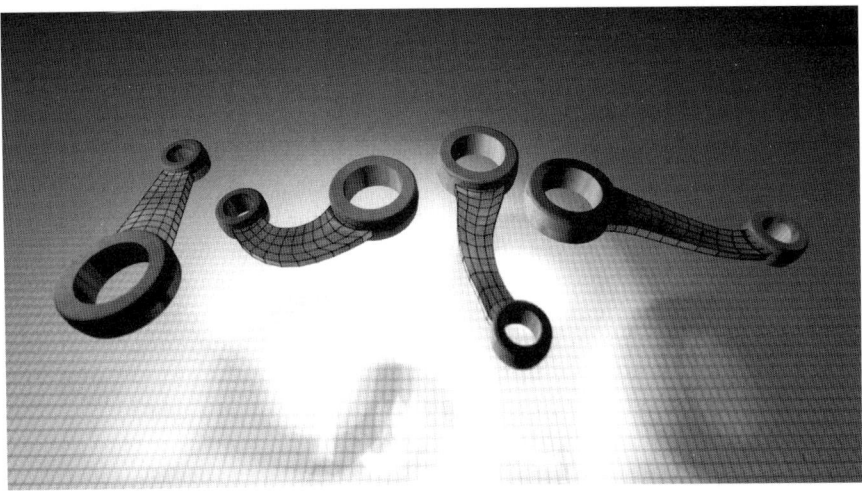

Fig. 2 Deformed configurations $\mathcal{B}_t$ for $t \in \{0, 2, 3.5, 5\}$.

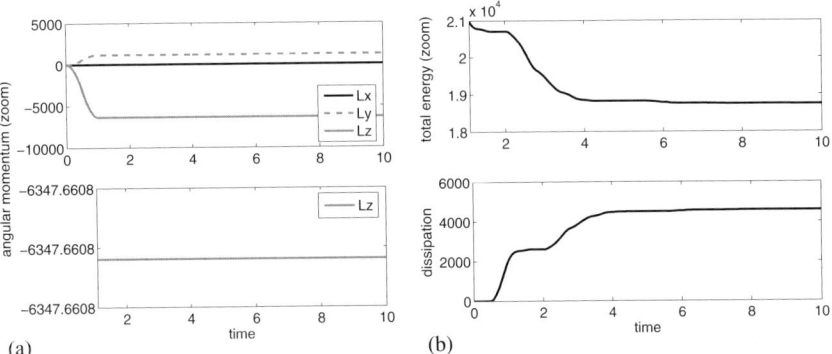

Fig. 3 (a) angular momentum; (b) total energy (zoom) & accumulated dissipation

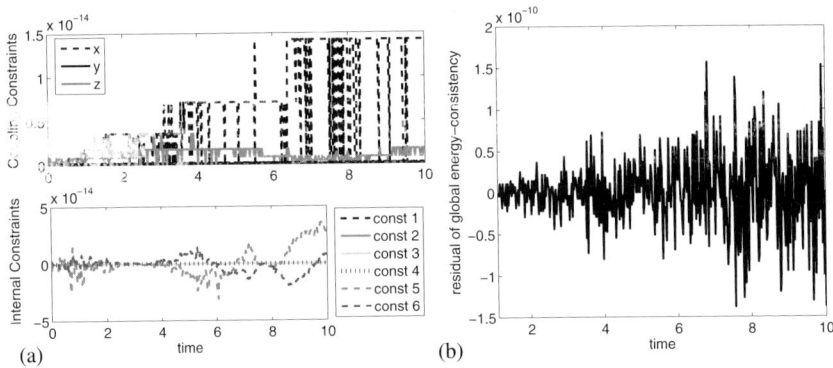

Fig. 4 Fulfilment of: (a) coupling constraints & internal constraints (rigid body 1); (b) residual of energy-consistency R.

To start the free flight, the control is equipped with an initial velocity. Furthermore, external loads act on both rigid bodies during a certain period, using a so-called 'hat function' in time. After unloading, the external loads vanish to expose the conservation properties of the time-stepping scheme. For the calculation, the elastic constants $\lambda = 3000$, $\mu = 1000$, the yield limit $Y_0 = 300$, the linear hardening modulus $H^{rd} = 300$, the mass densities $\rho^r = 3.0$, $\rho^d = 3.0$, and the time-step size $h_n = 0.02$ have been applied. A sequence of the motion can be regarded in Figure 2, and the conservation of angular momentum is pictured in Figure 3(a). Moreover, the total energy of the system decreases monotonically corresponding to a strictly non-negative plastic dissipation, see Figure 3(b). To demonstrate the rigorous energy-consistency of the integrator, we introduce the global dissipation increment ΔD and investigate the residual $R := H_{n+1} - H_n + \Delta D$, plotted in Figure 4(b). Obviously, energy-consistency is guaranteed within a numerically exact manner, as well as kinematic consistency that is pictured exemplarily for one of the rigid bodies in Figure 4(a). Summarising, we have outlined adequate concepts to model constrained inelastic systems in a completely consistent way, providing not only physically meaningful results but also particularly robust algorithms.

References

1. Armero F. (2006) Energy-dissipative momentum-conserving time-stepping algorithms for finite strain multiplicative plasticity. *Comput. Methods Appl. Mech. Engrg.* 195:4862–4889.
2. Armero F., Zambrana-Rojas C. (2007) Volume-preserving energy-momentum schemes for isochoric multiplicative plasticity. *Comput. Methods Appl. Mech. Engrg.* 196:4130–4159.
3. Armero F. (2007) Assumed strain finite element methods for conserving temporal integrations in non-linear solid dynamics. *Int. J. Numer. Meth. Engrg.*, doi: 10.1002/nme.2233.
4. Betsch P., Steinmann P. (2001) Conservation properties of a time FE method. Part II: Time-stepping schemes for non-linear elastodynamics. *Int. J. Numer. Meth. Engrg.* 50:1931–1955.
5. Betsch P., Steinmann P. (2002) Conservation properties of a time FE method. Part III: Mechanical systems with holonomic constraints. *Int. J. Numer. Meth. Engrg.* 53:2271–2304.
6. Betsch P., Uhlar S. (2007) Energy-momentum conserving integration of multibody dynamics. *Multibody System Dynamics* 17:243–289.
7. Biakeu G., Jezequel L. (2003) Simplified crash models using plastic hinges and the large curvature description. *Multibody System Dynamics* 9:25–37.
8. Gerstmayr J. (2004) The absolute coordinate formulation with elasto-plastic deformations. *Multibody System Dynamics* 12:363–383.
9. Gross M., Betsch P., Steinmann P. (2005) Conservation properties of a time FE method. Part IV: Higher order energy and momentum conserving schemes. *Int. J. Numer. Meth. Engrg.* 63:1849–1897.
10. Meng X.N., Laursen T.A. (2002) Energy consistent algorithms for dynamic finite deformation plasticity. *Comput. Methods Appl. Mech. Engrg.* 191:1639–1675.
11. Mohr R., Menzel A., Steinmann P. (2006) Galerkin-based time integrators for geometrically nonlinear elasto-plastodynamics – Challenges in modeling and visualization. In: Hagen H., Kerren A., Dannenmann P. (Eds), *Visualization of Large and Unstructured Data Sets*, GI-Edition Lecture Notes in Informatics (LNI) S-4, pp. 185–194.
12. Mohr R., Menzel A., Steinmann P. (2008) A consistent time FE-method for large strain elasto-plasto-dynamics. *Comput. Methods Appl. Mech. Engrg.*, doi: 10.1016/j.cma.2008.02.002.
13. Newmark N.M. (1959) A method of computation for structural dynamics. *ASCE J. Engrg. Mech. Div.* 85:67–94.

14. Noels L., Stainier L., Ponthot J.P. (2006) An energy momentum conserving algorithm using the variational formulation of visco-plastic updates. *Int. J. Numer. Meth. Engrg.* 65:904–942.
15. Noels L., Stainier L., Ponthot J.P. (2008) A first-order energy-dissipative momentum-conserving scheme for elasto-plasticity using the variational updates formulation. *Comput. Methods Appl. Mech. Engrg.* 197:706–726.
16. Simo J.C. (1998) Numerical analysis and simulation of plasticity. In: Ciarlet P.G., Lions J.L. (Eds), *Handbook of Numerical Analysis*, Vol. 6, pp. 183–499. Elsevier Science, Amsterdam.
17. Sugiyama H., Shabana A.A. (2004) Application of plasticity theory and absolute nodal coordinate formulation to flexible multibody system dynamics. *ASME J. Mech. Design* 126:478–487.

The Potential for SPH Modelling of Solid Deformation and Fracture

Paul W. Cleary and Rajarshi Das

*CSIRO Mathematical and Information Sciences, Private Bag 33,
Clayton South 3169, Australia
E-mail: paul.cleary@csiro.au*

Abstract. The advantages of using mesh free simulation methods such as SPH are demonstrated for elastic and elastoplastic deformation and for brittle fracture. The accuracy of SPH for deformation problems is demonstrated using a uniaxial test. An industrial example of forging is shown.

Key words: elastic solid, elastoplastic, fracture, damage, Smoothed Particle Hydrodynamics.

1 Introduction

Smoothed Particle Hydrodynamics (SPH) is a mesh-free or particle based Lagrangian method for solving systems of partial differential equations. Typically, it is used to solve Navier–Stokes equations for the flow of incompressible fluids (see [8, 2] for many examples) and for gas motions in astrophysical applications [8]. The earliest modelling of elastic solids using SPH was by Libersky and Petschek [6] and Wingate and Fisher [11]. Gray et al. [4] presented an extension of this early work including a method for overcoming the tensile instability that otherwise leads to numerical fracture. Over the past ten years the method has also been adopted for solving problems involving extreme solid deformation, such as the industrial forming processes of extrusion and forging [1].

Using only particles with no prescribed geometric linkages (such as in a mesh or a grid) allows high deformations to be dealt with easily in cases where finite element methods would either fail and/or require expensive and diffusive re-meshing. SPH is also able to predict complex free surface, including fragmentation without the need for any explicit surface tracking methods. SPH also has a natural ability to track material history. Fracture can also be predicted using a damage parameter for each particle, where the damage can evolve depending on the flow history [5, 9].

2 SPH Method for Solid Deformation

2.1 Elastic Solid Governing Equation

Following Monaghan [8] the interpolated value of a function A and its derivative at any position **r** can be expressed using SPH smoothing as:

$$A(\mathbf{r}) = \sum_b m_b \frac{A_b}{\rho_b} W(\mathbf{r} - \mathbf{r}_b, h)) \text{ and } \nabla A(\mathbf{r}) = \sum_b m_b \frac{A_b}{\rho_b} \nabla W(\mathbf{r} - \mathbf{r}_b, h) \quad (1)$$

where m_b and r_b are the mass and density of particle b and the sum is over all particles b within a radius $2h$ of **r**. Here $W(\mathbf{r},h)$ is a C^2 spline based interpolation kernel with radius $2h$, that approximates the shape of a Gaussian function but has compact support. Using these interpolation formulae and suitable finite difference approximations for second order derivatives, one can convert parabolic PDEs into ODEs for the motion of the particles and the rates of change of their properties.

The SPH continuity equation [8] is:

$$\frac{d\rho_a}{dt} = \sum_b m_b (\mathbf{v}_a - \mathbf{v}_b) \bullet \nabla W_{ab} \quad (2)$$

where ρ_a is the density of particle a with velocity $\mathbf{v}_a$ and m_b is the mass of particle b. $\mathbf{r}_{ab} = \mathbf{r}_a - \mathbf{r}_b$ is the position vector from particle b to particle a and $W_{ab} = W(\mathbf{r}_{ab}, h)$ is the interpolation kernel evaluated at distance $|\mathbf{r}_{ab}|$. This form is Galilean invariant and has good numerical conservation properties.

The governing or flow equation used for elastic, elastoplastic and brittle elastic deformation of the solids is:

$$\frac{d\mathbf{v}}{dt} = \frac{1}{\rho} \nabla \bullet \boldsymbol{\sigma} + \mathbf{g} \quad (3)$$

g denotes the body force and $\boldsymbol{\sigma}$ is the stress tensor which can be written as:

$$\boldsymbol{\sigma} = -P\mathbf{I} + \mathbf{S} \quad (4)$$

where P is the pressure, **S** is the deviatoric stress, and **I** is the identity matrix. We use a neo-Hookian elasticity model with a bulk modulus K and shear modulus G. The deviatoric stress **S** evolution is given by [5]:

$$\frac{dS^{ij}}{dt} = 2G\left(\dot{\varepsilon}^{ij} - \frac{1}{3}\delta^{ij}\dot{\varepsilon}^{kk}\right) + S^{ik}\Omega^{jk} + \Omega^{ik}S^{kj} \quad (5)$$

where the strain rate and Jaumann rotation tensors are:

$$\dot{\varepsilon}^{ij} = \frac{1}{2}\left(\frac{\partial v^i}{\partial x^j} + \frac{\partial v^j}{\partial x^i}\right) \text{ and } \Omega^{ij} = \frac{1}{2}\left(\frac{\partial v^i}{\partial x^j} - \frac{\partial v^j}{\partial x^i}\right) \quad (6)$$

SPH is a compressible method and uses the equation of state:

$$P = c^2(\rho - \rho_0) \tag{7}$$

where ρ_0 is the reference density and c is the speed of sound given by:

$$c^2 = K/\rho_0$$

2.2 Von Mises Plasticity with Linear Isotropic Hardening

The radial return plasticity model of Wilkins [10] is used to model elastoplastic deformation. A trial deviatoric stress S_{Tr}^{ij} is calculated assuming an initial elastic response. The increment of plastic strain is:

$$\Delta \varepsilon^p = \frac{\sigma_{vm} - \sigma_y}{3G + H} \tag{8}$$

where σ_{vm} is the von Mises stress and σ_y is the current yield stress. The plastic strain at the ith time step is then calculated as:

$$\varepsilon_i^p = \varepsilon_{i-1}^p + \Delta \varepsilon^p \tag{9}$$

The yield stress increment $\Delta \sigma_y$ at each time step is calculated as:

$$\Delta \sigma_y = H \Delta \varepsilon^p \tag{10}$$

where H is the hardening modulus. The deviatoric stress S^{ij} at the end of a time step is given by:

$$S^{ij} = r_s S_{Tr}^{ij} \tag{11}$$

where r_s is the radial scale factor given by:

$$r_s = \sigma_y / \sigma_{vm} \tag{12}$$

2.3 Damage Modelling

Brittle fracture of elastic solids plays an important role in many industrial processes such as in oil and gas recovery, mining, manufacturing processes such as casting, extrusion and forging, high speed impact, failure of mechanical components and in extreme geophysical events.

The Grady–Kipp model [3] is used for the prediction of rock damage (volume averaged cracking of rock) based on the local stress history and flaw distribution. Assuming a constant crack growth speed C_g, damage $D(t)$ evolution is given by the integral equation:

$$D(t) = \frac{4}{3} \pi C_g^3 \int_0^t n'(\varepsilon) \dot{\varepsilon} (1-D)(t-\tau)^3 d\tau \qquad (13)$$

where ε is the tensile strain, n' is the change in number of flaws, t is the current time, τ is the past times for history integration.

The approximate differential form (from [3]) is:

$$\frac{dD^{1/3}}{dt} = \frac{(m+3)}{3} \alpha^{1/3} \varepsilon^{m/3} \qquad (14)$$

α is a constant function of material parameters k, m, and Cg, given by:

$$\alpha = \frac{8\pi C_g^3 k}{(m+1)(m+2)(m+3)} \qquad (15)$$

To generalise the one-dimensional Grady–Kipp model to three dimensions Melosh et al. [7] introduced an effective tensile strain:

$$\varepsilon = \sigma_{\max} \bigg/ \left(K + \frac{4}{3}G\right) \qquad (16)$$

where $\sigma_{\max}$ is the maximum positive principal stress.

This evolution equation (14) is used to predict the damage state of the material represented by each SPH particle. The tensile component of the stress at each SPH particle is then scaled by its damage level. A material with no damage fully transmits tensile forces whereas a fully damaged piece of material cannot transmit any tensile force leading to a partial crack. Contiguous cracked material across a body leads to fragmentation.

3 Evaluation of Solution Accuracy for Uniaxial Compression

The accuracy and stability of SPH modelling of stress waves is examined using a uniaxial tester in which a rectangular rock specimen is vertically compressed between a fixed bottom plate and a piston descending from above at 1.5 mm/s. Figure 1 shows the SPH prediction of the von Mises stress during early wave propagation. Elastic waves initiate from the piston and propagate downwards and reflect from the rigid bottom plate. These reflected waves propagate back upwards and interfere with newly generated incident waves from the top. This creates a wave pattern by superposition of incident and reflected waves. The process continues leading to an increasingly complex stress pattern. Over time the spatial variations of stress decline leading to a uniform stress state with linearly increasing magnitude. Figure 2 compares the very early transient SPH response at the center of the specimen with a Finite Element (FE) solution obtained using an implicit dynamic analysis.

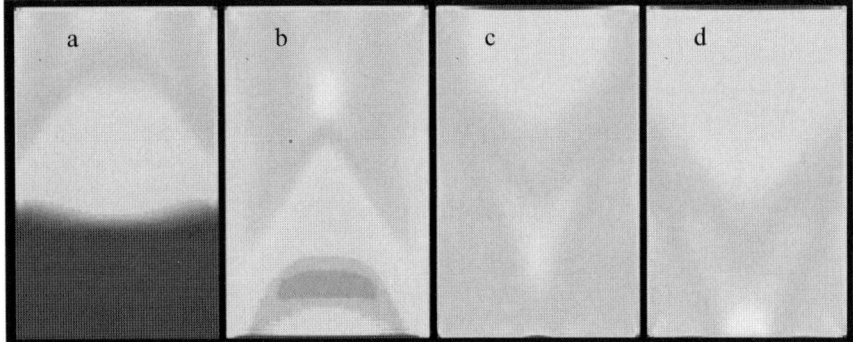

Fig. 1 Von Mises stress during early stress wave propagation for uniaxial compression; (a) 40 μs, (b) 60 μs, (d) 90 μs, and (d) 110 μs.

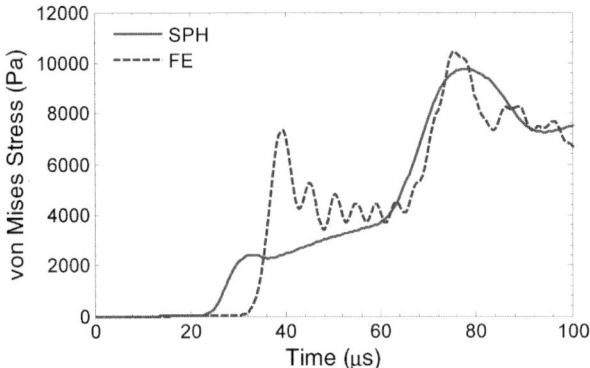

Fig. 2 Comparison of initial stress waves with FEM at the center of the specimen.

The specimen was meshed with quadratic isoparametric (plane stress) quadrilateral elements. This element type is commonly used for modelling 2D elasto-dynamic problems. The FE mesh was much finer (0.5 mm) than the SPH resolution (1 mm) so that the FE solution could be considered to be a reference solution. The two solutions show quite similar behaviour with the SPH solution being smoother and the FE solution having Gibbs oscillations after the initial stress rise. The initial stress rise is sharper for the FE. The small differences between the two solutions decline with time as the stress approaches a uniform state. The order of accuracy is found to be 1.3. Although the function approximations used in this SPH formulation are formally second order accurate, the boundaries degrade the solution somewhat reducing the order of accuracy. Overall, the SPH elastic stress solution has good accuracy, good stability properties and acceptable convergence for elastic solid deformations.

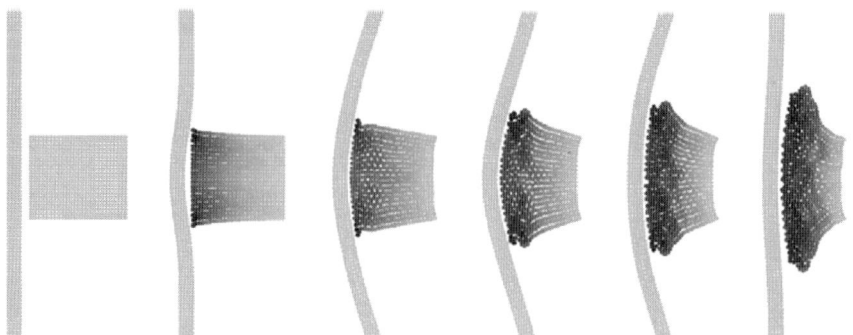

Fig. 3 Elastoplastic deformation of a rectangular projectile (moving to the left) impacting on a thin elastic wall.

4 Elastoplastic Collision

Figure 3 shows the elastoplastic deformation of a soft rectangular projectile (moving to the left) impacting on a thin elastic wall. Light grey shows material with no plastic strain and dark grey shows the maximum current plastic strain. Upon initial contact with the wall, there is some initial plastic deformation at the leading edge of the projectile with peak deformation (of 23%) occurring at its corners. As the projectile pushes against the wall and the wall bows to the left, the plastic failure extends into the projectile in a triangular zone from the corners. The peak strain is now 60%. In the fourth frame the wall has reached its maximum extension and has started moving back to the right. The wall and the projectile are now moving in opposite directions and the stresses in the projectile rise sharply. Significant plastic deformation now occurs at the front of the projectile with significant transverse flow. The peak strain is now 90%. The returning wall accelerates to the right and the rate of deformation increases leading to a significant flattening of the projectile and a near doubling of its width with peak strains reaching 155% at the time of separation from the wall.

The impact of an elastic projectile on a thin elastoplastic wall was simulated. Upon contact the wall bows to the left. Once the von Mises stress exceeds the yield stress plastic deformation starts to occur. This continues until all the kinetic energy of the projectile is absorbed by the wall. The projectile is then pushed back to the right by the elastic unloading of the wall. Figure 4 shows the final shape of the wall for four different yield stresses. For the strongest material (left) the plastic strain is concentrated near the attachment points of the wall and produces ductile necking. For a softer material, necking also occurs at the corners of the projectile where stress is concentrated. As the material is made softer, the plastic strain needed to stop the projectile increases and the thinned sections of the wall lengthen. In the final case the entire wall has been stretched to more than double its length with peak plastic strains of 150%. The SPH method shows remarkable stability for these very high deformation problems.

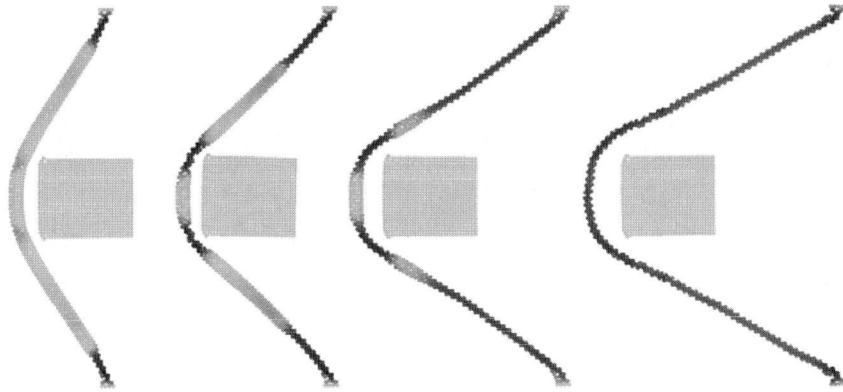

Fig. 4 Elastoplastic deformation of a thin elastic wall when impacted by a rectangular projectile (moving from the right). Yield stress increases from left to right.

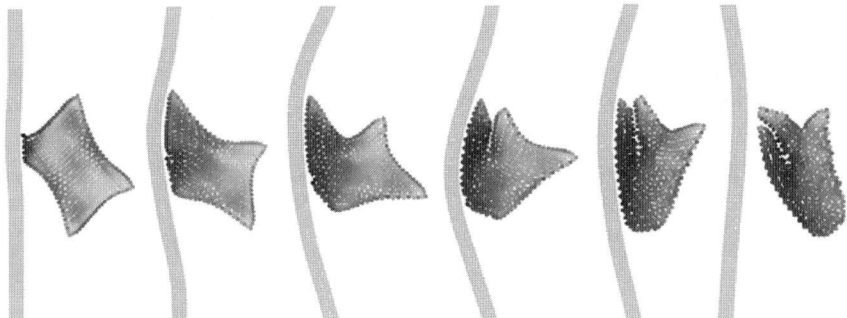

Fig. 5 Extreme plastic deformation of a high speed rapidly spinning projectile.

Figure 5 shows the collision of a very soft elastoplastic projectile, with an initial speed of 100 m/s and spin of 1000 rad/s, with an elastic wall. The deformations observed are extreme (225%) with folding of the body back onto itself. Such self collision is automatically resolved by SPH.

5 Forging

Forging is a popular process for manufacturing complex shaped industrial components. Here we demonstrate the ability of SPH to simulate the plastic deformation of a metal work piece driven by the motion of two moving pistons and two fixed pistons. A cylindrical blank is represented by 32000 particles which are coloured in vertical strata to allow the deformation to be tracked. Figure 6 shows the progress of the forging. At 1 s, three bulges are observed in the work piece. By 2 s, material

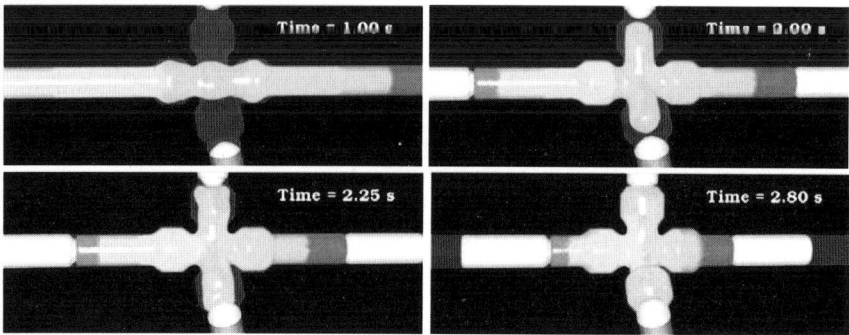

Fig. 6 Multi-die forging of an industrial component.

begins to fill the two vertical arms of the die. The wider diameter sections of the horizontal arms are now almost filled. At 2.25 s, the material motion in the vertical arms is halted and the metal is forced sideways towards the side walls and the die is almost full. Forging is completed at 2.8 s.

6 Brittle Fracture during Impact and Compression

Figure 7 shows the normal impact of a 100 mm circular rock on a rigid surface. Undamaged material is shaded light grey and fully damaged material is near black. The rock is sandstone with bulk modulus 12.2 GPa, shear modulus 2.67 GPa and density 2300 kg/m^3. The Weibull damage parameters k and m were 1.0×10^{22} and 9 respectively (taken from [7]). The impact speed was 75 m/s. The model used 2,029 particles. As the body compresses vertically during impact, high tensile stresses are generated in its middle. These lead to damage in four branches radiating from the center. Two are close together and directed up and the other two are pointing down. Damage continues leading to fracture along these paths. Two primary fragments are created on the sides with a small fragment at the top and a cloud of heavily fractured fine material below.

Figure 8 shows the brittle fracture of a square rock moving at 100 m/s on a rigid plate. The material and damage properties were same as those of the circular rock. The square rock has sides of 100 mm and the rock consisted of 2,514 particles. Damage was initiated from the vertex that collides with the surface, which is subjected to a very high stress due to the sharp geometry. High tensile stresses are also generated in the middle of the rock. Damage grows in both locations and becomes connected leading to primary fracture from the contact point through the center and up to the opposite corner. Secondary fragmentation is then observed including some delamination along the top and bottom surfaces.

The Potential for SPH Modelling of Solid Deformation and Fracture 295

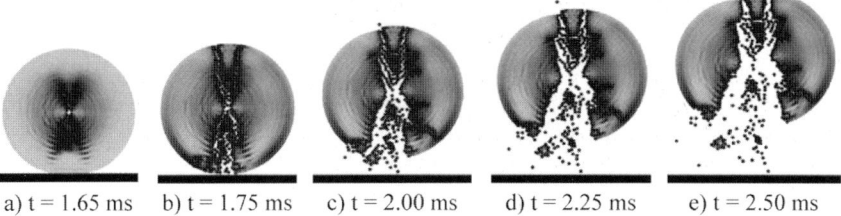

a) t = 1.65 ms   b) t = 1.75 ms   c) t = 2.00 ms   d) t = 2.25 ms   e) t = 2.50 ms

Fig. 7 Brittle fracture of a circular rock impacting on a rigid surface.

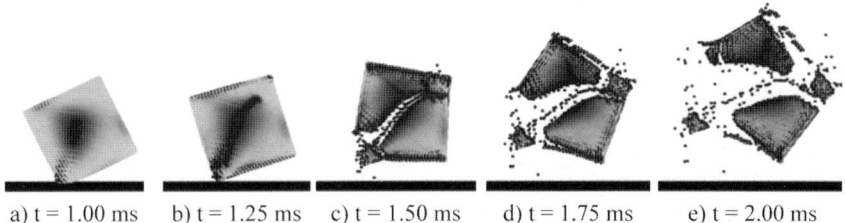

a) t = 1.00 ms   b) t = 1.25 ms   c) t = 1.50 ms   d) t = 1.75 ms   e) t = 2.00 ms

Fig. 8 Brittle fracture of a square rock impacting on a rigid surface.

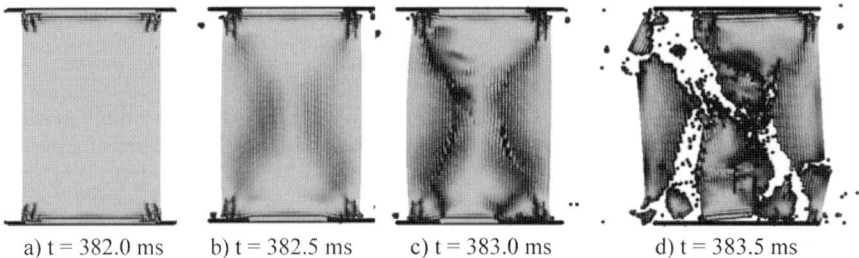

a) t = 382.0 ms   b) t = 382.5 ms   c) t = 383.0 ms   d) t = 383.5 ms

Fig. 9 Brittle fracture during uniaxial compression.

Figure 9 shows the brittle fracture of a uniaxially loaded rectangular rock specimen. This is the same test configuration with the same material properties as in the previous uniaxial test. The specimen domain was discretised with 12,040 particles of size 1 mm. Uniaxial compression generates lateral tension. Early damage occurs starting from the four corners where the stresses are very high. Damage propagates slowly downwards and slightly inwards creating multiple cracks. Just after 0.382 s, the growth of these cracks abruptly accelerates and they propagate diagonally towards the middle of the sample leading to catastrophic failure.

7 Conclusions

SPH is able to solve elastic deformation problems with accuracy comparable to finite elements and demonstrates excellent stability and adequate convergence. The method is highly suited to elastoplastic deformation problems resulting from compression or impact. It has powerful abilities to model material free surfaces, extreme deformation including self-collision and automatically carries stress and strain history and material dependent information. Brittle fracture of rocks during impact and compression has been demonstrated and with good predictions for fragmentation patterns.

References

1. Cleary, P.W., Prakash, M., and Ha, J., 2006. Novel applications of smoothed particle hydrodynamics (SPH) in metal forming. *J. Mat. Proc. Tech.* **177**: 41–48.
2. Cleary, P.W., Prakash, M., Ha, J., Stokes, A.N., and Scott, S., 2007. Smooth particle hydrodynamics: Status and future potential. *Progr. Comput. Fluid Dynam.* **7**: 70–90.
3. Grady, D.E. and Kipp, M.E., 1980. Continuum modelling of explosive fracture in oil shale. *Int. J. Rock Mech. and Mining Sci. & Geomech. Abst.* **17**: 147–157.
4. Gray, J.P., Monaghan J.J., and Swift, R.P., 2001. SPH elastic dynamics. *Comput. Meth. Appl. Mech. Engrg.* **190**: 6641–6662.
5. Gray, J.P. and Monaghan, J.J., 2004. Numerical modelling of stress fields and fracture around magma chambers. *J. Vulcanology Geothermal Res.* **135**: 259–283.
6. Libersky, L.D. and Petschek, A.G., 1990. Smooth particle hydrodynamics with strength of materials. In *Advances in the Free-Lagrange Method*, Trease and Crowley (Eds.), Springer, Berlin.
7. Melosh, H.J., Ryan, E.V., and Asphaug, E., 1992. Dynamic fragmentation in impacts: Hydrocode simulation of laboratory impacts. *J. Geophys. Res.* **97**: 14735–14759.
8. Monaghan, J.J., 2005 Smoothed particle hydrodynamics. *Rep. Prog. Phys.* **68**: 1703–1759.
9. Randles, P.W., Carney, T.C., Libersky, L.D., Renick, J.D., and Petschek, A.G., 1995. Calculation of oblique impact and fracture of tungsten cubes using smoothed particle hydrodynamics. *Int. J. Impact Engrg.* **17**: 661–672.
10. Wilkins, J.L., 1964. Calculation of elastic-plastic flow, *Meth. Comput. Phys.* **8**: 211–263.
11. Wingate, C.A. and Fisher, H.N., 1993. Strength modeling in SPHC. Los Alamos National Laboratory, Report No. LA-UR-93-3942.

Effect of Material Parameters in the Izod Test for Polymers

Viggo Tvergaard[1] and Alan Needleman[2]

[1]Department of Mechanical Engineering, Solid Mechanics,
Technical University of Denmark, DK-2800 Kgs. Lyngby, Denmark
E-mail: viggo@mek.dtu.dk
[2]Division of Engineering, Brown University, 182 Hope Street,
Providence, RI 02912, USA
E-mail: needle@brown.edu

Abstract. Numerical analyses of the Izod impact test for polymers are used to compare the effect of different sets of material parameters. Full 3D finite strain transient analyses are carried out using a nonlinear polymer constitutive relation. The predicted differences in the overall load-deflection response and in the evolution of the field quantities in the notch region are discussed.

Key words: Izod test, viscoplasticity, polymer, computer simulation.

1 Introduction

The evolution of stress- and strain fields in Izod specimens is analyzed. The constitutive relation used is mainly adopted from Wu and Van der Giessen [1] with a set of reference material parameters chosen so that the model qualitatively represents the polycarbonate response in Mulliken and Boyce [2]. Other materials are considered by varying single material parameters. For the reference material the effect of various specimen widths has previously been analyzed by Tvergaard and Needleman [3]. The analyses are continued up to very large strains, well beyond the stage where network locking initiates at the notch-tip, so at the end of the computations a region of very high network stiffness has developed around the notch. The analyses are carried out for an intermediate specimen width in the range of widths recommended in ASTM D256, and the effect of varying values of material parameters on the deformation response is studied, both for the force displacement curves and for the stress distributions that develop in the specimens. Material failure is not modeled in the analyses here.

2 Problem Formulation

The formulation follows that in [3] where further details and additional references are given.

B.D. Reddy (ed.), IUTAM Symposium on Theoretical, Modelling and Computational Aspects of Inelastic Media, 297–306.
© Springer Science+Business Media B.V. 2008

2.1 Initial/Boundary Value Problem

A convected coordinate Lagrangian formulation is used and full transient calculations are carried out. The finite element calculations are based on the dynamic principle of virtual work written as

$$\int_V \tau^{ij} \delta E_{ij} \, dV = \int_S T^i \delta u_i \, dS - \int_V \rho \frac{\partial^2 u^i}{\partial t^2} \delta u_i \, dV \tag{1}$$

with

$$T^i = (\tau^{ij} + \tau^{kj} u^i_{,k}) v_j \tag{2}$$

$$E_{ij} = \frac{1}{2}(u_{i,j} + u_{j,i} + u^k_{,i} u_{k,j}) \tag{3}$$

where τ^{ij} are the contravariant components of Kirchhoff stress on the deformed convected coordinate net ($\tau^{ij} = J\sigma^{ij}$, with σ^{ij} being the contravariant components of the Cauchy or true stress and J the ratio of current to reference volume), v_j and u_j are the covariant components of the reference surface normal and displacement vectors, respectively, ρ is the mass density, V and S are the volume and surface of the body in the reference configuration, and $(\;)_{,i}$ denotes covariant differentiation in the reference frame. All field quantities are considered to be functions of convected coordinates, y^i, and time, t.

The Izod specimen analyzed is sketched in Figure 1 and has overall length $L = 63.5$ mm and depth $b = 12.7$ mm (the x, y and z directions correspond to the y^1, y^2 and y^3 directions respectively). The origin of the coordinate system is taken so that $-L/2 \le y^3 \le L/2$, $0 \le y^1 \le b$ and $-w/2 \le y^2 \le w/2$. The notch depth is 2.54 mm, the notch radius is 0.25 mm and the notch angle is 22.5°. Here, the specimens are taken to have the width w equal to $b/2$. Symmetry about the midplane is assumed so that half the specimen is analyzed ($0 \le y^2 \le w/2$), with symmetry conditions imposed on $y^2 = 0$ and traction free conditions on $y^2 = w/2$.

The loading is applied by a prescribed velocity, $V(t)$, along a portion of the notched edge $y^1 = 0$ with

$$\dot{u}_1(0, y^2, y^3) = V(t) \quad \text{on } y^3 = C \tag{4}$$

where $C = 22$ mm and

$$V(t) = \begin{cases} V_1 t/t_r & \text{for } t < t_r \\ V_1 & \text{for } t > t_r \end{cases} \tag{5}$$

with $V_1 = 3.5$ m/s and $t_r = 0.3$ ms.

The lower half of the specimen is held in grips so that

$$u_i(0, y^2, y^3) = 0 \quad u_i(b, y^2, y^3) = 0 \quad \text{for } y^3 < 0 \tag{6}$$

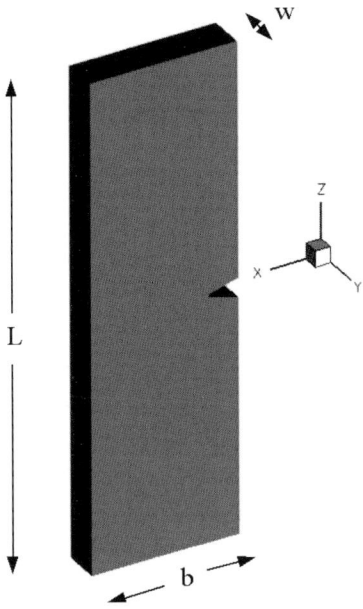

Fig. 1 Sketch of the Izod specimen.

At sufficiently large deformations the back side of the specimen (the side along $y^1 = b$) can come into contact with the top of the grip, which is taken to be planar at $y_3 = 0$. When contact occurs, frictionless sliding is assumed, as specified by the requirement

$$u_3(b, y_2, y_3) + y_3 \geq 0 \quad \text{for } y_3 > 0 \tag{7}$$

Twenty node brick elements are used with eight point integration for the force term, the left hand side of equation (1), and twenty-seven point integration for the mass matrix (which gave more accurate elastic wave speeds). The same mesh is used in all calculations.

2.2 Constitutive Relation

The constitutive formulation is taken from [1, 2, 4–6].

The rate of deformation tensor is written as

$$\boldsymbol{D} = \boldsymbol{D}^e + \boldsymbol{D}^p = \boldsymbol{L}^{-1} \cdot \hat{\boldsymbol{\sigma}} + \boldsymbol{D}^p = \boldsymbol{L}^{-1} \cdot \hat{\boldsymbol{\sigma}} + \dot{\gamma}^p \boldsymbol{p} \tag{8}$$

where $\boldsymbol{L}$ is the isotropic tensor of moduli with elastic constants Young's modulus E and Poisson's ratio ν and

$$\dot{\gamma}^p = \dot{\gamma}_0 \exp\left[-\frac{\Delta G}{kT}\left\{1 - \left(\frac{\tau}{s - \alpha(\mathrm{tr}\sigma)/3}\right)^m\right\}\right] - \dot{\gamma}_0 \exp\left[-\frac{\Delta G}{kT}\right] \quad (9)$$

Here, m, $\dot{\gamma}_0$, ΔG and α are material constants, T is the temperature (Kelvin) and k is Boltzmann's constant $k = 1.38 \times 10^{-23}$ J/K. The last term in equation (9) is included so that $\dot{\gamma}^p = 0$ when $\tau = 0$. Also,

$$p = \frac{\sigma' - b'}{\sqrt{2\tau}} \qquad \tau = \sqrt{\frac{1}{2}(\sigma' - b') : (\sigma' - b')} \quad (10)$$

with ( )′ denoting deviatoric quantities.

The hardness s in equation (9) is taken to have the initial value s_0 and to evolve as

$$\dot{s} = h\left(1 - \frac{s}{s_{ss}}\right)\dot{\gamma}^p \quad (11)$$

with h and s_{ss} material constants.

For the eight chain model, [4],

$$b = \frac{1}{3}C_R\sqrt{N}\frac{\beta_c}{\lambda_c}B \quad (12)$$

where $B = F \cdot F^T$, (F is the deformation gradient), C_R and $\sqrt{N}$ are specified material constants and

$$\lambda_c^2 = \frac{1}{3}\mathrm{tr}B \qquad \beta_c = \mathcal{L}^{-1}\left(\frac{\lambda_c}{\sqrt{N}}\right) \quad (13)$$

with

$$\mathcal{L}(x) = \coth(x) - \frac{1}{x} \quad (14)$$

The value of $\sqrt{N}$ defines a limit stretch such that when λ_c approaches $\sqrt{N}$ the hardening rate grows very large, modeling increased network stiffness. The parameter C_R gives the amplitude of this stiffening effect so that for $C_R = 0$ there is no network stiffening.

3 Numerical Results

The previous calculations of Tvergaard and Needleman [3] made use of material parameters that qualitatively represent the polycarbonate response in Mulliken and Boyce [2]. The elastic constants were taken as $E = 1.815$ GPa and $\nu = 0.38$, and the density ρ was taken to be 1300 kg/m^3. The values of the plastic constitutive parameters were $C_R = 12.6$ MPa, $N = 2.3$, $h = 50$ MPa, $s_{ss} = 22.71$ MPa and the initial value of s was 81.65 MPa. Furthermore, $\Delta G = 3.744 \times 10^{-19}$ J, $\alpha = 0.08$, $m = 0.2$, $\dot{\gamma}_0 = 8940$ s^{-1}, and the temperature T fixed at 298 K so

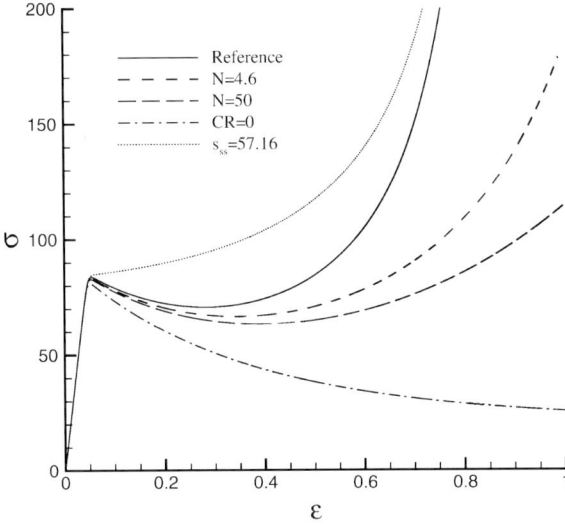

Fig. 2 Uniaxial stress-strain curves at strain rate 1/s in tension.

that $\Delta G/(kT) = 91.08$. The material specified by these parameters is denoted the reference material in the following, and the corresponding stress-strain curve for slow tension (strain rate 1/s) is included in Figure 2. Here, σ is the absolute value of the Cauchy (or true) stress and ϵ is the absolute value of the logarithmic strain. Stress-strain curves for both tension and compression at two different strain rates are shown in [3] for this reference material.

To study the effect of the material parameters five different materials are considered here, with the five stress-strain curves shown in Figure 2. In each material only one parameter is varied relative to the reference material. Thus, for the top curve in Figure 2 the steady-state value of the hardness has been increased to $0.7 s_0$, or $s_{ss} = 57.16$ MPa. The curve below the reference curve has the value of N multiplied by two, i.e. $N = 4.6$ and the curve below that has a much higher value, $N = 50$. Finally, the bottom curve in Figure 1 has $C_R = 0$ MPa.

In [3] Izod specimens of different width were analyzed, ranging from a square cross-section specimen with $w = 12.7$ mm over the half width specimen, $w = 6.35$ mm, to the thinnest specimen, $w = 3.0$ mm, usually applied. The present studies for different materials are carried out for the half width specimen, $w = 6.35$ mm. Figure 2 shows the corresponding normalized force displacement curves for the Izod specimens corresponding to the five different materials illustrated in Figure 1. The top curve is that for $s_{ss} = 57.16$ MPa, next comes the reference curve, then the curve for $N = 4.6$. The second lowest curve is that for $N = 50$, which ends rather early, at $U/b = 0.519$, and finally the curve for $C_R = 0$ MPa, which can hardly be distinguished from the other curves, since it ends early at $U/b = 0.205$. The reference curve and the curve for $N = 4.6$ show a force peak, before the force starts

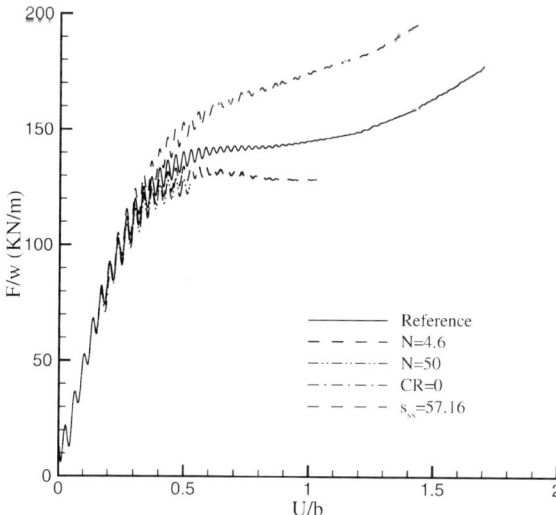

Fig. 3 Normalized force displacement curves for Izod specimens made of different materials.

to increase again, but this is not seen on the top curve, and the two lower curves have not reached that stage.

Figure 3 shows the distribution of plastic strain on the deformed specimen geometry for the calculation with $N = 4.6$ at $U/b = 1.015$, i.e. at the end of the curve shown in Figure 2. The deformation mode of the entire Izod specimen is shown, with the largest strains at the notch and also high strains opposite to the notch where bending strains are concentrated due to the clamping of the lower part of the specimen. In Figure 3, and in the subsequent figures the plots on the left show the specimen surface ($y^2 = w/2$), while the plots on the right show the symmetry plane ($y^2 = 0$). A band of deformation connecting these two regions of plastic strain concentration is seen at $y^2 = w/2$ (Figure 3a) but is less pronounced at the specimen midsection $y^2 = 0$ (Figure 3b). Somewhat elevated strains are also seen near the top of the specimen, where the pendulum has hit.

Figures 4 and 5 focus on the field distributions in the notch vicinity for the calculation with $N = 4.6$ at $U/b = 1.015$. The plastic strain concentrations from Figure 3 are more clearly illustrated in Figures 4a and 4b. Figures 4c and 4d show that the mean normal stress, $\sigma_m = \text{tr}(\sigma/3)$, is largest at the notch root and the region of large positive σ_m is directly ahead of the notch, rather than in the band where the plastic strain γ^p is concentrated. It is noted that in metal plasticity the maximum σ_m would occur some distance inside the material directly ahead of the notch. This difference is a consequence of the very high hardening response of the polymer at large strains. The distribution of the effective stress quantity τ, defined in equation (10), at $U/b = 1.015$ is shown in Figures 5a and 5b. The largest values of this effective shear stress occur in the deformation band above the notch, but not at the notch surface where the plastic strain is greatest. The hardness s (see equa-

Effect of Material Parameters in the Izod Test for Polymers 303

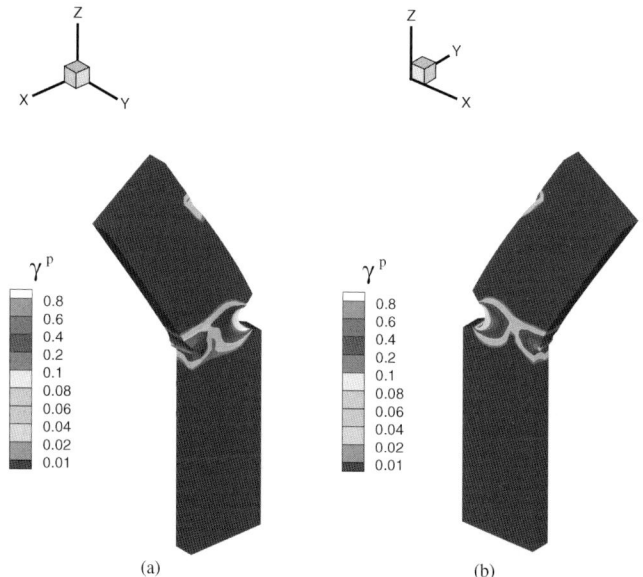

Fig. 4 The distribution of γ^p in the full specimen for the calculation with $N = 4.6$ ($U/b = 1.015$).

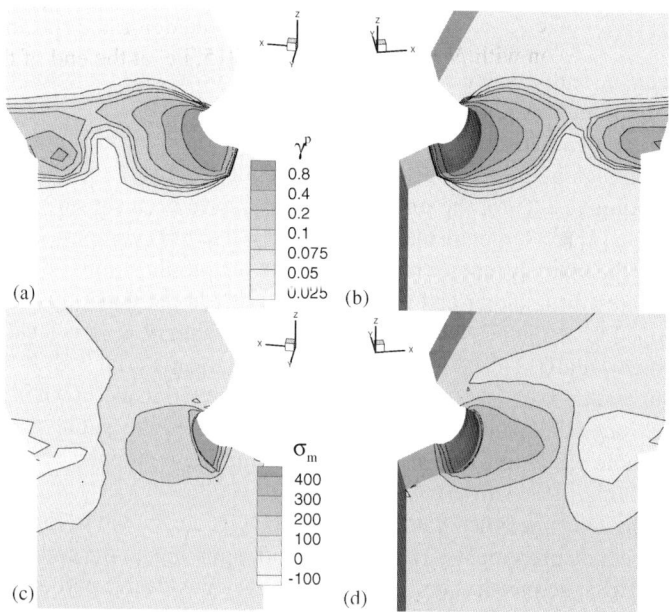

Fig. 5 The distribution of field quantities in the notch vicinity for the calculation with $N = 4.6$ at $U = 12.89$ mm (U/b=1.015). (a) and (b) γ^p. (c) and (d) σ_m.

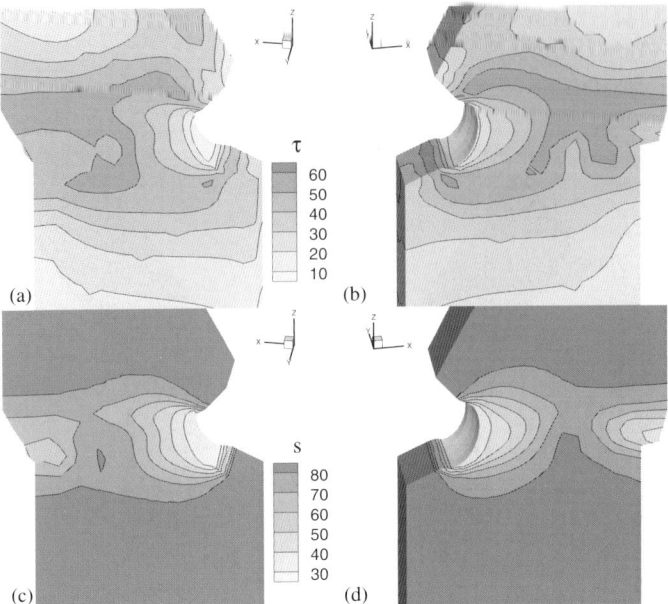

Fig. 6 The distribution of field quantities in the notch vicinity for the calculation with $N = 4.6$ at $U = 12.89$ mm (U/b=1.015). (a) and (b) τ. (c) and (d) s

tion (11)) has the initial value $s_0 = 81.65$ MPa and decreases monotonically with plastic deformation towards the steady state value 22.71 MPa. The smallest value of s occurs on the notch surface which is where the plastic strain is greatest. Away from the notch region, the state variable s has its initial value.

Corresponding results for the reference material have been shown by Tvergaard and Needleman [3] at a larger displacement, $U/b = 1.20$. By comparison of the two sets of results the main difference observed is that the plastic strains at the notch tip in Figure 4b are a little larger and that the values of s at the notch tip in Figure 5d have developed to a smaller value, even though the value of U/b is smaller in the case of Figures 4 and 5. This agrees with the fact that the material with $N = 4.6$ has a larger limit stretch $\lambda_{max} = \sqrt{N}$ than the reference material, as shown in Figure 1, and therefore once the material at the notch tip has passed the initial stress peak it will deform more easily to a larger strain before straining is stopped by the effect of network stiffening.

Plots for the Izod specimen with $s_{ss} = 57.16$ MPa are shown in Figures 6 to 8, now at the larger displacement $U/b = 1.454$, corresponding to the end of the curve in Figure 2. The specimen has been bent so far that an extensive part of the surface at the back side of the bending region has come in contact with the grip. Figures 7a and 7b show distributions of γ^p similar to those found before, but the maximum strain values at the notch tip are smaller than found in Figure 4 or for the reference material, even though U/b is larger. This is an effect of the higher value of s_{ss} which results in earlier hardening as seen in Figure 1. Corresponding to this, the value of

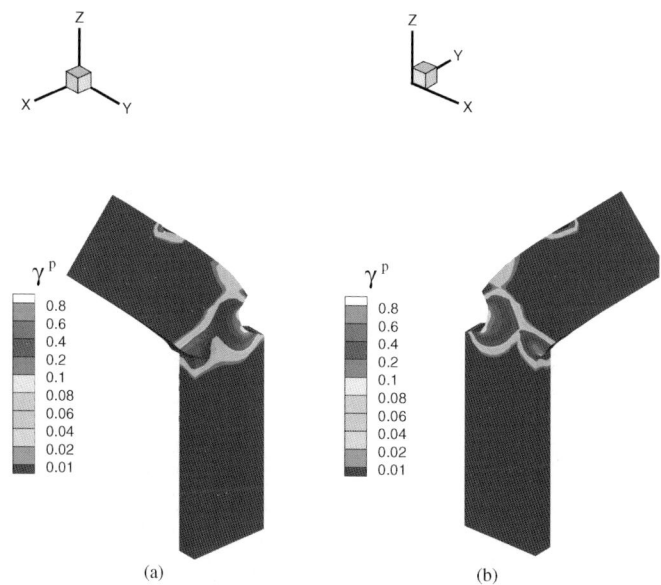

Fig. 7 The distribution of γ^p in the full specimen for the calculation with $s_{ss} = 57.16$ MPa at $U = 18.47$ mm ($U/b = 1.454$).

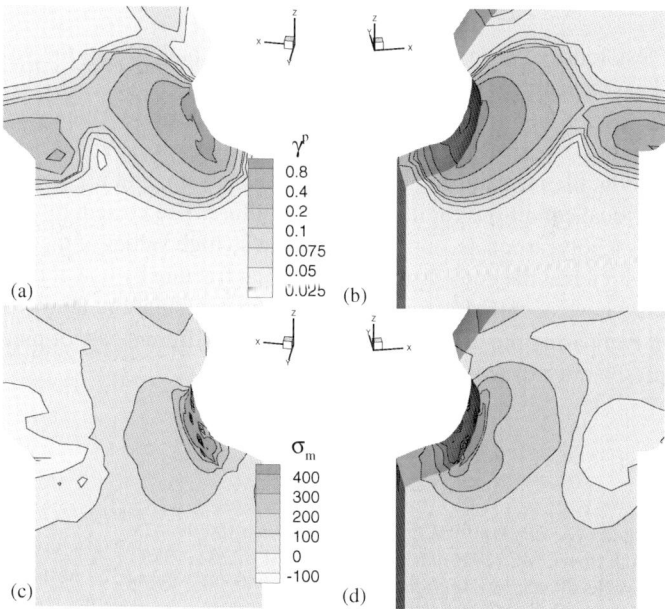

Fig. 8 The distribution of field quantities in the notch vicinity for the calculation with $s_{ss} = 57.16$ MPa at $U = 18.47$ mm ($U/b = 1.454$). (a) and (b) γ^p. (c) and (d) σ_m.

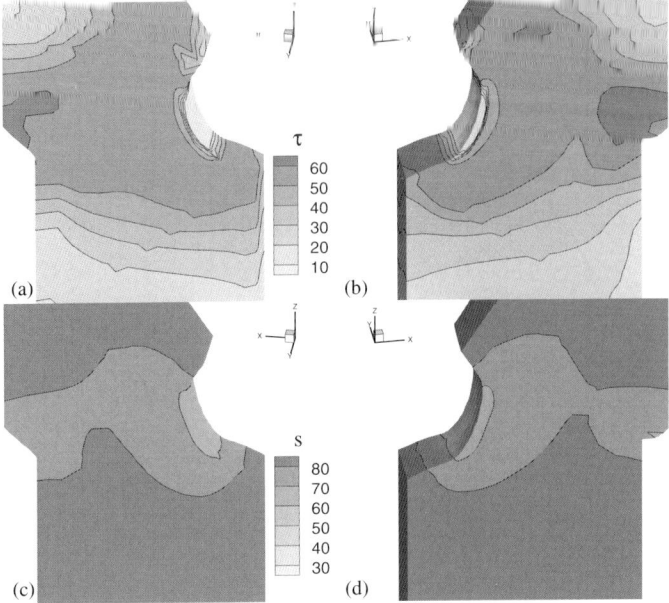

Fig. 9 The distribution of field quantities in the notch vicinity for the calculation with $s_{ss} = 57.16$ MPa at $U = 18.47$ mm ($U/b = 1.454$). (a) and (b) τ. (c) and (d) s.

s is much less reduced at the notch tip in Figures 8c and 8d and the values of the effective shear stress τ are larger in this region.

Material failure has not been accounted for in the present analyses. Fracture mechanisms for polycarbonate have been studied by Gearing and Anand [7], who considered a ductile mechanism and a brittle mechanism, and used experiments to determine critical conditions for these types of failure. The present reference material represents a polycarbonate, and it appears that so high values of the mean normal stress or the stretch at the notch tip are reached that fracture by one of the two mechanisms would have occurred before the end of the computation. However, here the focus is on comparing the deformation behavior for different sets of material parameters representative of polymers.

References

1. Wu PD, Van der Giessen E (1996) *Eur J Mech A/Solids* 15:799–823.
2. Mulliken AD, Boyce MC (2006) *Int J Solids Struct* 43: 1331–1356.
3. Tvergaard V, Needleman A (2008) *Int J Solids Struct*, 45:3951–3966.
4. Arruda EM, Boyce MC (1993) *Int J Plast* 9: 697–720.
5. Boyce MC, Parks DM, Argon AS (1988) *Mech Mater* 19:193–212.
6. Boyce MC, Arruda EM (1990) *Polym Engng Sci* 30:1288–1298.
7. Gearing BP, Anand L (2004) *Int J Solids Struct* 41:827–845.

Experimental and Computational Aspects

The Response of "Large" Square Tubes (Width/Thickness Ratio > 45) to Opposite Lateral Blast Loads Followed by Dynamic Axial Load

S. Chung Kim Yuen and G.N. Nurick

Blast Impact and Survivability Research Unit (BISRU), Department of Mechanical Engineering, University of Cape Town, Private Bag, Rondebosch 7701, South Africa
E-mail: steeve.chungkimyuen@uct.ac.za

Abstract. Experiments and Finite Element analyses are carried out to investigate the response of a square tube with width/thickness ratio (C/H) > 45 to two opposite lateral loads followed by dynamic axial load. The localised blast loads on opposite sides of 76 × 76 × 1.6 mm square tubes at mid-length create imperfections (triggers) that change the geometry and the material properties of the tube in the mid-section. The effects of the imperfections on the energy absorption characteristics of the tubular structure are investigated by means of the dynamic axial load. Similar studies have been carried out for tubes with C/H ratio of 33 [1–3]. In contrast to the tubes with C/H ratio of 33 where the lobe formation are regular in shape and size, the tubes with C/H ratio > 45 exhibit irregular lobe formation.

The Finite Element package ABAQUS/Explicit v6.5-6 is used to construct a 1/2 symmetry model using shell and continuum elements to simulate the tube response to the two loading conditions. The hydro-dynamic code AUTODYN is used to characterise the localised blast pressure spatial history. The Finite element simulations show satisfactory correlation with experiments for both crushed shapes.

Key words: crashworthiness, energy absorbers, tube crushing, triggers, imperfections.

1 Introduction

Studies on the behaviour of thin-walled structures to impact loading were pioneered in the 1960s by Pugsley and Macaulay [4] and Alexander [5] and have been ongoing ever since. The continued interest on the axial crushing behaviour of these structures has been overviewed by numerous authors – Reid [6], Alghamdi [7], Jones [8] and Chung Kim Yuen and Nurick [9].

These extruded thin-walled structures can convert high input kinetic energy into strain energy by large plastic deformation when subjected to high axial impact loads. The thin-walled structures fail in two distinct collapse modes, either Euler buckling or progressive buckling mode under axial load. While these collapse modes are highly influenced by the material and geometries (length, cross-section and wall thickness) of the thin-walled structure other factors such as loading conditions, boundary conditions, triggers (imperfections) and fillers also have an effect

B.D. Reddy (ed.), IUTAM Symposium on Theoretical, Modelling and Computational Aspects of Inelastic Media, 309–317.
© Springer Science+Business Media B.V. 2008

on the tube response. For crashworthiness applications, the idea is to absorb a large amount of kinetic energy in a controllable and predictable manner or at a predetermined rate. The progressive buckling mode hence is more desirable than the Euler buckling mode because of the large geometric deformations and various interactions between different deformation modes such as bending and stretching.

When a tubular structure is crushed in progressive buckling mode, the initial peak force is much greater than the subsequent peak force developed. For crashworthiness applications, this initial peak force is highly undesirable. The use of triggers to initiate a specific collapse mode, stabilising the collapse process and reduce the peak load magnitude as explored by Thornton and Magee [10] has become very "common". These triggers, can be either material or/and geometric modifications to the structure. Chung Kim Yuen and Nurick [9] presented an overview on studies investigating the effect of different types of triggers on the crushing characteristics of tubular structures.

Imperfections, such as pre-triggers, cut-outs and dents have to be pre-induced, consequently weakening the structure. Explosives, as use for air bags deployment in case of car crashes, can be applied to generate imperfections in tubular structures (car chassis) almost instantaneously. The main objective of the experiments is to investigate if such concept is achievable. The use of blast-induced triggers would create geometric and material imperfections.

Hitherto, most of the studies report on tubes with width to thickness (C/H) ratio less than 40. In contrast to the tubes with C/H ratio of 33 where the lobe formation are regular in shape and size, the tubes with C/H ratio $>$ 45 exhibit irregular lobe formation. This paper reports on the results of an experimental and numerical investigation into the response of extruded square tubes with C/H ratio $>$ 45 to two opposite lateral blast loads (to create triggers) followed by dynamic axial impact load. Comparisons are made with tubes that have C/H ratio of 33.

2 Typical Experiments

Two equal masses of plastic explosive (PE4) of the same load diameters centrally positioned on opposite sides of the 76 × 76 × 1.6 mm square tube are used to create the imperfections. Two different load diameters (25 and 37.5 mm) are used. By differing the masses and load diameters of the explosives only two types of imperfections are induced in the tubes; simple mode 1, where the sides of the tube deform in such a way that they form two non-touching domes, and capping domes, where the sides of the tube plastically deform with a disc-shaped fragment blown out in the central area. For example in Figure 2, the diameter of the load of explosive is 25 mm and the mass ranges from 0 g to 5.2 g (from right to left) resulting in imperfections which vary from non touching domes to capping in the tube with C/H ratio $>$ 45.

In cases of tubes with C/H ratio of 33 a third type of imperfection is observed; rebound domes whereby the sides of the tubes plastically deform to such an extent

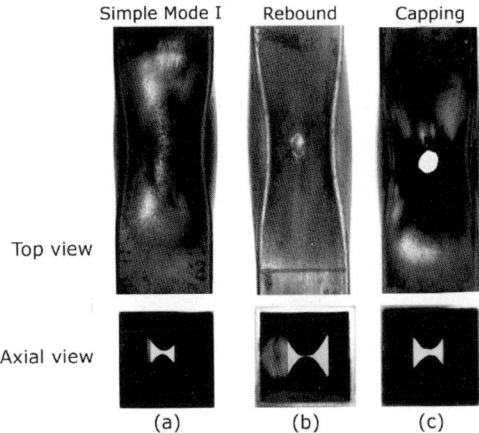

Fig. 1 Photograph showing the different modes of imperfections.

that sides make contact and rebound. In some cases, partial tearing is observed. The different types of imperfections are shown in Figure 1 [1, 2], for tubes with C/H ratio of 33.

The tubes with the blast-induced imperfections are clamped at a length of 50 mm onto the anvil and subjected to a dynamic axial impact load by means of a known drop mass from a specified height.

Figure 2 shows the final collapse shapes of square tubes ($C/H > 45$) with and without induced localised blast imperfection to dynamic impact of 329 kg mass dropped from nominally 3.3m (impact velocity 8.1 m/s). In the case of the "as-received" tube, specimen P012, the tube buckles progressively from the impact end. It is observed that the size of the lobes formed, in specimen P012, is irregular. Specimen T004, with small domes, appears to collapse progressively and symmetrically from the non-impact end. The three other tubes (T006, T007 and T033) show skew fold formations that seem to change the mode of buckling from progressive to Euler. When the tubes are loaded axially, the tubes T006, T007 and T033 crushed in such a way that the opposing sides of the tube touch each other. Tubes T007 and T033 show the same global response as the square tubes with C/H ratio of 33 with rebound imperfections [1, 2]. The crushed distance of these two tubes is less than tubes T004 and T006 which is subjected to a lesser charge load. This would suggest that an increasing mass of explosive does not necessarily result in increasing crushed distance.

It is difficult to compare the effects of the blast induced imperfection and crushed response of tube with C/H ratio of 33 and tube with C/H ratio > 45 because of scaling issues, range of damage and energy absorption capacity. However, in Figure3 a photo of a series of drop test carried out on 50 mm square tubes ($C/H = 33$) without imperfections and with induced geometric imperfections induced by 25 mm blast loads is shown for qualitative comparison. The "as-received" tubes with C/H

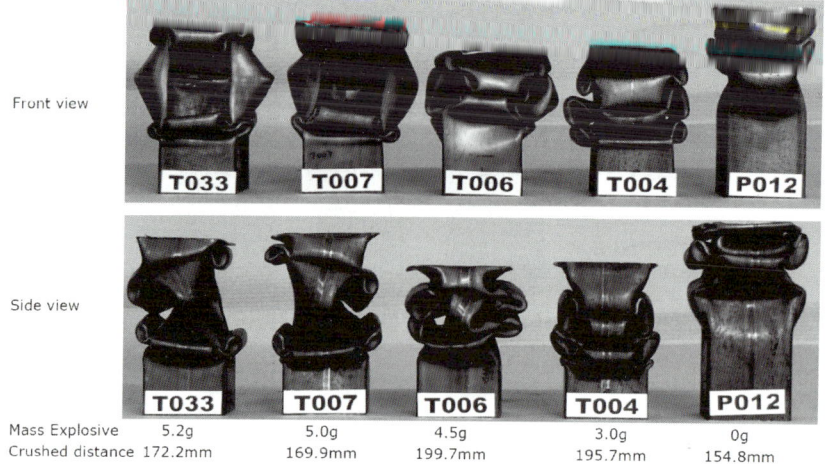

Fig. 2 Photographs showing dynamic axial crushed square tubes ($C/H > 45$) without and with blast imperfection: load diameter 37.5 mm, drop mass 329 kg, nominal drop height 3.3 m.

Fig. 3 Photographs showing dynamic axial crushed square tubes ($C/H = 33$) with and without imperfections caused by explosive of load diameter of 25 mm.

of 33 crush with regular lobe size. For tubes with the blast induced imperfections, there is a difference in the size of the lobe at the imperfection, similar to tubes with $C/H > 45$, as expected. Asymmetric in imperfections causes skew buckling but less pronounced than in tubes with $C/H > 45$. Possible reasons for the differences in the crushing behaviours of the $C/H = 33$ and $C/H > 45$ are the aspect ratios (tube width to tube length) of the structures and the stress wave propagation in the tubular structures.

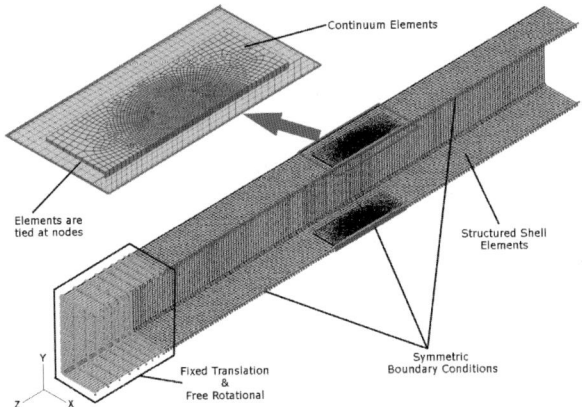

Fig. 4 1/2 Tube FE model showing boundary conditions and assigned element type.

3 Finite Element Formulation

AutoDYN is used to model the blast; the resulting pressure profile is then applied to the general-purpose finite element code; ABAQUS/*Explicit* v6.5-6 which incorporates non-linear geometry, strain rate sensitivity and temperature effects is used to model the tube response to the blast loads followed by an unloading phase during which the tube deforms elastically as a result of inertia effects and a dynamic axial load. Any residual stresses created by the simulation of the blasting are kept in the model when the unloaded phase and dynamic loads are applied. A half-symmetry model, shown in Figure 4, using a combination of four-noded shell elements (S4R) and 8-noded linear brick, 3-d continuum element with reduced integration and hourglass control (C3D8R) is used.

4 Material Properties

Stress-strain data, obtained from standard uni-axial tensile tests extracted from the tube profile, are converted into true stress and logarithmic plastic strain and corrected using the MLR_σ function proposed by Mirone [11].

The model is assumed to response adiabatically to the blast loads due to the very short duration of the blast (± 2 μseconds) and the subsequent inertia (± 640 μseconds) allowing very little heat to be transferred from the structure. The strain rate effects on the material properties are incorporated using the Cowper–Symonds relationship:

$$\frac{\sigma_y^1}{\sigma_0} = 1 + \left(\frac{\dot{\varepsilon}}{\dot{\varepsilon}_0}\right)^{1/\eta}, \qquad (1)$$

Table 1 Typical values of r_b, r_p and k obtained for different load diameters

Load Diameter (mm)	r_b (mm)	r_p (mm)	k
25	11	23	307
37.5	18	27	255

where σ_y^1 denotes the dynamic yield stress; $\dot{\varepsilon}$ the strain rate; $\dot{\varepsilon}_0$ and η are material constants; $\dot{\varepsilon}_0 = 844$ s^{-1} and $\eta = 2.207$ values for common South African mild steel obtained from Marais et al. [12].

Temperature effects are incorporated using the variations of Young's modulus and yield stress as functions of temperature as proposed by Masui et al. [13] and applied by Chung Kim Yuen et al. [14–17] and Balden and Nurick [18].

5 Modelling the Blast

The pressure-time history characteristics are obtained by simulating the typical localised explosive blast on a circular plate using the hydrodynamic code AUTODYN 2D as carried out by Balden and Nurick [18]. An axi-symmetric AUTODYN model, comprising of two Eulerian spaces (air and explosive) interacting with a Lagrangian mesh occupied by the deformable plate is used. The air media is assumed to behave as an ideal gas and the explosive is modelled using the JWL EOS for PE4 as defined in the AUTODYN material library.

The resulting plot of the normalised maximum pressure at the gauge points along the plate radius from the explosive centre shows a pressure profile that resembles a constant pulse coinciding with the burn radius (as defined by Nurick and co-workers [19, 20]) and subsequently decays in the assumed exponential manner and is described by equation (2). This blast pressure, $P(r)$, is applied to the opposite sides of the square tube to simulate the blast load.

$$\begin{aligned} P(r) &= P_0 & \text{for } 0 \leq r \leq r_b \\ P(r) &= P_0 e^{-k(r-r_b)} & \text{for } r_b \leq r \leq r_p \end{aligned} \quad (2)$$

The burn radius, r_b, the exponential decay constant, k, and the zero pressure radius, r_p, are extracted from AUTODYN simulations. r_b, r_p and k are found to be a function of the explosive radius and not mass. Table 1 shows the different values obtained for the different load diameters used.

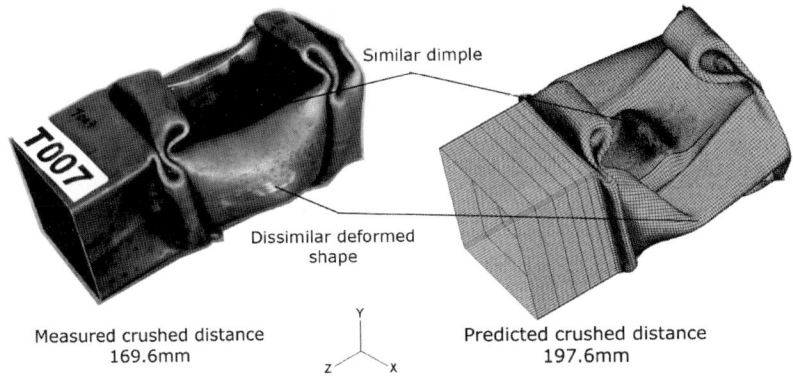

Fig. 5 Predicted collapse mode of 75 mm square tube with blast-induced Mode I imperfections (drop mass 329 kg, drop height 3.35 m, charge mass 5 g each, load diameter 37.5 mm).

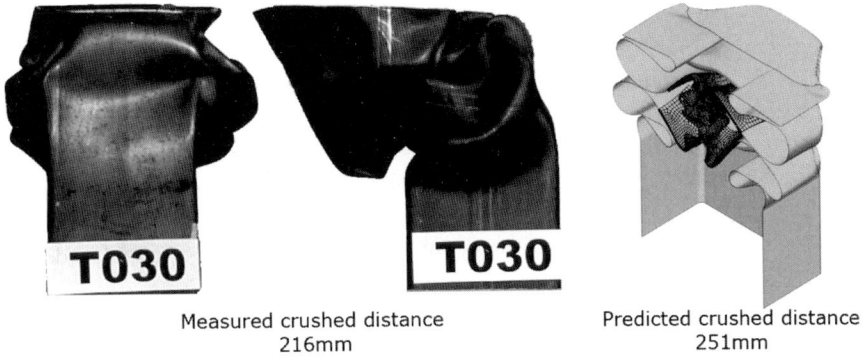

Fig. 6 Predicted collapse mode of 75 mm square tube with blast-induced capping imperfections (drop mass 329 kg, drop height 3.35 m, charge mass 5 g each (load diameter 25 mm).

6 Numerical Predictions

A typical collapse shape for a tube with dome imperfections is shown in Figure 6. The imperfections are induced using two blast loads of 5 g in mass (equivalent impulse 9.5 Ns) and 37.5 mm in diameter on opposing faces of the tube. The "imperfect" tube is crushed by a 329 kg mass dropped from a height of 3.3 m (impact velocity of 8.1 m/s). The finite element simulation over predicts the crushed distance by 17%. The crushed shape is well predicted with lobe formed at either side of the imperfection. It should be noted that while the experimental result display skew lobe formation (which could be a result of experimental scatter) the numerical model shows symmetric lobe formations.

The predicted folded shape of a 75 mm square tube with blast induced imperfections impacted by a 329 kg rigid mass from a drop height of 3.3 m (impact velocity of 8.1 m/s) is shown in Figure 6. Prior to the dynamic axial load the square tube is subjected to two localised blast loads of 5 g of PE4 (equivalent impulse 9.5 Ns), 25 mm in diameter, to create capping imperfections. The numerical simulation under-predicts capping – capping is only predicted at impulses of 12 Ns. In the experiment, the nominal diameter of the caps is 19.9 mm. The caps are not symmetric despite same loading conditions. The latter resulted in askew crushing. The finite element model, however, predicts perfect symmetry, as expected because of the boundary conditions, with a crushed distance of 251 mm.

7 Conclusions

While the lobe formation for tube with $C/H > 45$ is similar to that with tubes with C/H ratio of 33 the size of the lobe differs. For tubes with C/H ratio > 45, the size of the lobe differs from lobe to lobe and is erratic unlike tubes with C/H ratio of 33 whereby lobe sizes are of similar sizes. Blast-induced triggers at mid-points of the extruded tube affect the final crushed shape. Asymmetry in imperfections leads the tube to crush in a predominantly Euler mixed crushing mode which is undesirable for crash-worthiness applications and not predictable by the finite element model. The numerical predictions and experiments of axial loading of square tubes with blast-induced imperfections correlate reasonably for overall buckling shape.

References

1. Chung Kim Yuen, S. and Nurick, G.N., Modelling of axial loading of square tubes with blast-induced imperfections, in *Proceedings of 9th International Symposium on Plasticity and Impact Mechanics*, IMPLAST 2007, Bochum, Germany, pp. 553–560, 2007.
2. Chung Kim Yuen, S. and Nurick, G.N., The crushing charateristics of square tubes with blast-induced imperfections – Part I: Experiments, *J. Appl. Mech.*, 2008 (in review).
3. Chung Kim Yuen, S. and Nurick, G.N., The crushing charateristics of square tubes with blast-induced imperfections – Part II: Numerical simulations, *J. Appl. Mech.*, 2008 (in review).
4. Pugsley, A.G. and Macaulay, M., The large scale crumpling of thin cylindrical columns, *Quarterly J. Mech. Appl. Math.* **13**(1), 1–9, 1960.
5. Alexander, J.M., An approximate analysis of the collapse of thin cylindrical columns, *Quarterly J. Mech. Appl. Math.* **13**(1), 10–15, 1960.
6. Reid, S.R., Plastic deformation mechanisms in axial compressed metal tubes used as impact energy absorbers, *Int. J. Mech. Sci.* **35**(12), 1035–1052, 1993.
7. Alghamdi, A.A.A., Collapsible impact energy absorbers: An overview, *Thin-Walled Structures* **39**(2), 189–213, 2001.
8. Jones, N., Several phenomena in structural impact and structural crashworthiness, *Eur. J. Mech. A/Solids* **22**(5), 693–707, 2003.
9. Chung Kim Yuen, S. and Nurick, G.N., The energy absorbing characteristics of tubular structures with geometric and material modifications: An overview, *Appl. Mech. Reviews* **61**(2), 020802, 2008, doi:10.1115/1.2885138.
10. Thornton, P.H. and Magee, C.L., The interplay of geometric and materials variables in energy absorption, *Trans. ASME, J. Eng. Matl. Tech.* **99**(2), 114–120, 1977.

11. Mirone, G., A new model for the elastoplastic characteriztion and the stress-strain determination on the necking section of a tensile specimen, *Int. J. Solids Structures* **41**(13), 3545–3564, 2004.
12. Marais, S.T., Tait, R.B., Cloete, T.J. and Nurick, G.N., Material testing at high strain rate using the split Hopkinson pressure bar, *Lat. Am. J. Solids Structures* **1**(3), 319–338, 2004.
13. Masui, T., Nunokawa, T. and Hiramatsu, T., Shape correction of hot rolled steel using an on line leveller, *J. Japan Soc. Technol. Plasticity* **28**(312), 81–87, 1987.
14. Chung Kim Yuen, S. and Nurick, G.N., Deformation and tearing of uniformly blast-loaded quadrangular stiffened plates, in *Proceedings International Conference on Structural Engineering, Mech. and Comp. (SEMC)*, Cape Town, South Africa, pp. 1029–1036, 2001.
15. Chung Kim Yuen, S. and Nurick, G.N., Modelling the deformation and tearing of thin and thick plates subjected to localised blast loads, in *Proceedings 8th International Symposium on Plasticity and Impact Mechanics*, IMPLAST 2003, New Delhi, India, pp. 729–739, 2003.
16. Chung Kim Yuen, S. and Nurick, G.N., Experimental and Numerical studies on the response of quadrangular stiffened plates – Part I: Subjected to uniform blast load, *Int. J. Imp. Eng.* **31**(1), 55–83, 2005.
17. Langdon, G.S., Chung Kim Yuen, S. and Nurick, G.N., Experimental and numerical studies on the response of quadrangular stiffened plates – Part II: Subjected to localised load, *Int. J. Imp. Eng.* **31**(1), 85–111, 2005.
18. Balden, V.H. and Nurick, G.N., Numerical simulation of the post failure motion of steel plates subjected to blast loading, *Int. J. Imp. Eng.* **32**(1–4), 14–34, 2005.
19. Nurick, G.N. and Radford, A.M., Deformation and tearing of clamped circular plates subjected to localised central blast loads, in *Recent Developments in Computational and Applied Mechanics: A Volume in Honour of John B. Martin*, International Centre for Numerical Methods in Engineering (CIMNE), Barcelona, Spain, pp. 276–301, 1997.
20. Chung Kim Yuen, S. and Nurick, G.N., The significance of the thickness of a plate when subjected to localised blast load, in *Proceedings 16th International Symposium on Military Aspects of Blast and Shock*, (MABS 16), Oxford, UK, pp. 491–499, 2000.

Modelling the Behaviour of Fibre-Metal Laminates subjected to Localised Blast Loading

D. Karagiozova[1,2], G.S. Langdon[1] and G.N. Nurick[1]

[1] Blast Impact and Survivability Research Unit (BISRU), Department of Mechanical Engineering, University of Cape Town, Private Bag, Rondebosch 7701, South Africa
[2] Institute of Mechanics, Bulgarian Academy of Sciences, Acad. G. Bonchev Street, Block 4, Sofia 1113, Bulgaria
E-mail: d.karagiozova@imbm.bas.bg

Abstract. The modelling particulars of the response of Fibre-Metal Laminates (FML) to localised blast loading are discussed, particularly considering the debonding failure at the composite-metal interface. Attention is paid to the though-thickness transient deformation process in order to interpret the deformation mechanism due to highly localised pressure pulses. The study is based on previously reported experimental results on FML panels comprising different numbers of aluminium alloy layers and different thickness blocks of GFPP material. Good agreement between the experimental results and numerical predictions is demonstrated. A brief comparison between the response of a relatively thin FML panel and a monolithic aluminium alloy plate is presented.

Key words: fibre-metal laminate, panel, localised blast loading, debonding, modelling.

1 Introduction

The analysis of the response of various sandwich structures to intensive dynamic loading – impact and blast – has been attracting increasing attention during the past few years due to the improved resistance of these structures in comparison to the conventional monolithic members. While a large number of studies were published on the modelling of sandwich and laminated structures under high and low velocity impact [3, 9, 12, 13], the number of studies which examine the response of these structures to blast loading, has increased only recently [14]. However, the analyses of blast-resisting structures, which are reported in the open literature, are focused mostly on a type of structure consisting of two plates sandwiching a core material. The complex behaviour of laminated structural elements subjected to a blast loading has been examined mainly experimentally [7, 8].

Fibre-metal laminates (FMLs) are hybrid materials based on stacked arrangements of metal alloy and reinforced composite layers. When FMLs are subjected to localised blast loading, high transverse velocities are induced which cause large inelastic deformations and damage. The latter appears mainly as debonding between the composite and metal layers and, for relatively large impulses, as fibre breakage and petalling of the metal layers.

The purpose of modelling the response of FML panels is to analyse the transient deformation processes in order to clarify the mechanism of deformation and failure. The response to a localised blast loading is analysed in order to estimate the enhancement, if any, of the blast resistance due to the different structure and thickness of the panels. A second objective is to analyse the performance of these hybrid materials when compared to monolithic panels made of a conventional ductile metal (aluminium alloy).

2 Modelling Particulars

A number of challenges had to be tackled in order to model the blast loading response of FML panels when using ABAQUS/Explicit. This section discusses the modelling particulars and presents a brief comparison of the simulation and experimental results. Further details of the deformation phenomena in FML panels are available in [6].

2.1 Defining the Load

An explosion in air causes blast loading, which is highly intensive short duration loading that is difficult to measure and hence is difficult to prescribe in a Lagrangian code such as ABAQUS/Explicit. The route commonly taken, for example as in [1], when modelling monolithic metal panels response to localised blast loading, is to idealise the blast loading as a pressure loading related to the impulse calculated for each experiment. The pressure loading is a function of both time and distance from plate centre, but is assumed to be impulsive, meaning that the exact variation of the loading in time is unimportant. For simplicity, the pressure–time function is assumed to be a rectangular pulse, with an instantaneous rise time and peak pressure P_0 [1].

However, the compressibility of the sandwich structures is found to cause a significant fluid-structure interaction under blast [4]. The beneficial effects of fluid-structure interaction in reducing the impulsive loads produced on a structure as a result of an explosion were recognized early by Taylor [10]. The basic concept is that the motion of the structure relieves the pressure acting on it, thus reducing the transmitted impulse and, as a consequence, the effects of the blast. The amount of momentum acquired by the structure will, as a result, depend on its inertia. As the limit of a plate of infinite mass is approached, the plate hardly moves and the pressure history experienced by the plate corresponds to the stagnation values produced upon wave reflection on a fixed boundary. In this case, the impulse transferred to the plate is maximum. In the opposite limit of a plate of negligible mass, the plate is rapidly accelerated, which reduces the reflected pressure. Therefore only a small portion of the available momentum is transferred to the plate. This behaviour suggests that sandwich-plate construction with thin front face sheets may also provide opportunities for blast mitigation in the case of strong air blasts. A similar effect has been observed in the response of FML panels, particularly for those having thick composite layers when the transverse compressibility leads to increased pulse duration thus reducing the maximum pressure acting on the front plate.

To gain insight into the loading profile, both in time and space, the detonation and loading of the FML was simulated using AUTODYN, an explicit code for modelling non-linear dynamic behaviour of solids, liquids and gases. These simulations revealed that, due to the compressive composite core material, the pulse duration, which characterises the blast load, was larger than the pulse acting on a monolithic plate of equal mass and was well described by an exponentially decaying in time function. The spatial distribution of the loading was described with a reasonable accuracy by a constant pressure over some characteristic radius and an exponentially decaying function in the radial direction [1]. The pressure functions, $P(r, t)$, used in the present article are shown in Figure 1 and given by Equations (1, 2a, b)

$$P(r, t) = p_1(r) p_2(t) \quad (1)$$

where

$$p_1(r) = \begin{cases} P_0, & r \leq r_0 \\ P_0 \exp[-k(r - r_0)], & r_0 < r \leq r_b \\ 0, & r > r_b \end{cases}, \quad p_2(t) = \exp(-t/t_0). \quad (2a, b)$$

In Equation (2a), $r_0 = 15$ mm was the radius of the explosive disc used in the experiments [7], $k = 125$ m^{-1} was an exponential decay constant, which modelled the pressure distribution on the exposed area of the plate, $r_b < L/2$, where the length of the panel was $L = 300$ mm and t_0 was the characteristic decay time for the pulse. The total impulse was defined as

$$I = 2\pi \int_0^\infty \int_0^{r_b} P(r, t) dr dt. \quad (3)$$

FML panels comprising four layers of aluminium alloy and three blocks with up to eight plies of a woven glass fibres polypropylene composite (Twintex) were analysed. Due to the symmetry of the problem one quarter of the panel was modelled. An area with a refined mesh of size $R_m \times R_m$ was used at the central part of the panel due to expected large deformation gradients caused by the localised blast.

2.2 Strain Rate Effects in the FML Panels

The high intensity localised blast can cause a rapid local compression in the transverse direction of the woven material. The through-thickness compression modulus

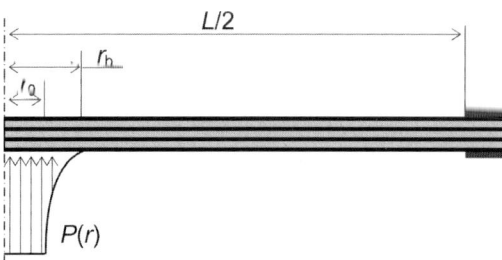

Fig. 1 Spatial distribution of the idealised pressure loading on a FML panel used in ABAQUS/Explicit.

Table 1 Characteristics of the hybrid material.

Al 2024-O	ρ kg/m^3	A MPa	B MPa	n	C	$\dot{\varepsilon}_0$	E GPa	Johnson–Cook model		
	2700	85	325	0.4	0.0083	0.001	73.4			
Twintex	ρ kg/m^3	E_{11} GPa	E_{22} GPa	E_{33} GPa	ν_{12}	ν_{13}	ν_{23}	G_{12} GPa	G_{13} GPa	G_{23} GPa
	1800	13	13	4.8	0.1	0.3	0.3	1.72	1.72	1.69

varies with strain rate, and increases significantly under a rapid dynamic load and reaches about 4.8 GPa at strain-rates in the order of 10^3 s^{-1}. The characteristics of the hybrid material used in the numerical simulations are given in Table 1.

2.3 Modelling Debonding Failure

Delamination within a block of GFPP (glass fibre polypropylene) occurred very rarely in the blast tests reported in [7, 8] and so this failure mode was not included as a possibility within the modelling. Because of computational constraints, the GFPP (Twintex) was modelled as an elastic orthotropic material since the detailed structure of the woven material was not considered.

The observed debonding at the interface of the GFPP and aluminium sheets was modelled using the potential of cohesive interface elements [2] available in ABAQUS. The cohesive tractions were related to the relative displacements of the cohesive surfaces by a constitutive law that simulated the accumulation of damage through progressive decohesion in the process zone. The initiation and progression of damage were explicitly incorporated in the formulation of the element. This technique was used in the present study to model the initiation and development of debonding between the aluminium layers and woven fabric blocks. A thickness of 50 mm was assumed for the cohesive interface.

Table 2 Damage characteristics [11].

Elastic properties			Damage initiation			Damage evolution		
E_{11} GPa	G_{12} GPa	G_{13} GPa	σ_N MPa	σ_T MPa	σ_S MPa	G_n Jm^{-2}	G_s Jm^{-2}	G_t Jm^{-2}
2.05	0.72	0.72	180	300	300	2000	3000	3000

The stress-based quadratic criterion was used to assess the beginning of the decohesion stage. Damage was assumed to initiate when a quadratic interaction function involving the nominal stress ratios reaches a value of one. This criterion can be represented as

$$\left\{\frac{\sigma_n}{\sigma_N}\right\}^2 + \left\{\frac{\sigma_s}{\sigma_S}\right\}^2 + \left\{\frac{\sigma_t}{\sigma_T}\right\}^2 = 1 \qquad (4)$$

where σ_N is the maximum nominal stress in the normal-only mode, σ_S and σ_T are maximum nominal stress in the first and second shear direction, respectively. The evolution of damage was controlled by the energy dissipated as work of separation during the decohesion process. The dependence of the fracture energy on the mode mixity was defined by the following linear interaction criterion

$$\left\{\frac{G_n}{G_N}\right\} + \left\{\frac{G_s}{G_S}\right\} + \left\{\frac{G_t}{G_T}\right\} = 1 \qquad (5)$$

where G_N is the normal mode fracture energy, G_S and G_T are the shear mode fracture energy for failure in the first and second shear direction, respectively G_n, G_s and G_t refer to the work done by the traction and its conjugate relative displacement in the normal, the first, and the second shear directions, respectively. The values of G_N, G_S and G_T were specifically verified by the available experimental results for low-velocity impact delamination initiation [11]

3D models of the panels were created with 8-node elements (C3D8R) for the composite and aluminium layers while 8-node cohesive interface elements (COH3D8) were used for the interface between these two layers. The damage parameters used in the numerical simulations are given in Table 2.

2.4 Mesh Sensitivity Analysis

The simulation of progressive debonding using cohesive elements posed numerical difficulties related to the proper definition of the cohesive layer stiffness, the requirement of extremely refined meshes, and the convergence difficulties associated with problems involving softening constitutive models. The element size in the plane of the panel was varied between 1.5 × 1.5 mm and 0.5 × 0.5 mm in the refined mesh area while a constant element size of 1.5 × 1.5 mm was used outside this area. It was

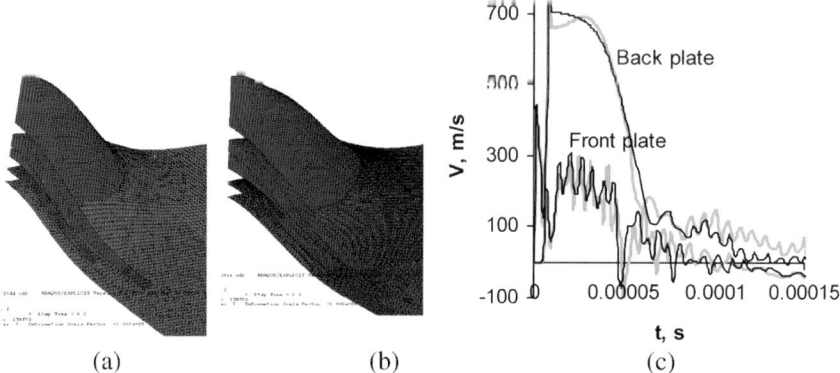

Fig. 2 (a, b) Debonding areas for models M1 and M2, respectively; (c) Resulting transverse velocities at the mid-span of the front and back plates corresponding to models M1 (—) and M2 (- - -).

observed during the simulations that the induced transverse velocities of the centre of the panel were the most sensitive variables to the element size variation. Satisfactory accuracy of the numerical results was obtained with a model using 1×1 mm with area of 60×60 mm (Model M1). A comparison with the results using the latter model and a more refined model having element size 0.5×0.5 mm within the refined area of 120×120 mm (Model M2) is shown in Figure 2. It is evident that there were marginal differences in terms of the debonding areas and characteristic velocities while the computational cost increased about four times when using model M2.

3 Comparisons with Experiments

The challenges involved in the modelling of FML panels subjected to localised blast loading were mainly related to the fluid-structure interaction, (which influenced the load definition) and representing the strain-rate sensitivity of the composite core material when using the standard elastic material models available in ABAQUS/Explicit.

Due to the fluid-structure interaction, the load duration, characterised by t_0 was varied with the variation of the thickness of the GFPP blocks. The particular values of this parameter were obtained from the model verification and a detailed discussion on the range of t_0 is given in [6]. The characteristic constant t_0 increased from 6 µs for the two-ply composite blocks to about 10.5 µs for the six-ply composite blocks for the analysed FML geometries. The use of a constant high value transverse elastic modulus ($E_{33} = 4.8$ GPa) was justified by the fact that the highest strain rates (of order 10^3 s^{-1}) occur during the very short compression phase of the GFPP blocks governed by this elastic modulus. The compression properties of the

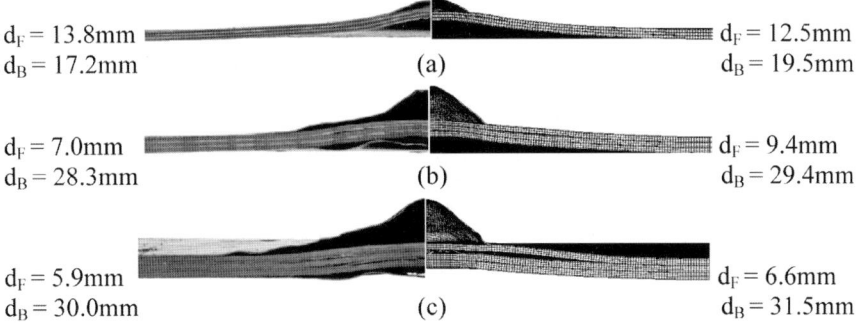

$d_F = 13.8$mm, $d_B = 17.2$mm (a) $d_F = 12.5$mm, $d_B = 19.5$mm

$d_F = 7.0$mm, $d_B = 28.3$mm (b) $d_F = 9.4$mm, $d_B = 29.4$mm

$d_F = 5.9$mm, $d_B = 30.0$mm (c) $d_F = 6.6$mm, $d_B = 31.5$mm

Fig. 3 Cross-sections of the deformed FML panels, numerical simulations (right-hand side) compared with the experimental results (left-hand side) [7] (a) A4T32, $I = 7.25$ Ns; (b) A4T34, $I = 11.84$ Ns; (c) A4T36, $I = 17.25$ Ns.

woven material did not play a significant role in the deformation of the FML panel after the compression phase of deformation.

A comparison between photographs of deformed FML panels and the simulation results are shown in Figure 3 for three typical panels each with four layers of aluminium alloy and three Twintex blocks of different thicknesses. The experimentally and numerically obtained values of the mid-span deflections of the front, d_F, and back plates, d_B, of the panels are presented on the left- and right-hand side of Figure 3.

Reasonable correlation of mid-point displacements was observed and some of the failure mechanisms evident in the experiments were detected in the simulation predictions, namely: large plastic deformation, debonding of the back face and multiple debonding of the internal aluminium layers. Buckling of the internal aluminium layer in panel A4T36, as shown in Figure 3c, was also captured by the simulation. The large local plastic deformation of the front face (Figure 3c – experiment), which were possibly caused by the local softening of the aluminium alloy due to the large temperature gradient, was not apparent in the simulation predictions since no temperature effects were taken into account by the constitutive model of the aluminium alloy. The exact shape of the debonding of the back plate was also not completely captured. The largest debonding area was observed experimentally between the back plate (fourth layer) and the attached composite block while the most significant debonding was predicted by the model between the third alumimium layer and the third composite block. However, the energies associated with debonding predicted by the simulations were similar to those estimated by Langdon et al. [7] from the experimental data since the total debonding area obtained from the numerical simulations was similar to the corresponding experimentally observed one. It should be noted that the energy dissipation due to debonding mechanism was a very small proportion of the total energy induced by the localised blast. The plastic deformations, which developed in the back plate of the model of the panel, were governed

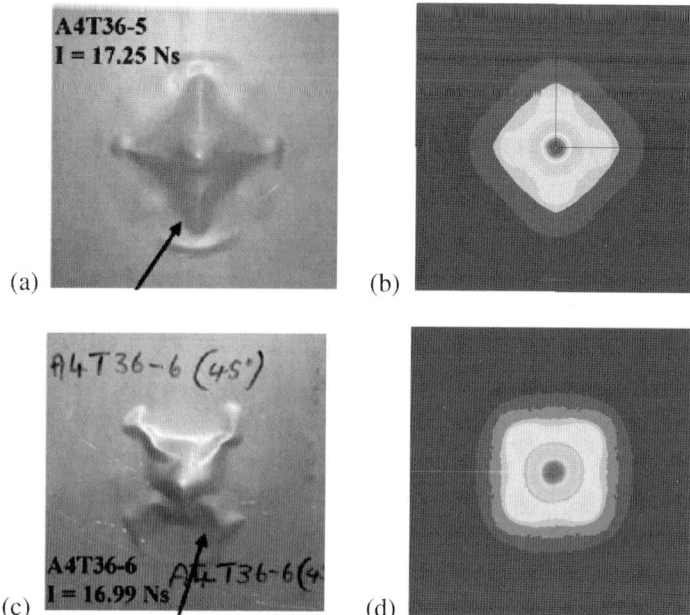

Fig. 4 Deformation of the back plates of FMLs; (a, b) Twintex fibres oriented at 90°/0° to panel edges, experiment [7] and equivalent plastic strains, respectively; (c, d) Twintex fibres oriented at 45°, experiment [7] and equivalent plastic strains, respectively.

by the orientation of the Twintex fibres with respect to the panel edges as shown in Figure 4 for two panel geometries.

4 Comparison between the Response of a FML Panel and Equivalent Monolithic Plate

Numerical simulations of monolithic aluminium plates were compared with the simulations of equal mass FML panels when subjected to identical blast loading. It was observed larger mid-point deflections and considerable damage were observed in the simulated response of the monolithic plate (shown in Figure 5b) while smaller deflections without damage occurred in the tested FML panel (shown in Figure 5a).

5 Conclusions

The results from the numerical simulations show that the existing tools and material models available in ABAQUS/ Explicit 6.6-4 (and higher versions) allow an

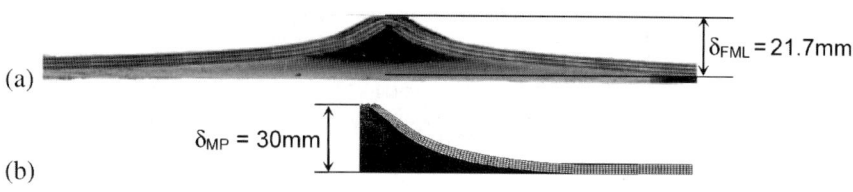

Fig. 5 Comparison between an FML panel (a) (see [8]) and aluminium alloy monolithic plate of equal mass (b) subjected to blast loading with $I = 9.06$ Ns, $t_0 = 7$ μs.

adequate modelling and with a sufficient accuracy the deformation of FML panels subjected to localised blast, which causes debonding between the layers of the hybrid material. Damage inside the GFPP blocks, however, cannot be modelled directly using the damage models available in ABAQUS due to the essentially 3D deformation process involving large transverse compression and strain-rate sensitivity.

It is shown that FML panels comprising layers of aluminium alloy and relatively thin composite blocks can tolerate larger localised blast loading compared to monolithic plates of equivalent mass made of aluminium alloy.

Acknowledgements

The authors are indebted to Dr S. Chung Kim Yuen and Mr V. Balden for modelling the blast loading using AUTODYN.

References

1. Chung Kim Yuen S, Nurick GN (2005) Experimental and numerical studies on the response of quadrangular stiffened plates. Part I: subjected to uniform blast load. *Int J Impact Eng* **31**: 55–83.
2. Espinosa HD, Dwivedi S, Lu H-C (2000) Modeling impact induced delamination of woven fiber reinforced composites with contact/cohesive laws. *Comput Methods Appl Mech Eng* **183**: 259–290.
3. Hoo Fatt MS, Lin C, Revilock Jr. DM, Hopkins DA (2003) Ballistic impact of GLARETM fibre metal laminates. *Compos Struct* **61**: 73–88.
4. Hōle J, Librescu L (2008) Recent results on the effect of the transverse core compressibility on the static and dynamic response of sandwich structures. Composites, Part B: *Engineering* **38**: 108–119.
5. Johnson AF, Holzapfel M (2006) Influence of delamination on impact damage in composite structures. *Composites Science and Technology* **66**: 807–815.
6. Karagiozova D, Langdon GS, Nurick GN, Chung Kim Yuen S (2008) The response of fibre-metal laminates to localised blast loading – A numerical perspective. *Int J Solids Struct*, submitted.

7. Langdon GS, Lemanski SL, Nurick GN, Simmons MD, Cantwell WJ, Schleyer GK (2007) Behaviour of fibre-metal laminates subjected to localised blast loading: Part I – Experimental observations and failure analysis. *Int J Impact Eng* **34**: 1202–1222.
8. Lemanski SL, Nurick GN, Langdon GS, Simmons MD, Cantwell WJ, Schleyer GK (2007) Behaviour of fibre-metal laminates subjected to localised blast loading: Part II – Quantitative analysis. *Int J Impact Eng* **34**: 1223–1245.
9. Reyes G, Cantwell WJ (2004) The high velocity impact response of composite and FML-reinforced sandwich structures. *Comp Sci Tech* **64**: 35–54.
10. Taylor GI (1963) The pressure and impulse of submarine explosion waves on plates. In Batchelor GK (Ed.), *Aerodynamics and the Mechanics of Projectiles and Explosions*, The Scientific Papers of Sir Geoffrey Ingram Taylor, Vol. III, Cambridge University Press, Cambridge, England, pp. 287–303.
11. Villanueva GR (2001) *Processing and Characterisation of the Mechanical Properties of Novel Fibre-Metal Laminates*. Ph.D. Thesis, The University of Liverpool.
12. Vlot A (1996) Impact loading on fibre metal laminates. *Int J Impact Eng* **18**: 291–307.
13. Vogelesang LB, Vlot A (2000) Development of fibre metal laminates for advanced aerospace structures. *J Materials Processing Technol* **103**: 1–5.
14. Zhu F, Lu G (2007) A review of blast and impact of metallic and sandwich structures. *EJSE Special Issue: Loading on Structures*, pp. 92–101.

On the Measurement and Evaluation of the Width of Portevin–Le Chatelier Deformation Bands with Application to AA5083-H116 Aluminium Alloy

A. Benallal[1], T. Berstad[2], T. Børvik[2], O. Hopperstad[2] and R. Nogueira de Codes[1]

[1]*LMT-Cachan, ENS Cachan/CNRS/Université Paris 6,*
PRES UniverSud Paris, Cachan, France
E-mail: benallal@lmt.ens-cachan.fr
[2]*Structural Impact Laboratory – SIMLab, Centre for Research-based Innovation (CRI), Norwegian University of Science and Technology, Trondheim, Norway*
E-mail: odd.hopperstad@ntnu.no

Abstract. This paper is devoted to experimental measurement, theoretical prediction and numerical simulation of the width of deformation bands observed in various materials exhibiting dynamic strain ageing and Portevin–Le Chatelier effect.

Key words: PLC deformation bands, dynamic strain ageing, thermography.

1 Introduction

The issue of the experimental measurement, the theoretical prediction and the numerical simulation of the width of the deformation bands associated to the Portevin–Le Chatelier phenomenon are considered in this paper. First, experimental characterization of these deformation bands is undertaken through infrared thermography for AA5083-H116 aluminium alloy. Then a theoretical framework is proposed to predict the width of the bands. Finally, nonlinear numerical simulations are used to evaluate the width of the bands. In a recent paper [3], the authors proposed a three-dimensional linear stability analysis for the predictions of the PLC bands characteristics for AA5083-H116 aluminium alloy. This approach was used with the elastic-viscoplastic McCormick model [1] including dynamic strain ageing. The results were applied to different tension tests carried out at different overall strain rates and it was shown there that the approach was able to predict qualitatively the onset of instabilities due to dynamic strain ageing and the orientations of the associated deformation bands. The onset of instability was seen to not fit with experimental results, and the discrepancy of the results may be attributed either to the parameter identification or to the constitutive model itself (see e.g. [7]). Regarding the orientations of the bands, the approach predicts for the tension test a wide range of potential normals and both the numerical simulations and experimental observations were shown to fall within this range. However, the thickness of the deformation bands was left completely unspecified. It is commonly argued in the literature that the width of the bands is set by geometrical properties such as the thickness of the specimen itself. The linear stability analysis is therefore extended here for a finite thickness plate including boundary conditions allowing the width to be estimated.

2 Experimental Observations

AA5083-H116 aluminium alloy exhibits irregular plastic flow in a given range of strain rates and temperature [6]. In a uniaxial tension test for instance, this irregular flow results in inhomogeneous deformation with different types of localization bands. Various tests have been conducted on smooth and notched specimen where the spatio-temporal features were analyzed and measured. For details, we refer to [3]. We display in Figure 1 typical results obtained in the case of a tension test on a flat smooth specimen (shown in Figure 1f) with 3 mm thickness, 15 mm width and 80 mm gage length. The specimen was cut from a 5 mm thick rolled plate and its axial loading direction aligned with the sheet rolling direction. The test was carried out at room temperature in a servo-hydraulic material testing system (MTS model 810) with a 10 kN load cell, in displacement control with clamp velocity adjusted to the desired strain rate. Both force and displacement data were recorded digitally. Further two different techniques were used to observe and eventually characterize the PLC bands spatially and temporally, namely Digital Image Correlation (DIC) and Digital Infrared Thermography (DIT). The gage length of the flat specimen was imaged with a fast CCD camera (model Ultima APX-RS) on one side and with an infrared camera (model JADE 570M) on the other side. The imaged zones for DIC and DIT are shown in Figure 1f. Prior to the tests, one side of the specimen was decorated with finely sprayed black and white paints to enhance the image contrast and the other was decorated with a fully black paint in order to enhance its emissivity. For Digital Image correlation, recorded digital images had a 128×540 pixels size. Images were recorded at a shutter speed of 125 frames per second. A cumulative strain map can be obtained by comparing each current deformation image with the initial image while an incremental strain map can be computed by comparing the image at the current load step with the image recorded just before the current load increment. These maps are computed from the recorded data by the software Correli developed by Hild and co-workers [8] where principles of Digital Image Correlation can be found. Figure 1a shows the load-displacement response where the serrations are easily observed. Figures 1b and 1c represent the strain and temperature histories at the center of the specimen respectively, obtained by Digital Image Correlation and Digital Infrared Thermography. The two curves display the stair-like behaviour characteristic of the development and evolution of deformation bands. The acquired thermal information clearly shows in Figure 1d where and when bands nucleate and how they propagate. Figure 1d actually shows a number of these bands appearing, propagating and disappearing. It displays the temperature change versus time from the beginning of the test until the time 9.3 s, just before the fracture of the specimen. Figure 1e shows in the time interval corresponding to the seventh band in Figure 1d. The thickness of the bands is most easily obtained from the pictures shown in Figure 1e. Results obtained for tension tests at two different overall strain rates on 2 mm thick specimens are provided in Figure 3 where the evolution of the bandwidth with strain is displayed. One can see in this figure that the width of the PLC bands is greater than the thickness of the specimen.

3 Theoretical Prediction of the Width of Deformation Bands

This unstable plastic flow observed for AA5083-H116 and different other alloys is usually understood as the consequences of solute-dislocation interaction at the microscopic level as suggested by Cottrel [4, 5].The reason for the serrations in the stress-strain curve is negative steady-state strain-rate sensitivity. The negative strain-rate dependence of the flow stress is attributed to dynamic strain ageing (DSA) associated with conditions when point defects can diffuse towards mobile dislocations and temporarily arrest them. This phenomenon is also referred to as the Portevin–Le Chatelier (PLC) effect. McCormick [1] has proposed an elastic-viscoplastic model for metals exhibiting dynamic strain ageing. This model is considered here. In our former stability analysis, the wavenumber was indeterminate. For completness we recall the expression of the yield function in the McCormick model, that is

$$f(\sigma, p, t_a) = \sqrt{\frac{3}{2}\mathbf{s}:\mathbf{s}} - \sigma_y(t_a) - R(p) \tag{1}$$

where $\mathbf{s}$ is the stress deviator, p is the accumulated plastic strain, $R(p)$ is the strain hardening, t_a is the average waiting time (used to describe dynamic strain ageing) and σ_y is the yield stress. The strain hardening is defined by

$$R(p) = Q_1(1 - \exp(-C_1 p)) + Q_2(1 - \exp(-C_2 p)) \tag{2}$$

where Q_1, Q_2 and C_1, C_2 are hardening parameters. The yield stress, which accounts for dynamic strain ageing through the average waiting time, is taken as

$$\sigma_y(t_a) = \sigma_0 + SH\left[1 - \exp\left(-\left\{\frac{t_a}{t_d}\right\}^\alpha\right)\right] \tag{3}$$

where σ_0 is the yield stress for $t_a = 0$, S determines the instantaneous strain-rate sensitivity and H, t_d and α are parameters governing dynamic strain ageing. The evolution of the average waiting time reads

$$\dot{t}_a = 1 - \frac{t_a}{\Omega}\dot{p} \tag{4}$$

where Ω is a parameter.

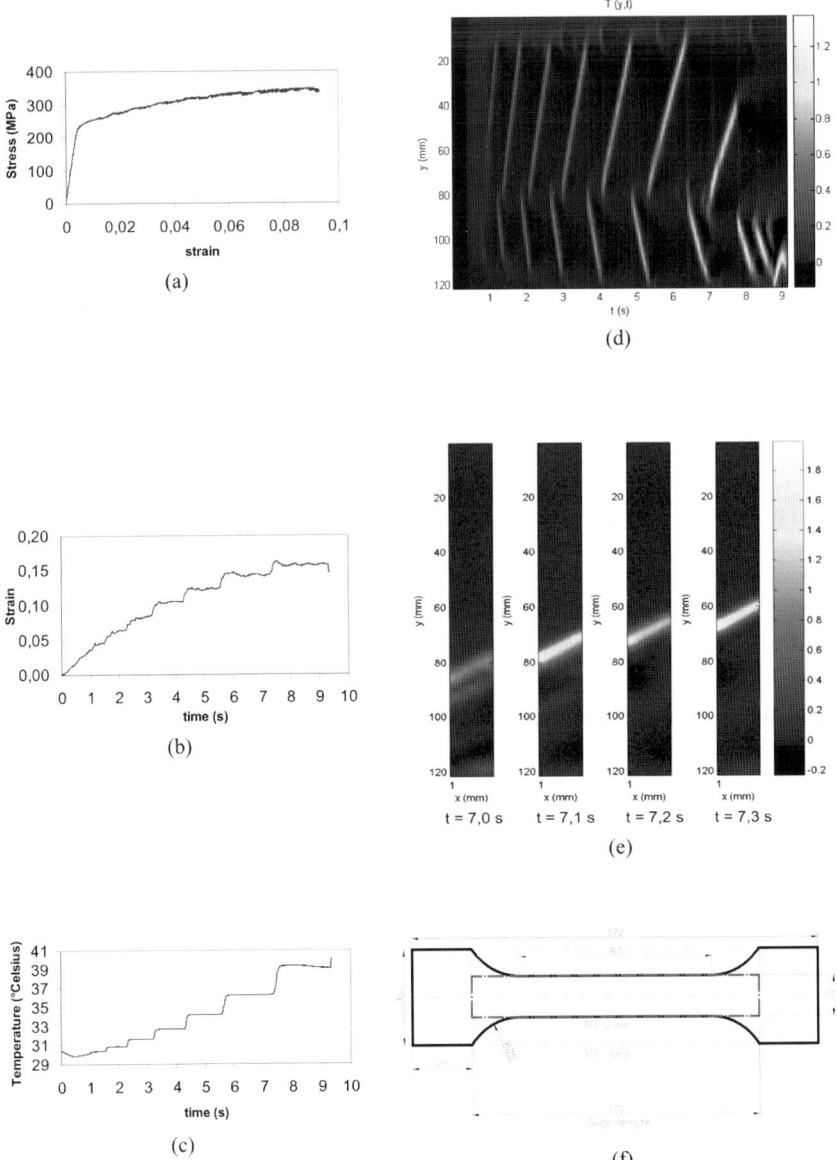

Fig. 1 AA5083-H118 tension test on a flat specimen at an overall strain rate 10^{-2} s^{-1}. (a) Load-displacement response; (b) strain history at the center of the specimen measured by Digital Image Correlation; (c) temperature history at the centre of the specimen measured by Infrared thermography; (d) visualisation of the bands during the test; (e) band evolution showing the orientation and the thickness; (f) specimen used in the tests. The imaged zones for image correlation and infrared thermography are displayed in bold and broken lines.

The linear perturbation approach used to detect instabilities due to dynamic strain ageing and negative strain-rate sensitivity is similar to the one performed in [3]. The analysis is carried out assuming small strains, for an infinite homogeneously deformed plate with uniform thickness $2e$ and with uniform properties. It is assumed therefore that during the tension tests to be considered later, this situation indeed prevails before the emergence of the PLC effect. The solution associated to the homogeneous path is denoted by $\mathbf{X}_0$ and at a given stage of the deformation path, a small perturbation $\delta \mathbf{X}$ is applied to this homogeneous path. As the applied perturbation is small, the perturbed motion $\mathbf{X}_0 + \delta \mathbf{X}$ is linearized around the homogeneous solution $\mathbf{X}_0$. When seeking perturbation fields in the form $\delta \mathbf{X} = \tilde{\mathbf{X}}(\mathbf{x})e^{\eta t}$, the stress perturbation $\tilde{\sigma}$ was shown to be related to the strain perturbation $\tilde{\epsilon}$ through the relation (see [3])

$$\tilde{\sigma} = \mathbf{H}(\eta) : \tilde{\epsilon} \tag{5}$$

where the moduli $\mathbf{H}$ for the McCormick elasto-viscoplastic model are given in the Appendix. Reporting (5) in the equilibrium equations and seeking solutions in the form $\delta \mathbf{u} = \mathbf{v}(z) \exp[i\xi(\mathbf{n}\cdot\mathbf{x}+\eta t)]$ for the displacement perturbation field (z referring to the through-thickness variable), one obtains

$$-\xi^2[\mathbf{n}.\mathbf{H}.\mathbf{n}].\mathbf{v} + i\xi[\mathbf{n}.\mathbf{H}.\mathbf{m} + \mathbf{m}.\mathbf{H}.\mathbf{n}].\frac{\partial \mathbf{v}}{\partial z} + [\mathbf{m}.\mathbf{H}.\mathbf{m}].\frac{\partial^2 \mathbf{v}}{\partial z^2} = 0 \tag{6}$$

where η is related to the growth rate of the perturbation, ξ is the mode wave number and $\mathbf{n}$ is a unit vector in the plane of the plate while $\mathbf{m}$ is the normal to the plate.

This linear differential equation has solutions of the form

$$\mathbf{v} = \mathbf{g} \exp(i\xi\tau z) \tag{7}$$

where $\mathbf{g}$ is the amplitude and τ describing the exponential decay and oscillating behaviour of the mode. Substitution in equation (6) gives again

$$[\mathbf{N}.\mathbf{H}.\mathbf{N}].\mathbf{g} = 0 \tag{8}$$

and a nontrivial solution exists then if and only if

$$\det[\mathbf{N}.\mathbf{H}.\mathbf{N}] = 0 \tag{9}$$

where we have defined

$$\mathbf{N} = \mathbf{n} + \tau \mathbf{m} \tag{10}$$

For $\mathbf{n}$ and $\mathbf{m}$ fixed Equation (9) is a sixth order equation in τ. The general solution is therefore

$$\mathbf{v} = \sum_{s=1}^{s=6} A_s \mathbf{g}_s \exp(i\xi\tau_s z) \tag{11}$$

The boundary conditions take the form

$$i\xi[\mathbf{m}.\mathbf{H}.\mathbf{n}].\mathbf{v} + [\mathbf{m}.\mathbf{H}.\mathbf{m}].\frac{\partial \mathbf{v}}{\partial z} = \mathbf{0}, z = \pm e \quad (12)$$

and give after substitution of (11) the following conditions

$$\sum_{s=1}^{s=6} A_s[\mathbf{m}.\mathbf{H}.\mathbf{N}_s].\mathbf{g}_s \exp(i\xi \tau_s e) = \mathbf{0} \quad (13)$$

$$\sum_{s=1}^{s=6} A_s[\mathbf{m}.\mathbf{H}.\mathbf{N}_s].\mathbf{g}_s \exp(-i\xi \tau_s e) = \mathbf{0} \quad (14)$$

Equations (13) and (14) form a linear system for the amplitudes A_s and nontrivial solutions are available if and only if its determinant vanishes. This condition represents a relation between the normal $\mathbf{n}$, the combination ξe and the growth rate $\bar{\eta}$. This relation has to be solved during a loading process and the critical conditions are obtained at the first instant of this loading process when this relation gives real normal $\mathbf{n}_c$ associated to positive ξ_c and rate of growth η_c with positive real part. We have then

$$e\xi_c = G(\mathbf{n}_c, \eta_c) \quad (15)$$

and the width of the deformation band w_b is therefore obtained by half of the wavelength of the critical mode π/ξ_c so that

$$w_b = \frac{\pi}{\xi_c} = \frac{\pi}{G(\mathbf{n}_c, \eta_c)} e \quad (16)$$

which shows indeed that the width of the band is proportional to the thickness of the plate.

4 Nonlinear Simulations and Evaluation of the Bands Width

Characterization and material identification for AA5083-H116 alloy have been performed earlier and can be found in [9]. The obtained parameters are summarized in Table 1. The numerical study was performed using the explicit solver of LS-DYNA. The model was made using 8-node constant-stress solid elements with a one-by-one Gauss quadrature rule in the centre of the brick, and viscous-based hourglass control was applied to avoid zero energy modes. The loading was always applied smoothly

Table 1 Constitutive parameters for AA5083-H118 aluminium alloy at room temperature.

E	ν	σ_0	C_1	C_2	Q_1	Q_2	S	$\dot{p}_0$	H	Ω	t_d	α
MPa		MPa	–	–	MPa	MPa	MPa	s^{-1}	–	–	s	
70000	0.3	165.4	1385.8	15.9	8.8	219.3	2.23	10^{-8}	27.9	10^{-4}	0.02	0.336

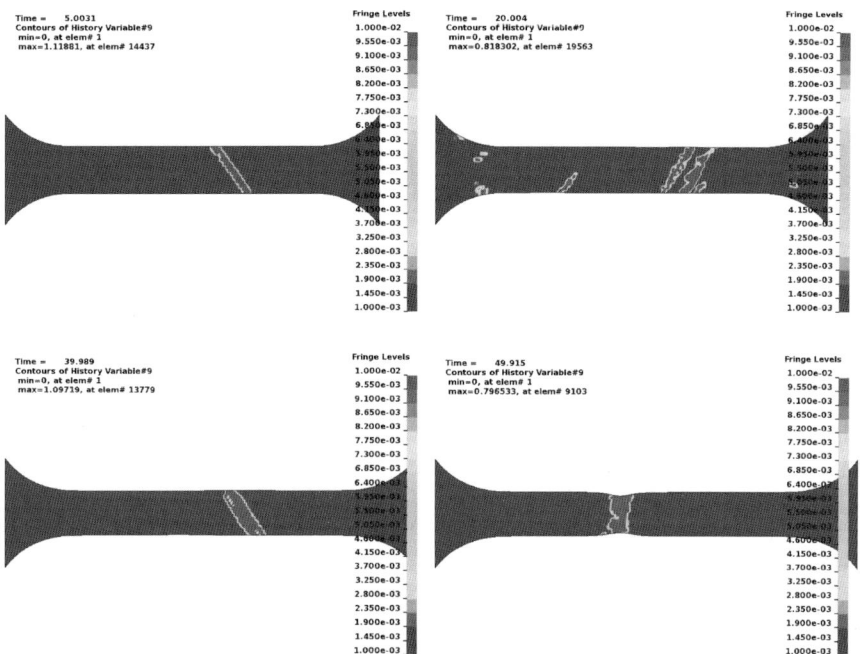

Fig. 2 3D nonlinear finite element simulation of a tension test at overall strain rate 3.3×10^{-3} showing fringes of plastic strain rate and deformation bands along the gage length of the specimen at different times from the beginning of the test untill the end. The band is trapped in the necked area. Calculations were carried out in large strain.

to avoid introducing spurious high frequency noise in the simulations. The geometry of the model was identical to the geometry used in the corresponding experimental tests (see Figure 1f). The whole gage length of the specimen was modeled, using an element size of $0.5 \times 0.5 \times 0.5$ mm^3. This gave 4 elements over the thickness, 30 elements over the width and a total of 24000 elements for the whole specimen. Examples of PLC bands in the simulation of the test at a strain rate of 3.3×10^{-3} s^{-1} are provided in Figure 2. It is seen that there may be one or several bands occurring simultaneously along the specimen. A numerical strain gage located at any position allows the local nominal strain versus time during straining to be registered. Based on these curves, an estimate of band width can be obtained. By assuming that all plastic deformation takes place inside the propagating band and that the strain is approximately uniform within the band, the band width w_b is connected to the strain rate $\dot{\epsilon}_b$ inside the band, the imposed strain rate $\dot{\epsilon}$ and the gage length L_g (if one assumes that the band travels all along the specimen gage length) by the relation

$$w_b = L_g \frac{\dot{\epsilon}}{\dot{\epsilon}_b} \qquad (17)$$

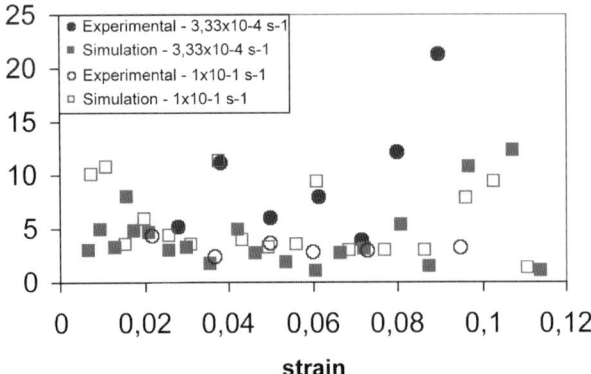

Fig. 3 Band widths measured by infrared thermography and simulated by finite element for two tests at overall strain rates 3.3×10^{-4} and 10^{-1} s^{-1}

Using this relation, the band width may be calculated. Finally, the average strain rate $\dot{\epsilon}_b$ inside the PLC band is given by the slope of the strain increments as shown in Figure 1b and the corresponding simulated one. Figure 3 shows the band width from experiments and simulations. In the simulations the band width is generally between 2 and 6 mm, while it is somewhat larger in the experiments. The scatter in the data prohibits more detailed comparisons.

5 Conclusions

In this paper, we have studied the width of the deformation bands that develop in presence of dynamic strain ageing and Portevin–Le Chatelier effect in some metallic alloys. Thermography analysis is a very efficient tool for the observation of these bands from their nucleation to their development and their propagation all along the specimen. It also allows the determination of various characterisitics of the bands and in particular their widths. Further experimental investigation is needed in order to evaluate the heat conduction consequences in the measured widths. We have also proposed here a theoretical framework to calculate the width of the deformation bands and the results obtained need still to be compared to the experimental results. Inclusion of thermo-mechanical effects and heat conduction in this theoretical framework is under investigation. Finally, numerical simulations with the Mc-Cormick's elastic-viscoplastic model including dynamic strain ageing have been carried out. These numerical simulations show indeed that the computed widths scale with the thickness specimen and that the scaling factor in formula (16) is in general greater than 1 as observed in the experiments. The agreement between experimental and numerical simulations is rather good except maybe at the end of the test.

Direct Evaluation of Limits in Plasticity and Creep Deformation*

Alan R.S. Ponter

University of Leicester, University Road, Leicester LE1 7RH, United Kingdom
E-mail: asp@le.ac.uk

Abstract. The paper discusses the Linear Matching Method for the direct evaluation of shakedown and ratchet limits for an elastic perfectly plastic body subjected to cyclic load and temperature histories. The method is reviewed and a new development is described for locating the ratchet limit by evaluating load states where a small ratchet strain per cycle of loading occurs. The method is applied to the Bree problem.

Key words: plasticity, direct methods, cyclic loading, shakedown, ratchetting.

1 Introduction

The paper is concerned with a recent development in the numerical analysis of Direct Methods in elastic plastic analysis. By this we refer to methods where the objective is the evaluation of a scalar parameter that corresponds to magnitude of a load or temperature history applied to an elastic plastic body so that a predetermined restriction on the deformation history is satisfied. The most well known and extensively investigated version of this problem is the evaluation of the shakedown limit of an elastic perfectly plastic body for cyclic loading histories where the limit corresponds to the maximum load level where the steady cyclic state involves no cyclic increase in plastic strain, i.e. the body reaches a steady state where purely elastic behaviour occurs against a background of a constant residual stress.

The origin of the interest in such problems lies in structural design and life assessment of structures where there is a need to characterise structural behaviour against uncertainties in the details of the loading history and material data. The particular interest of this author is the developments of methods for the UK life assessment method R5 [5].

For shakedown limits and the special case of limit analysis, the primary emphasis in the literature has been on the use of programming methods, applied to the upper and lower bound shakedown theorems. Significant progress has been made in re-

---

* This paper is dedicated to the memory of John Martin.

cent years, and this has recently been summarised in a review article [6]. At the same time an alternative approach has been developed, beginning with a number of ad-hoc methods developed to allow the use of a standard finite element algorithm. These are iterative methods where each iteration consists of a linear finite element solution where the material coefficients vary spatially, i.e. the standard solutions adopted in all finite element codes for materially non-linear problems. Out of this approach has developed a method that may be regarded as a particular type of upper bound programming method, the Linear Matching Method for which, in many circumstances, strict convergence proofs may be constructed.

In this paper a brief summary of the Linear Matching Method is given for an elastic perfectly plastic solid where the objective is to evaluate the two significant load limits, the shakedown limit and the ratchet limit. The body is assumed to be subject to a cyclic history of load $\lambda P(x_i, t)$ and temperature $\lambda \theta(x_i, t)$, where λ is a (positive) scale parameter acting on a prescribed history of load $P(x_i, t)$, operating on part S_T of the surface S and $\theta(x_i, t)$ a history of temperature acting in the volume V. A typical cycle in the steady state will be discussed for $0 \leq t \leq \Delta t$. The objective is to evaluate the values of λ corresponding to shakedown and ratchetting.

Consider such a body composed of a solid under conditions of small strains where the total strain is the sum of a linearly elastic and perfectly plastic component

$$\dot{\varepsilon}_{ij} = \dot{\varepsilon}_{ij}^e + \dot{\varepsilon}_{ij}^p \qquad (1)$$

where the plastic strains are associated with a yield condition $f(\sigma_{ij}) \leq 0$

$$\dot{\varepsilon}_{ij}^p = \dot{\alpha} \frac{\partial f}{\partial \sigma_{ij}}, \quad f = 0, \qquad (2)$$

The steady cyclic state of stress is given by

$$\sigma_{ij}(x_i, t) = \lambda \hat{\sigma}_{ij}(x_i, t) + \bar{\rho}_{ij}(x_i) + \rho_{ij}^r(x_i, t) \qquad (3)$$

The linear elastic solution (i.e. $\dot{\varepsilon}_{ij}^p = 0$) is denoted by $\lambda \hat{\sigma}_{ij}$. $\bar{\rho}_{ij}$ denotes a constant residual stress field in equilibrium with zero surface traction on S_T and corresponds to the residual state of stress at the beginning and end of the cycle. The history ρ_{ij}^r is the change in the residual stress during the cycle and satisfies

$$\rho_{ij}^r(x_i, 0) = \rho_{ij}^r(x_i, \Delta t) = 0 \qquad (4)$$

The significant limits are now given by:

1. The shakedown limit $\lambda = \lambda_s$, so that for $\lambda \leq \lambda_s$, in the steady state, the behaviour is purely elastic. The residual stress is constant, i.e. $\rho_{ij}^r = 0$.
2. The ratchet limit $\lambda = \lambda_R$, so that for $\lambda \leq \lambda_R$, in the steady state, there is no growth in inelastic displacement, although plastic strains occur in some part of the volume, summing to a zero increment over a cycle.

In the following two sections, methods are described that produce convergent solutions to these problems.

2 A Minimum Theory

The methods are based on a general minimum theorem for cyclic loading [4, 9]. For a typical cycle $0 \leq t \leq \Delta t$ consider the functional

$$I(\dot{\varepsilon}_{ij}^c, \lambda) = \int_0^{\Delta t} \int_V (\sigma_{ij}^c - (\lambda \hat{\sigma}_{ij} + \rho_{ij}^c)) \dot{\varepsilon}_{ij}^c \, dt \, dV \tag{5}$$

where $\dot{\varepsilon}_{ij}^c$ is kinematically admissible in the sense that the accumulated strain over the cycle;

$$\Delta \varepsilon_{ij}^c = \int_0^{\Delta t} \dot{\varepsilon}_{ij}^c \, dt \tag{6}$$

is compatible with a displacement field Δu_{ij}^c which satisfies the displacement boundary conditions. The stress state σ_{ij}^c is associated with $\dot{\varepsilon}_{ij}^c$ at yield. We now apply two additional conditions which place a restriction on the magnitude of $\dot{\varepsilon}_{ij}^c$:

1. Corresponding to $\dot{\varepsilon}_{ij}^c$ we define a cyclic history of residual stress $\rho_{ij}^c(x_i, t)$ which satisfies the relationship

$$\dot{\varepsilon}_{ij}^{cc} = C_{ijkl} \dot{\rho}_{ij}^c + \dot{\varepsilon}_{ij}^c \tag{7}$$

where $\dot{\varepsilon}_{ij}^{cc}$ is a compatible strain rate. Note that $\rho_{ij}^c(0) = \rho_{ij}^c(\Delta t) = 0$

2. Corresponding to $\rho_{ij}^c(x_i, t)$ we place a restriction on the absolute magnitude of $\dot{\varepsilon}_{ij}^c$ by requiring that there exists a *constant* residual stress field $\bar{\rho}_{ij}$ so that the composite stress history;

$$\sigma_{ij}^* = \lambda \hat{\sigma}_{ij} + \bar{\rho}_{ij} + \rho_{ij}^c \tag{8}$$

satisfies the yield condition $f(\sigma_{ij}^*) \leq 0$ for $0 \leq t \leq \Delta t$.

For any kinematically admissible $\dot{\varepsilon}_{ij}^c$ and prescribed λ

$$I(\dot{\varepsilon}_{ij}^c, \lambda) \geq 0 \tag{9}$$

and $I(\dot{\varepsilon}_{ij}^c, \lambda) = 0$ when $\dot{\varepsilon}_{ij}^c = \dot{\varepsilon}_{ij}^{pr}$ the exact cyclic solution. The upper bound shakedown theorem is recovered when $\dot{\varepsilon}_{ij}^c$ is so small that ρ_{ij}^c is negligible compared with $\lambda \hat{\sigma}_{ij}$ and (5) and (9) yields

$$\lambda_s \leq \frac{\int_0^{\Delta t} \int_V \sigma_{ij}^c \dot{\varepsilon}_{ij}^c \, dV \, dt}{\int_0^{\Delta t} \int_V \hat{\sigma}_{ij} \dot{\varepsilon}_{ij}^c \, dV \, dT} = \lambda_{UB} \tag{10}$$

The Linear Matching Method consists of the solution of a sequence of linear problems with a constitutive equation, for the von Mises yield condition, of

$$\dot{\varepsilon}_{ij}^f = \frac{3}{2\mu^i}(\lambda_{UB}^i \hat{\sigma}_{ij}' + \bar{\rho}_{ij}'^f), \quad \dot{\varepsilon}_{kk}^f = 0 \tag{11}$$

and μ^i is a shear modulus determined from the previous solution $\dot{\varepsilon}_{ij}^i$ by matching to the yield surface,

$$\mu^i \bar{\bar{\varepsilon}}(\dot{\varepsilon}_{ij}^i) = \sigma_y \tag{12}$$

for the von Mises yield condition. A solution to this problem becomes possible when equation (11) is integrated over the cycle and invoicing the compatibility of $\Delta \varepsilon_{ij}^f$ and equilibrium of $\bar{\rho}_{ij}^f$. The superscripts f refers to the current solution whereas i refers to the previous solution. Hence λ_{UB}^i refers to upper bound (10) evaluated from a previous solution $\dot{\varepsilon}_{ij}^i$. This sequence of solutions defines a sequence of monotonically reducing values of the function I and, at the same time, of the load parameter [7, 8], i.e.

$$\lambda_{UB}^f \leq \lambda_{UB}^i \tag{13}$$

where equality only applies when successive solutions are identical and, at the same time, satisfy the conditions for the shakedown limit λ_s. Hence the method may be described analytically without recourse to numerical approximation. For numerical purposes, the strain rate history is replaced by a sequence of strain increments and the compatibility condition is represented by the class of displacements defined by a finite element mesh. In this case the upper bounds converge to the least upper bound defined by the finite element mesh.

3 Evaluation of the Ratchet Limit

The evaluation of the ratchet limit falls outside the range of methods that rely upon programming methods as no bounding theorems exist that relate to the ratchet limit directly. The limit correspond to the load where the exact solution, in the steady states, changes from one where no accumulated strain occurs over a cycle to one where the accumulation corresponds to a displacement on which the loads do work. However, it is possible to, numerically, evaluate the boundary in some cases. These methods rely upon the evaluation of the changing residual stress $\rho_{ij}^c(x_i, t)$ by the minimization of increments of the functional I for an increment of strain $d\varepsilon_{ij}^c$ at time t, assuming $\rho_{ij}^c(x_i, t)$ is known;

$$dI(d\varepsilon_{ij}^c, \lambda) = \int_V (\sigma_{ij}^c - (\lambda \hat{\sigma}_{ij} + \rho_{ij}^c)) d\varepsilon_{ij}^c dV \tag{14}$$

Again we can match $d\varepsilon_{ij}^c$ to a plastic strain associated with the yield surface for a linear relationship between $d\varepsilon_{ij}^c$ and $d\rho_{ij}^c$ yielding a shear modulus;

$$\bar{\mu} = \sigma_y/\bar{\varepsilon}(d\varepsilon_{ij}^i) \tag{15}$$

where, as before $d\varepsilon_{ij}^i$ is a previously evaluated estimate. Taking into account that the increment of total strain is made up of both an elastic and plastic component, we arrive at the following equation for an improved estimate of $d\varepsilon_{ij}^c$,

$$d\varepsilon_{ij}^{fT} = \frac{3}{2\mu}d\rho_{ij}^{f'} + \frac{3}{2\bar{\mu}}(\lambda\hat{\sigma}_{ij} + \rho_{ij}^c + d\rho_{ij}^f)' \text{ and } d\varepsilon_{kk}^f = 0. \tag{16}$$

The approach here exactly matches that of the shakedown method and is a simple matter to demonstrate that

$$dI(d\varepsilon_{ij}^f) \leq dI(d\varepsilon_{ij}^i). \tag{17}$$

However the application of this algorithm through the entire cycle requires prior knowledge of the initial residual stress $\bar{\rho}_{ij}$. There exists, therefore, two interconnecting numerical strategies:

(a) If we possess an estimate of ρ_{ij}^c throughout the cycle from estimates of $d\rho_{ij}^c$, it is possible to evaluate estimates of $\bar{\rho}_{ij}$ by applying the shakedown algorithm, outlined above, where the elastic solution is augmented by ρ_{ij}^c.
(b) If we possess an estimate of $\bar{\rho}_{ij}$, then we are able to generate estimates of $d\rho_{ij}^c$ by the method described above.

3.1 A Simplified Method

In one particular case, a strictly convergent method may be derived from these results. The method was derived for problems where the variable loading is dominated by a change of temperature between two extremes. In this case it is permissible to make two assumptions; the plastic strains are produced at only two instants in the cycle; and for values of λ near to but greater than the shakedown limit, the load state will be in the reverse plasticity regime within the ratchet limit. In this case there is only a single residual stress increment $\Delta\rho_{ij}^c$ to evaluate but two equation (16). Hence it is possible to eliminate $\bar{\rho}_{ij}$ and we obtain a strictly converging process for these incremental values for fixedλ. The proximity of the load point to the ratchet limit may then be evaluated by applying an additional constant load with it's own load parameter. The application of the shakedown method where the elastic solution is augmented by the evaluated values of $\Delta\rho_{ij}^c$ yields the ratchet boundary [1]. This limited application corresponds to the method normally applied in the Life Assessment method R5 [5]. Based on this simple assumption, a set of methods have been devised for the R5 method [2, 3] where examples of both shakedown and ratchet solutions are given.

3.2 A Direct Method

The following describes a first attempt to provide a direct evaluation of the ratchet limit, making use of the algorithms described above. The additional ingredient that makes this possible is a reinterpretation of the general minimum theorem (5) and (9).

We look for a value of λ where the exact solution satisfies a certain finite strain constraint. For example, we may require that the accumulated effective plastic strain at a particular point in a structure has a specific small positive value. In this case we seek a value of λ that exceeds the ratchet limit λ_R by a small amount. There exists a subclass of kinematically admissible strain rate histories that satisfy this constraint which we denote by $\dot{\varepsilon}_{ij}^{cc}$ (or increments $d\varepsilon_{ij}^{cc}$ for numerical solutions). The minimum theorem may be applied to this subclass of histories where $\lambda = \lambda^{cc}$ now only applies to those values where the exact solution satisfies this condition. Hence the following upper bound may be derived from (5) and (9);

$$\lambda^{cc} \le \frac{\int_0^{\Delta t} \int_V (\sigma_{ij}^c - \rho_{ij}^{cc})\dot{\varepsilon}_{ij}^{cc} dV dt}{\int_0^{\Delta t} \int_V \hat{\sigma}_{ij} \dot{\varepsilon}_{ij}^{cc} dV dT} = \lambda_{UB}^{cc} \qquad (18)$$

with a corresponding form in the incremental case. The member of the class $\dot{\varepsilon}_{ij}^{cc}$ that minimizes the upper bound λ_{UB}^{cc} provides the exact solution λ^{cc}.

The application of this theorem within a numerical scheme poses the particular problem of finding a procedure, of the type discussed above, were we remain within this predefined class $\dot{\varepsilon}_{ij}^{cc}$. By linear scaling, any kinematically admissible strain rate history may be changed to one that satisfied the required condition. However, the resulting solution generally falls outside the class. The following suggested procedure attempts to make the best use of known methods but lacks a strict convergence proof.

A numerical procedure consisting of four stages is proposed based on the procedures (a) and (b) discussed in Section 3.1.

1. A value of $\bar{\rho}_{ij}$ is obtained by applying procedure (a).
2. Procedure (b) is then applied to produce a sequence of $d\varepsilon_{ij}^c$ throughout the cycle.
3. The accumulated strain over the cycle is evaluated and a linear scaling factor is found that returns the strain increment history to the defined class, with a simultaneous scaling of $d\rho_{ij}^c$.
4. The upper bound (18) provides a new estimate of λ_{UB}^{cc}.

The process then reverts to stage 1. The process is begun with an arbitrary constant choice of the shear moduli and assuming the $d\rho_{ij}^c$ are zero for the first application of stage 1.

Although each stage corresponds to a numerical process that improves the solution, in the sense of reducing either I or dI, no strict convergence proof seems possible at the present time. The primary barrier to a convergence proof is the fact

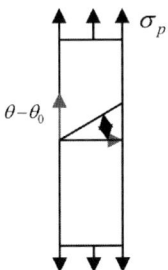

Fig. 1 Bree problem.

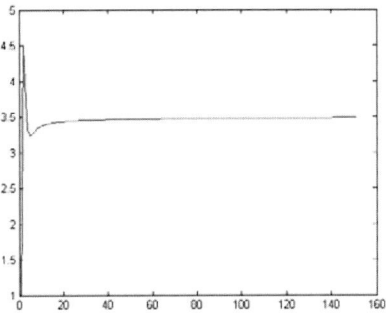

Fig. 2 Variation of λ_{UB}^{cc} with number of iterations for $\Delta \bar{\varepsilon}_p = \sigma_y/E$ and $\sigma_p/\sigma_t = 4$.

that the sequences of plastic strain increments $d\varepsilon_{ij}^c$ so generated are not necessarily kinematically admissible. However, the method has been applied to simple problems and consistently converges although not necessarily monotonically as the following example demonstrates.

Consider the plane stress problem shown in Figure 1, the simplest form of the Bree problem [11]. A plate is subjected to an axial stress σ_p and a varying linear through thickness temperature gradient, characterized by a maximum thermo elastic stress σ_t, the variation being between this value and zero stress. The plate is restrained against in-plane bending. Hence compatibility implies constant axial strain and equilibrium that axial stress integrates to the total applied stress.

Figure 3 shows a set of converged solutions for various ratios of σ_p/σ_t and for $\Delta \bar{\varepsilon}_p = 0.001 \sigma_y/E$, a very small ratchet rate. Convergence occurred for all solutions. A convergence sequence is shown in Figure 2 for a more severe case where $\Delta \bar{\varepsilon}_p = \sigma_y/E$ and convergence is significantly slower. Convergence is smooth but does not monotonically reduce. All solutions converged to the analytic solution of Bree [11], taking into account integration errors. A complete description of these solutions and others is described elsewhere [10].

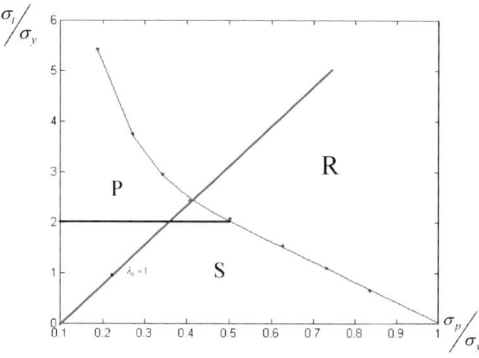

Fig. 3 A sequence of converged solutions for $\Delta \bar{\varepsilon}_p = 0.001 \sigma_y / E$.

References

1. Chen, H.F and Ponter, A.R.S. (2001), A method for the evaluation of a ratchet limit and the amplitude of plastic strain for bodies subjected to cyclic loading, *European Journal of Mechanics A/Solids* **20**, 555–572.
2. Chen, H.F., Ponter, A.R.S. and Ainsworth, R.A. (2006), The linear matching method applied to the high temperature life assessment of structures. Part 1, Assessments involving constant residual stress fields, *International Journal of Pressure Vessels and Piping* **83**, 123–135.
3. Chen, H.F., Ponter, A.R.S. and Ainsworth, R.A. (2006), The linear matching method applied to the high temperature life assessment of structures. Part 2, Assessments beyond shakedown involving changing residual stress fields, *International Journal of Pressure Vessels and Piping* **83**, 136–147.
4. Gokhfeld, D.A. and Cherniavsky, D.F. (1980), *Limit Analysis of Structures at Thermal Cycling*, Sijthoff & Noordhoff, Alphen a/d Rijn, the Netherlands.
5. Goodall, I.W., Goodman, A.M., Chell, G.C., Ainsworth R.A., and Williams, J.A. (1991), R5: An assessment procedure for the high temperature response of structures, Nuclear Electric Ltd., Report, Barnwood, Gloucester.
6. Maier, G., Pastor, J., Ponter, A.R.S., and Weichert, D. (2003), Direct methods for limit and shakedown analysis, in *Comprehensive Structural Integrity – Vol. 3*, R. de Borst and H.A. Mang (Eds.), Elsevier Science, Oxford, 48 pages.
7. Ponter, A.R.S., Fuschi, P. and Engelhardt, M. (2000), Limit analysis for a general class of yield conditions, *European Journal of Mechanics, A/Solids*, **19**, 401–422.
8. Ponter, A.R.S. and Engelhardt, M. (2000), Shakedown limits for a general yield condition: Implementation and examples for a Von Mises yield condition, *European Journal of Mechanics, A/Solids* **19**, 423–446.
9. Ponter, A.R.S. and Chen H.F. (2001), A minimum theorem for cyclic loading in excess of shakedown, with applications to the evaluation of a ratchet limit, *European Journal of Mechanics A/Solids* **20**, 539–554.
10. Ponter, A.R.S. (2008), A direct method for ratchet limits, to appear.
11. Bree, J. (1987), Elasto-plastic behaviour of thin tubes subjected to internal pressure and intermittent high heat fluxes with application to fast-nuclear-reactor fuel elements, *Journal of Strain Analysis* **2**, 226–238.

On Recent Progress in Shakedown Analysis and Applications to Large-Scale Problems*

D. Weichert, A. Hachemi, S. Mouhtamid and A.D. Nguyen

Institute of General Mechanics, RWTH-Aachen, 52056 Aachen, Germany
E-mail: weichert@iam.rwth-aachen.de

Abstract. Shakedown and Limit Analysis have proven to be powerful tools to determine limit states of mechanical structures operating beyond the elastic limit under monotonous or variable thermo-mechanical loads. In this paper, recent results obtained by using the lower-bound theorem of shakedown analysis are presented focusing on new methods for solving large-scale problems by using a selective algorithm. Illustrative examples from mechanical and pavement engineering are presented.

Key words: shakedown analysis, direct methods, optimisation, plasticity, failure.

1 Introduction

Direct Methods, comprising Limit Analysis (LA) and Shakedown Analysis (SDA) are powerful tools to predict if under monotonous or variable thermo-mechanical loads structural failure may occur or not. Characteristic features are that this information is obtained directly, without solving an evolution problem and that in case of SDA, the loading path is not be known except for its bounding envelop. Reader interested in the foundations of "Direct Methods" are referred to [1–4].

From practical point of view, lower-bound theorems are of special interest, because they provide in principle safe bounds to the loading space. Nevertheless, their practical use has been for long time handicapped as they involve genuinely an optimisation procedure, which is in general non-linear and numerical solutions in case of large numbers of degrees of freedom suffered from time consuming calculations.

Over the last years, considerable progress has been made in developing problem-tailored numerical algorithms to overcome this problem. In particular the so called Interior Point Methods have proven of high efficiency [5, 6]. In this paper focus is laid on the use of the IPDCA algorithm (Interior Point Difference of Convex functions Algorithm) developed by the authors in cooperation with others [7] and a new

---

* Dedicated to the memory of Professor John B. Martin.

B.D. Reddy (ed.), IUTAM Symposium on Theoretical, Modelling and Computational Aspects of Inelastic Media, 349–359.
© Springer Science+Business Media B.V. 2008

selective algorithm, reducing the number of optimisation variables to the plastically active set located in the plastic process zone.

2 Melan's Theorem

The starting point is the static shakedown theorem [1]: An elastic-perfectly plastic body $\mathcal{B}$ shakes down if there exists a real number $\alpha > 1$ and a time independent field of self-equilibrated stresses ρ with

$$Div\rho = 0 \quad \text{in } V \tag{1}$$

$$\mathbf{n} \cdot \rho = 0 \quad \text{on } S_p \tag{2}$$

such that the superposition of the elastic stresses σ^E with ρ constitutes a safe state of stresses:

$$\alpha\sigma^E + \rho \subset C \tag{3}$$

Here, σ^E is the solution of a reference problem, differing from the original one only by the fact, that the material behaves purely elastically, C denotes the elastic domain, usually defined by the yield criterion, $\mathbf{n}$ denotes the outer normal vector on the surface of the body $\mathcal{B}$, V and S_p stand, respectively, for the volume of $\mathcal{B}$ and the part of the surface with prescribed tractions.

If relations (1–3) are valid in any point $\mathbf{x}$ of $\mathcal{B}$ and at any instant of time, then the total dissipation is bounded and $\mathcal{B}$ is said to shake down. In practice, the largest load factor, defined by $\alpha_{SD} = \max \alpha$ is searched for, giving the maximum lower bound to the load carrying capacity of $\mathcal{B}$. The resulting optimisation problem is convex, with α and ρ as variables and the yield criterion (3) as subsidiary condition which is, e.g., in the case of von-Mises-type yield conditions convex and non-linear.

It should be mentioned, that this theorem is valid in this form only for elastic-perfectly materials in the geometrically linear framework of continuum mechanics. We note that over the last decades many studies have been devoted to the extension of this theorem, in particular to more general material laws involving non-linear hardening, material damage, non-associated flow rules and composites as well as to the geometrically non-linear framework [8].

3 Numerical Procedure

Only few analytical solutions are available for technically relevant problems and mostly numerical methods are applied. They are in general achieved in two steps: (i) the construction of the purely elastic reference solution σ^E and (ii) the solution of the constrained optimisation problem.

The purely elastic reference solution can be constructed by using standard commercial or ad-hoc algorithms. Here, one has to decide, if to use a displacement based or a stress based approximation. Although very good results had been obtained in the past with ad-hoc stress based algorithms [8], preference is given now to displacement based finite element codes, because of their intensive use in practical engineering. This way, shakedown analysis can be considered as post-processing procedure to classical finite element calculation.

The second step consists in determining the lower bound shakedown factor α_{SD}. For this purpose, usually finite-element methods are used. The variables in this optimisation procedure are the load factor α and the components of the stress tensor at the Gauss points. The eigenstress-restrictions (1–2) are to be satisfied in a weak form [8]. Furthermore, the stress field is subjected to the yield condition (3) which varies according to the material law under consideration.

In discrete form, the purely elastic stresses σ^E satisfies the virtual work principle (4) with body forces $\mathbf{f}^*$ and surface tractions $\mathbf{p}^*$

$$\int_V \{\delta\boldsymbol{\varepsilon}(\mathbf{x})\}^T\{\sigma^E(\mathbf{x})\}dV = \int_{S_p} \{\delta\mathbf{u}\}^T\{\mathbf{p}*\}dS + \int_V \{\delta\mathbf{u}\}^T\{\mathbf{f}*\}dV \tag{4}$$

for any virtual displacement field $\delta\mathbf{u}$ and any virtual strain field $\delta\boldsymbol{\varepsilon}$ satisfying the compatibility condition. Similarly, the self-equilibrated residual stresses (1–2), fulfil the condition

$$\int_V \{\delta\boldsymbol{\varepsilon}(\mathbf{x})\}^T\{\boldsymbol{\rho}(\mathbf{x})\}dV = 0 \tag{5}$$

Using this relation and introducing the unknown residual stress vector $\boldsymbol{\rho}_i$ at each Gaussian point i, the equilibrium condition (5) is integrated numerically by using the well-known Gauss–Legendre technique. The integration has to be carried out over all Gaussian points NGE with their weighting factors w_i in the considered element

$$\int_V \{\delta\boldsymbol{\varepsilon}(\mathbf{x})\}^T\{\boldsymbol{\rho}(\mathbf{x})\}dV = \sum_{i=1}^{NGE} w_i |\mathbf{J}|_i \left(\sum_{k=1}^{NKE} \mathbf{B}_k \delta\mathbf{u}_k^e\right) \boldsymbol{\rho}_i \tag{6}$$

By summation of the contributions of all elements and by variation of the virtual node-displacements respecting the boundary conditions, one obtains the linear system of equations

$$\sum_{i=1}^{NG} \mathbf{C}_i \boldsymbol{\rho}_i = [\mathbf{C}]\{\boldsymbol{\rho}\} = \{\mathbf{0}\} \tag{7}$$

where NG denotes the total number of Gaussian points of the reference body, [**C**] is a constant equilibrium matrix, uniquely defined by the chosen discretisation and the boundary conditions and $\{\boldsymbol{\rho}\}$ is the global residual stress vector of the discretised reference body. The discrete formulation of the lower bound shakedown theory for the determination of the shakedown loading factor is then given by

$$a_{SD} = \max_{\rho} \alpha \tag{8}$$

with the subsidiary conditions

$$[C]\{\rho\} = \{0\} \tag{9}$$

$$F(\alpha \sigma_i^E(P_j) + \rho_i) \leq 0 \tag{10}$$

$$\forall i \in [1, NG] \quad \text{and} \quad \forall j \in [1, NV].$$

The yield criterion F has to be fulfilled at all Gaussian points $i \in [1, NG]$ and each load vertex $j \in [1, NV]$. The number of unknowns of the optimisation problem (8)–(10) is $N = 1 + NG \times NS$ corresponding to α and $\{\rho\}$. The number of constraints is $NV \times NG + NK$, where NS is the dimension of the stress vector at each Gaussian point and NK denotes the degrees of freedom of displacements of the discretised body. This problem can not be solved efficiently by classical algorithms of optimisation because in technical design applications the number of unknowns is in general very high. Therefore, special software has been developed for solving large-scale non linear optimisation problems. It has turned out, that interior-point or barrier methods provide an attractive possibility to handle such problems and constitute to the authors' opinion a major breakthrough for numerical shakedown analysis. The interior point difference of convex functions algorithm (IPDCA) used here is described as follows [7]:

Find a Karush–Kuhn–Tucker (KKT) point of the following non-linear programming problem:

$$(\mathcal{P}_I) \begin{cases} \min f(x) \\ Ax - b = 0 \\ c(x) \geq 0 \\ x \in \mathbb{R}^n \end{cases} \tag{11}$$

where f and $c : \mathbb{R}^n \to \mathbb{R}^{m_I}$ are two twice continuous differentiable functions and c is supposed to be concave. $A \in \mathbb{R}^{m_E \times n}$ a surjective matrix and $b \in \mathbb{R}^{m_E}$ a vector. We assume that we have constraints qualification in the sense presented in (8)–(10).

The first step in the chosen interior-point approach is to add slack variables to each of the inequality constraints in $(\mathcal{P}_I)$ and to add linear constraints in order to handle free variable x. The second step is to consider the problem with the barrier objective function. The problem $(\mathcal{P}_I)$ is then transformed to

$$(\mathcal{P}_\mu) \begin{cases} \min \bar{f}_\mu(w, x, y, z) \\ Ax - b = 0 \\ c(x) - w = 0 \\ x - y + z = 0 \\ w > 0, y > 0, z > 0 \end{cases} \tag{12}$$

with

$$\bar{f}_\mu(w,x,y,z) = f(x) - \mu \sum_{i=1}^{m_I} \log(w_i) - \mu \sum_{j=1}^{n} \log(y_j) - \mu \sum_{j=1}^{n} \log(z_j) \quad (13)$$

IPDCA requires a difference of convex functions (DC decomposition) of $f = g - h$, where g and h are two convex functions [6]. By linearising the concave component of the objective function, IPDCA solves approximately the problem

$$(\mathcal{DC}_k) \begin{cases} \min \bar{g}_k(x) \\ Ax - b = 0 \\ c(x) - w = 0 \\ x - y + z = 0 \\ w \geq 0, y \geq 0, z \geq 0 \end{cases} \quad (14)$$

where, $\bar{g}_k(x) = g(x) - \nabla^T h(x_k)x$.

The problem is then transformed to a sequence of problems with the logarithmic barrier function

$$(\mathcal{DC}_\mu) \begin{cases} \min g_\mu(w,x,y,z) - \nabla^T h(x_k)x \\ Ax - b = 0 \\ c(x) - w = 0 \\ x - y + z = 0 \end{cases} \quad (15)$$

with

$$g_\mu(w,x,y,z) = g(x) - \mu \sum_{i=1}^{m_I} \log(w_i) - \mu \sum_{j=1}^{n} \log(y_j) - \mu \sum_{j=1}^{n} \log(z_j)$$

$$h_\mu(w,x,y,z) = h(x).$$

To further enhance the calculation, a new selective algorithm has been developed, taking care that only the plastically active zones of the considered structure are involved in the optimisation procedure. Among several ad-hoc criteria for the selection of the active Gauss-points the following has turned out to be the most efficient one: A Gauss-point is identified as active, if in the considered iteration step of the optimisation procedure its equivalent eigenstress in the von Mises sense is superior to the maximum value of equivalent eigenstress detected in the entire structure, divided by a weighting factor β

$$F(\rho) \geq (F(\rho))_{max}/\beta \quad (16)$$

Because the active zones change during the process, the selective algorithm is disabled at several levels of calculation in order to guaranty that the calculations does not collapse in a subspace.

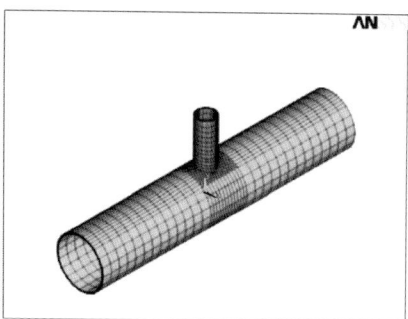

Fig. 1 FE-Mesh of pipe-junction.

Table 1 Geometrical characteristics.

	Pipe	Nozzle
Length (mm)	600	100
Internal Radius (mm)	53.55	18.6
Thickness (mm)	3.6	2.6

4 Examples

To show the efficiency, the robustness and the superiority of the interior point method with respect to standard optimisation code, several numerical results are presented. The input data for the optimisation procedure have been obtained with the commercial code ANSYS [9], where the lower bound direct method has been implemented as post-processor. In all examples SOLID45 elements are used, defined by eight nodes having three degrees of freedom at each node.

4.1 Pipe-Junction under Internal Pressure

As first example, a pipe-junction under internal pressure is considered. It is discretised by 8544 elements and 10874 nodes where the FE-mesh and essential dimensions of the pipe-junction are represented in Figure 1 and Table 1. The mechanical characteristics adopted in this analysis are as follows: Young's modulus $E = 2.1 \times 10^{+5}$ MPa; Poisson's ratio $\nu = 0.3$ and the yield stress $\sigma_Y = 300$ MPa. The results presented in Table 2 for the case of limit analysis, which can be considered as particular case of shakedown analysis, are compared with those obtained by the German design rules AD-Merkblatt [10], Cloud and Rodabaugh [11] and the incremental method [9].

Table 2 Limit pressure.

AD-Merkblatt	Cloud–Rodabaugh	Increment. Meth.	Present
$0.0427\ \sigma_Y$	$0.0511\ \sigma_Y$	$0.0456\ \sigma_Y$	$0.0450\ \sigma_Y$

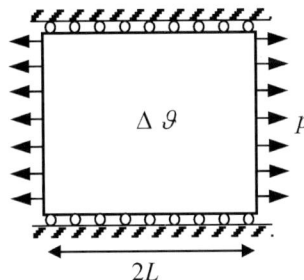

Fig. 2 Sheet under pressure and temperature variation.

4.2 Sheet under Thermo-mechanical Loading

We consider a sheet under pressure $p (0 \leq p \leq \mu_1^+ p_0;\ 0 \leq \mu_1^+ \leq 1$ and $p_0 = \sigma_Y)$ and temperature variation $\Delta\vartheta (0 \leq \Delta\vartheta \leq \mu_2^+ \Delta\vartheta_0;\ 0 \leq \mu_2^+ \leq 1$ and $\Delta\vartheta_0 = \sigma_Y/(E\alpha_\vartheta))$ (Figure 2). The mechanical characteristics used are as follows: Young's modulus $E = 1.6 \times 10^{+5}$ MPa; Poisson's ratio $\nu = 0.3$; the yield stress $\sigma_Y = 205$ MPa and the thermal expansion coefficient $\alpha_\vartheta = 0.2 \times 10^{-4} K^{-1}$. The obtained elastic, limit and shakedown domains are shown in Figure 3a for the case of perfectly plastic material, where the results given by Fuschi and Polizzotto [12] are represented. The results for shakedown analysis by taken into account limited and unlimited kinematical hardening are presented in Figure 3b, where the results obtained by Schwabe [13] are depicted.

4.3 Pavement with Locally Stationary Line Contact

The following numerical example concerns the resolution problem of repeated locally stationary rolling/sliding line contact by using the shakedown analysis [14]. As yield condition of pressure-sensitive material, the rounded Mohr-Coulomb criterion [15] is used to eliminate the singularity at the tip as well as at the edge intersections of the yield surface. This problem has been considered by Sharp and Booker [16]. Instead of a trapezoid distribution of contact pressure as in Sharp and Booker [16], a semi-cylindrical pressure distribution for both vertical and horizontal loading over the surface is assumed. The material properties used in the presented analysis are as follows: Young's modulus $E = 468$ MPa; Poisson's ratio $\nu = 0.4$; cohesion $c = 17.4$ KPa and different value of frictional angles $\varphi = 15°$, $30°$ and

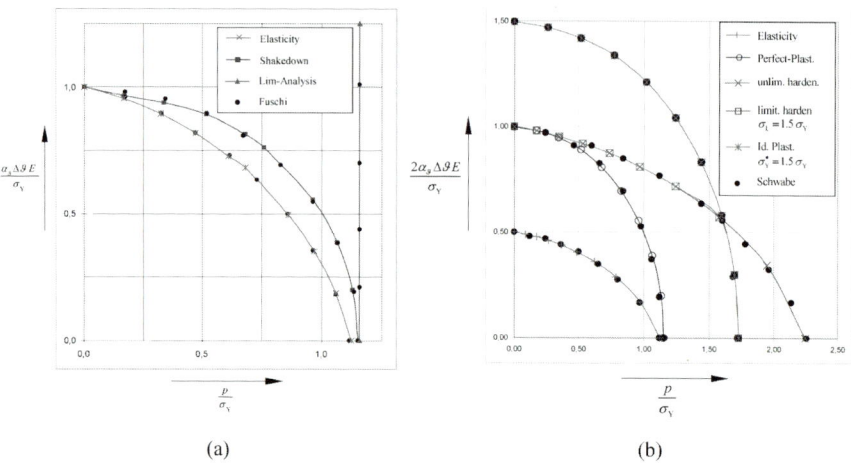

Fig. 3 Admissible loading domains.

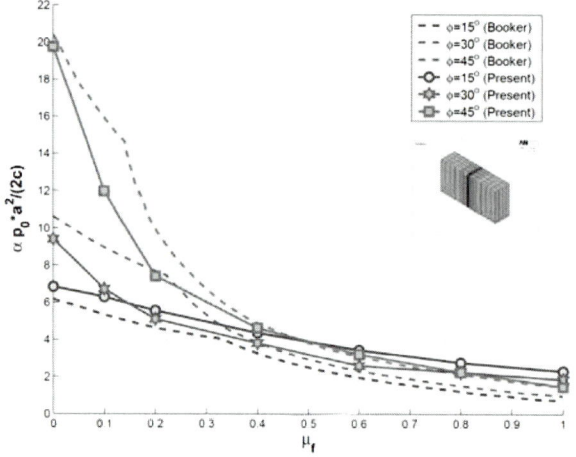

Fig. 4 Shakedown load factor versus angle of friction.

45°. Figure 4 depicts the variation of the shakedown load factor versus angle of friction μ_f in comparison with those of Sharp and Booker [16].

4.4 Flanged-Pipe under Internal Pressure and Axial Load

The shakedown problem of flanged-pipe under internal pressure p_0 and axial load Q_0 is considered (Figure 5) with the following material properties: Young's modu-

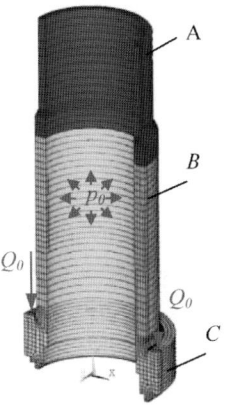

Fig. 5 Flanged-pipe.

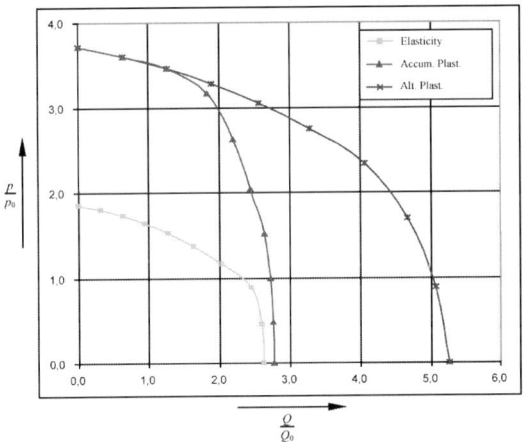

Fig. 6 Elastic and shakedown domains.

lus $E = 2.0 \times 10^{+5}$ MPa; Poisson's ratio $\nu = 0.3$; the yield stress $\sigma_Y = 200.63$ MPa. The FE-mesh of the flanged-pipe is presented in Figure 5. The length and the internal diameter are respectively $L = 386.93$ mm and $d_i = 120$ mm. The external diameters d_e of the parts A, B and C are respectively $d_e = 136.1$ mm, $d_e = 155.66$ mm and $d_e = 181$ mm. The elastic domain as well as the domains of failures due to respectively alternating plasticity and accumulated plastic deformation are presented in Figure 6.

Here, the previously mentioned selective algorithm has been applied. Table 3 gives an indication on the loading factor obtained by this method for differing values of β. By disabling the selective algorithm, the loading factor $\alpha_{SD} = 2.9005$ had

Table 3 Loading factor with selective algorithm.

β	Number of active Gaussian points	α_{SD}	CPU (s)
100	1470	2.9006	1140
80	1285	2.9010	1080
70	1180	2.9010	900

been obtained for 2120 Gauss-points ($NG = 2120$) with the CPU-time of 100 200 s. This time is roughly divided by the factor 100 when using the selective algorithm.

5 Conclusion

Compared to high-standard general codes for non-linear optimisation, a combination of IPDCA [6, 7] and Selective Optimisation allowed to reduce dramatically the necessary CPU time for lower bound Shakedown and Limit Analysis. The presented solutions of complex problems from different technical areas demonstrate the wide range of applicability of the method and its high potential for industrial design purposes, in particular as a post-processing method for conventional finite-element analyses.

References

1. Melan, E., Zur Plastizität des räumlichen Kontinuums. *Ing. Arch.* **9** (1938) 116–126.
2. Koiter, W.T., General theorems for elastic-plastic solids. In Sneddon, I.N., Hill, R. (Eds.), *Progress in Solid Mechanics*, North-Holland, Amsterdam, pp. 165–221 (1960).
3. König, J.A., *Shakedown of Elastic-Plastic Structures*, Elsevier, Amsterdam (1987).
4. Martin, G.B., *Plasticity, Fundamentals and General Results*, MIT Press, Cambridge, MA (1975).
5. Wächter, A., Biegler, L.T., *On the Implementation of a Primal-Dual Interior Point Filter Line Search Algorithm for Large-Scale Non-linear Programming*. IBM Research Report RC 23149, IBM T.J. Watson Research Center March, Yorktown Heights, NY, pp. 1–28 (2004).
6. Le Thi Haoi An, Pham Dinh Tao, The DC (difference of convex functions) Programming and DCA revisited with DC models of real world nonconvex optimization problems. *Ann. Oper Res.* **133** (2005) 23–46.
7. Akoa, F., Hachemi, A., Le Thi Hoai An, Mouhtamid, S., Pham Dinh Tao, Application of lower bound direct method to engineering structures. *J. Global Optim.* **37** (2007) 609–630.
8. Weichert, D., Maier, G., *Inelastic Behaviour of Structures under Variable Repeated Loads, Direct Analysis Methods*. CISM Courses and Lectures No. 432, International Centre for Mechanical Sciences, Springer, Wien/New York (2002).
9. *Ansys Release 8.0*, Ansys, Inc. Southpointe 275 Technology Drive Canonsburg, PA 15317.
10. Wagner, W., *Festigkeitsberechnungen im Apparate- und Rohrleitungsbau*. Vogel Buchverlag, Würzburg (1991).
11. Cloud, R.L., Rodabaugh, E.C., Approximate analysis of the plastic limit pressures of nozzles in cylindrical shells. *Trans. of the ASME, J. Engrg. Power* **4** (1968) 171–176.

12. Fuschi, P., Polizzotto, C., The shakedown load boundary of an elastic-perfectly plastic structure. *Meccanica* **30** (1995) 155–174.
13. Schwabe, F., *Einspieluntersuchungen von Verbundwerstoffen mit periodischer Mikrostruktur.* Doctor Thesis, RWTH-Aachen (2000).
14. Nguyen An Danh, Hachemi, A., Weichert, D., Application of the interior-point method to shakedown analysis of pavements. *Int. J. Numer. Meth. Engng.* (in print).
15. Aboudi, A.J., Sloan, S.W., A smooth hyperbolic approximation to the Mohr–Coulomb yield criterion. *Comput & Structures* **54** (1995) 427–441.
16. Sharp, R.W., Booker, J.R., Shakedown of pavements under moving surface loads. *J. Transp. Enrg.* **110** (1984) 1–14.

Viscoelasticity

Local and Global Regularity in Time Dependent Viscoplasticity

Hans-Dieter Alber and Sergiy Nesenenko

*Department of Mathematics, Darmstadt University of Technology,
Schlossgartenstr. 7, 64289 Darmstadt, Germany
E-mail: {alber, nesenenko}@mathematik.tu-darmstadt.de*

Abstract. In this note we announce the results obtained in [4] on local and global regularity for quasistatic initial-boundary value problems from viscoplasticity. The problems considered belong to a general class with monotone constitutive equations modelling inelastic materials showing kinematic hardening. A standard example is the Melan–Prager model. In [4] it is proved that the strain/stress/internal variable fields have $H^{1+1/3-\delta}/H^{1/3-\delta}/H^{1/3-\delta}$ regularity up to the boundary. We also show that in the case of generalized standard materials the same regularity can be obtained under weaker assumptions on the given data.

Key words: regularity, viscoplasticity, maximal monotone operator, difference quotient technique, interpolation, Melan–Prager model.

1 Introduction and Setting of the Problem

In this note we present interior and boundary regularity results for solutions of quasistatic initial-boundary value problems from viscoplasticity. The models we study use constitutive equations with internal variables to describe the deformation behavior of inelastic metals at small strain.

We consider constitutive equations of monotone type introduced in [1], which generalizes the class of generalized standard materials introduced by Halphen and Nguyen Quoc Son [12]. The class includes the well known models of Prandtl–Reuss, Norton–Hoff and Melan–Prager [15,16,20,22], to mention a few. In this work we deal only with models of monotone type, for which the associated free energy is a positive definite quadratic form. Materials showing linear kinematic hardening can be described by such models. This excludes the models of Prandtl–Reuss and Norton–Hoff, but includes the model of Melan–Prager. For a larger number of examples of constitutive equations used in engineering and for details on the monotone type class we refer to [1].

1.1 Setting of the Problem

Let $\Omega \subseteq \mathbb{R}^3$ be an open bounded set, the set of material points of the solid body. If not otherwise stated we assume that Ω has C^1-boundary. By T_e we denote a positive number (time of existence), which can be chosen arbitrarily large. $\mathcal{S}^n$ denotes the set of symmetric $n \times n$-matrices. Unknown are the displacement $u(x,t) \in \mathbb{R}^3$ of the material point x at time t, the Cauchy stress tensor $T(x,t) \in \mathcal{S}^3$ and the vector $z(x,t) \in \mathbb{R}^N$ of internal variables. The model equations of the problem are

$$-\operatorname{div}_x T(x,t) = b(x,t), \tag{1}$$

$$T(x,t) = \mathcal{D}[x]\big(\varepsilon(\nabla_x u(x,t)) - Bz(x,t)\big), \tag{2}$$

$$\partial_t z(x,t) \in g\big(x, -\nabla_z \psi(x, \varepsilon(\nabla_x u(x,t)), z(x,t))\big)$$
$$= g\big(x, B^T T(x,t) - L[x]z(x,t)\big), \tag{3}$$

which must be satisfied in $\Omega \times (0, T_e)$. The initial condition and the Dirichlet boundary condition are

$$z(x,0) = z^{(0)}(x), \quad x \in \Omega, \tag{4}$$

$$u(x,t) = \gamma(x,t), \quad (x,t) \in \partial\Omega \times [0, T_e]. \tag{5}$$

Here $\nabla_x u(x,t)$ denotes the 3×3-matrix of first order partial derivatives. The strain tensor is

$$\varepsilon(\nabla_x u(x,t)) = \frac{1}{2}\big(\nabla_x u(x,t) + (\nabla_x u(x,t))^T\big) \in \mathcal{S}^3,$$

with the transposed matrix $(\nabla_x u)^T$. For every $x \in \Omega$, the elasticity tensor $\mathcal{D}[x] : \mathcal{S}^3 \to \mathcal{S}^3$ is a linear, symmetric mapping, which is positive definite, uniformly with respect to x. The linear mapping $B : \mathbb{R}^N \to \mathcal{S}^3$ assigns to the vector $z(x,t)$ the plastic strain tensor $\varepsilon_p(x,t) = Bz(x,t)$. The free energy is

$$\psi(x, \varepsilon, z) = \frac{1}{2}\big(\mathcal{D}[x](\varepsilon - Bz)\big) \cdot (\varepsilon - Bz) + \frac{1}{2}(L[x]z) \cdot z, \tag{6}$$

where $L[x]$ denotes a symmetric $N \times N$-matrix, which is positive definite, uniformly with respect to $x \in \Omega$. The assumptions for $\mathcal{D}$ and L imply that ψ is a positive definite quadratic form with respect to (ε, z). Finally, we require that the nonlinear mapping $g : \Omega \times \mathbb{R}^N \to 2^{\mathbb{R}^N}$ satisfies

$$0 \in g(x,0), \tag{7}$$

$$0 \leq (z_1 - z_2, y_1 - y_2), \tag{8}$$

for all $x \in \Omega$, $z_i \in \mathbb{R}^N$, $y_i \in g(x, z_i)$, $i = 1, 2$. This means that g is monotone with respect to z. Given are the volume force $b(x,t) \in \mathbb{R}^3$, the boundary data $\gamma(x,t) \in \mathbb{R}^3$ and the initial data $z^{(0)}(x) \in \mathbb{R}^N$.

This completes the formulation of the initial-boundary value problem. (2) and (3) are the constitutive equations, which assign the stress $T(x, t)$ to the strain history $s \mapsto \varepsilon(\nabla_x u(x, s))$, $s \leq t$, and which model the viscoelastic material behavior of the solid body.

Under suitable regularity assumptions on the volume force and the boundary data we show that the time derivative $\partial_t u$ belongs to $L^\infty(0, \infty; H^1(\Omega))$ and the space derivatives $\nabla_x u$ to $L^\infty(0, \infty; H^1_{\text{loc}}(\Omega))$. Concerning derivatives at the boundary we prove that the tangential derivatives $\partial_\tau u$ belong to $L^\infty(0, \infty; H^1(\Omega))$, whereas for the normal derivatives we can only show a weaker result. Namely, we show that $\nabla_x u$ belongs to $L^\infty(0, \infty; H^{1/3-\delta}(\Omega))$ for every $\delta > 0$. The stress field T and the vector of internal variables z have the same regularity as the $\nabla_x u$-field.

For the time dependent problem to the Norton–Hoff and Prandtl–Reuss laws it was shown in [6] that the stress field T belongs to $L^\infty(0, \infty; H^1_{\text{loc}}(\Omega))$. In [10] this result is proved again using other methods and under different assumptions on the data. We are not aware of regularity results for the normal derivatives up to the boundary in the time dependent case.

Such results exist for time independent problems. In [21] it is shown for the stationary problem of elasto-plasticity with linear hardening in two space dimensions that the strain and stress fields belong to $H^2(\Omega)$ and $H^1(\Omega)$, respectively. For a stationary power-law model in the full three-dimensional case it is proved in [13, 14] that these fields belong to $H^{3/2-\delta}(\Omega)$ and $H^{1/2-\delta}(\Omega)$, whereas in [17] it is shown for a class of time discrete models, which includes a Cosserat model, that the displacement is in $H^2(\Omega)$ and the stress field in $H^1(\Omega)$. For local regularity results in the time independent case we refer to [5, 7, 9, 23–26] and to [11] for a survey on other results.

We consider coefficients and constitutive functions, which depend on x. Our results thus generalize and extend the local regularity results for constant coefficients in the time dependent case obtained in [18].

2 Regularity for Materials with Monotone Constitutive Equations

We use the following notations. For functions w defined on $\Omega \times [0, \infty)$ we denote by $w(t)$ the mapping $x \mapsto w(x, t)$, which is defined on Ω. The space $W^{m,p}(\Omega, \mathbb{R}^k)$ with $p \in [1, \infty]$ consists of all functions in $L^p(\Omega, \mathbb{R}^k)$ with weak derivatives in $L^p(\Omega, \mathbb{R}^k)$ up to order m. We set $H^m(\Omega, \mathbb{R}^k) = W^{m,2}(\Omega, \mathbb{R}^k)$. For the space of linear, symmetric mappings from a vector space V to itself we write $\mathcal{LS}(V, V)$.

The basis for our regularity results is the existence theorem for the initial-boundary value problem (1)–(5), which is proved in [2] in the case where the coefficient functions $\mathcal{D}$, L and the constitutive function g are independent of x. It is shown in [19] that the proof generalizes immediately to x-dependent coefficient and constitutive functions satisfying some natural conditions. In the statement of this general existence theorem given below we use that for fixed t the equations (1),

(2) and (5) together form an elliptic boundary value problem, the Dirichlet problem of linear elasticity theory. The data of this problem are $b(t)$, $z(t)$ and $\gamma(t)$. For $(b(t), z(t), \gamma(t)) \in L^2(\Omega) \times L^2(\Omega) \times H^1(\Omega)$ this problem has a unique weak solution $(u(t), T(t)) \in H^1(\Omega) \times L^2(\Omega)$. The existence theorem is

Theorem 1 (Existence). *Assume that the coefficient functions satisfy* $L \in L^\infty(\Omega, \mathcal{S}^N)$, $\mathcal{D} \in L^\infty(\Omega, \mathcal{LS}(\mathcal{S}^3, \mathcal{S}^3))$, *and that there is a constant $c > 0$ such that*

$$(\zeta, L[x]\zeta) \geq c|\zeta|^2, \quad (\sigma, \mathcal{D}[x]\sigma) \geq c|\sigma|^2, \quad \text{for all } x \in \Omega, \ \zeta \in \mathbb{R}^N, \ \sigma \in \mathcal{S}^3. \tag{9}$$

Let the mapping $g : \Omega \times \mathbb{R}^N \to 2^{\mathbb{R}^N}$ *satisfy the following three conditions:*

- $0 \in g(x, 0)$,
- $z \mapsto g(x, z) : \mathbb{R}^N \to 2^{\mathbb{R}^N}$ *is maximal monotone,*
- *the mapping* $x \mapsto j_\lambda(x, z) : \mathbb{R}^3 \to \mathbb{R}^N$ *is measurable for all* $\lambda > 0$, *where* $z \mapsto j_\lambda(\cdot, z)$ *is the inverse of* $z \mapsto z + \lambda g(\cdot, z)$.

Suppose that $b \in W^{2,1}(0, T_e; L^2(\Omega, \mathbb{R}^3))$ *and* $\gamma \in W^{2,1}(0, T_e; H^1(\Omega, \mathbb{R}^3))$. *Finally, assume that* $z^{(0)} \in L^2(\Omega, \mathbb{R}^N)$ *and that there exists* $\zeta \in L^2(\Omega, \mathbb{R}^N)$ *such that*

$$\zeta(x) \in g(x, B^T T^{(0)}(x) - L z^{(0)}(x)), \quad \text{a.e. in } \Omega, \tag{10}$$

with the weak solution $(u^{(0)}, T^{(0)}) \in H^1(\Omega) \times L^2(\Omega)$ *of the Dirichlet problem* (1), (2), (5) *of linear elasticity theory to the given data* $b(0)$, $z(0) = z^{(0)}$, $\gamma(0)$.

Then to every $T_e > 0$ *there is a unique solution*

$$(u, T) \in W^{1,\infty}(0, T_e; H^1(\Omega, \mathbb{R}^3)) \times W^{1,\infty}(0, T_e; L^2(\Omega, \mathcal{S}^3)), \tag{11}$$

$$z \in W^{1,\infty}(0, T_e; L^2(\Omega, \mathbb{R}^N)) \tag{12}$$

of the initial-boundary value problem (1)–(5).

Now we are in a position to state our main results.

2.1 Interior Regularity

Theorem 2 (Interior regularity). *Let all conditions of Theorem 1 be satisfied. Assume further that there are constants* C, C_1, C_2 *such that for every* $x \in \Omega$ *and every* $y \in \mathbb{R}^3$ *with* $x + y \in \Omega$, *for every* $z \in \mathbb{R}^N$ *and all* $\lambda > 0$ *the Yosida approximation* $z \mapsto g^\lambda(x, z)$ *of* $z \mapsto g(x, z)$ *and the mappings* $\mathcal{D}$, L *satisfy*

$$|g^\lambda(x+y, z) - g^\lambda(x, z)| \leq C|y||g^\lambda(x, z)|, \tag{13}$$

$$\|\mathcal{D}[x+y] - \mathcal{D}[x]\|_{\mathcal{LS}(\mathcal{S}^3, \mathcal{S}^3)} \leq C_1|y|, \tag{14}$$

$$\|L[x+y] - L[x]\|_{\mathcal{S}^N} \leq C_2|y|. \tag{15}$$

Suppose that $b \in W^{2,1}(0, T_e; L^2(\Omega, \mathbb{R}^3))$, $\gamma \in W^{2,1}(0, T_e; H^1(\Omega, \mathbb{R}^3))$ *and* $z^{(0)} \in H^1(\Omega, \mathbb{R}^N)$.

Then in addition to (11), (12), *the solution of the problem* (1)–(5) *satisfies*

$$(u, T) \in L^\infty(0, T_e; H^2_{\text{loc}}(\Omega, \mathbb{R}^3)) \times L^\infty(0, T_e; H^1_{\text{loc}}(\Omega, \mathcal{S}^3)), \tag{16}$$

$$z \in L^\infty(0, T_e; H^1_{\text{loc}}(\Omega, \mathbb{R}^N)). \tag{17}$$

Remark 1. If the function g is univalued, then (13) is equivalent to

$$|g(x+y, z) - g(x, z)| \leq C|y||g(x, z)|.$$

This follows directly from the relation $g^\lambda(y, z) = g(y, j_\lambda(y, z))$, which holds in this case. In general we only have $g^\lambda(y, z) \subseteq g(y, j_\lambda(y, z))$.

Of course, (14), (15) mean that $\mathcal{D}$ and L are Lipschitz continuous.

2.2 Boundary Regularity

At the boundary the tangential derivatives are as regular as all derivatives in the interior. This is shown by the next theorem.

Theorem 3 (Boundary regularity, tangential derivatives). *Let all conditions of Theorem 2 be satisfied. Assume additionally that* $\partial\Omega \in C^2$ *and* $\gamma \in W^{2,1}(0, T_e; H^2(\Omega, \mathbb{R}^3))$.

Then, for any vector field $\tau \in C^1(\overline{\Omega}, \mathbb{R}^3)$, *which is tangential at the boundary* $\partial\Omega$, *the solution of the problem* (1)–(5) *satisfies*

$$(\partial_\tau u, \partial_\tau T) \in L^\infty(0, T_e; H^1(\Omega, \mathbb{R}^3)) \times L^\infty(0, T_e; L^2(\Omega, \mathcal{S}^3)), \tag{18}$$

$$\partial_\tau z \in L^\infty(0, T_e; L^2(\Omega, \mathbb{R}^N)), \tag{19}$$

where ∂_τ *denotes derivation in the direction of the vector field.*

The next regularity result for normal derivatives at the boundary holds in Besov spaces. Here we recall only the definition of Besov spaces. A detailed exposition can be found in [8], for example.

For $h \in \mathbb{R}^n$ and an open set $\Omega \subseteq \mathbb{R}^n$ we define

$$\Omega_h = \bigcap_{j=0}^{1} \{x \in \Omega \mid x + jh \in \Omega\}.$$

Definition 1. Let $1 \leq p, \theta \leq \infty$, $s \geq 0$ and let $\ell \in \mathbb{N}$ with $\ell > s$. The function f belongs to the Besov (Nikol'skii–Besov) space $B^s_{p,\theta}(\Omega) = B^s_{p,\theta}(\Omega, \mathbb{R}^n)$ with order of smoothness s, if f is measurable on Ω and satisfies

$$\|f\|_{B^s_{p,\theta}(\Omega)} = \|f\|_{L^p(\Omega)} + \|f\|_{b^s_{p,\theta}(\Omega)} < \infty,$$

where

$$\|f\|_{b^s_{p,\theta}(\Omega)} = \begin{cases} \left(\int_{\mathbb{R}^n} \left(\frac{\|\Delta^\ell_h f\|_{L^p(\Omega_h)}}{|h|^s} \right)^\theta \frac{dh}{|h|^n} \right)^{1/\theta}, & \text{for } 1 \leq \theta < \infty, \\ \sup_{h \in \mathbb{R}^n, h \neq 0} \frac{\|\Delta^\ell_h f\|_{L^p(\Omega_h)}}{|h|^s}, & \text{for } \theta = \infty. \end{cases}$$

The ℓ-th order difference operator Δ^ℓ_h is defined by $\Delta_h f(x) = f(x+h) - f(x)$ and $\Delta^\ell_h f(x) = \Delta_h(\Delta^{\ell-1}_h f(x))$. Of course, the norm $\|f\|_{b^s_{p,\theta}(\Omega)}$ depends on the choice of ℓ, but for $\ell > s$ all norms are equivalent, cf. [8]. There exist other equivalent norms on the space $B^s_{p,\theta}(\Omega)$, but this one is the most convenient for our purposes.

Theorem 4 (Boundary regularity, all derivatives). *Under the conditions of Theorem 3 the solution of the problem* (1)–(5) *satisfies*

$$(u, T) \in L^\infty(0, T_e; B^{5/4}_{2,\infty}(\Omega, \mathbb{R}^3)) \times L^\infty(0, T_e; B^{1/4}_{2,\infty}(\Omega, \mathcal{S}^3)), \tag{20}$$

$$z \in L^\infty(0, T_e; B^{1/4}_{2,\infty}(\Omega, \mathbb{R}^N)). \tag{21}$$

For $H^s = B^s_{2,2}$ *and for* $\delta > 0$ *we have*

$$(u, T) \in L^\infty(0, T_e; H^{4/3-\delta}(\Omega, \mathbb{R}^3)) \times L^\infty(0, T_e; H^{1/3-\delta}(\Omega, \mathcal{S}^3)), \tag{22}$$

$$z \in L^\infty(0, T_e; H^{1/3-\delta}(\Omega, \mathbb{R}^N)). \tag{23}$$

3 Regularity for Generalized Standard Materials

For all our regularity results we need that the data b and γ have time derivatives of second order; the solutions obtained are in general only differentiable. We do not know whether this regularity requirement for the data is optimal and whether this loss of a derivative can be avoided. Yet, we show that in the special case when $g(x, z) = \partial_z \chi(x, z)$ with a convex function $z \mapsto \chi(x, z)$, that is for a generalized standard material, it suffices when the data have one time derivative. Indeed, to see that we first reduce the initial boundary value problem (1)–(5) to an evolution equation in a Hilbert space. To this end, we employ the standard reduction procedure from [2, 3] (see also [4]). Assume first that $b \in W^{1,2}(0, T_e; L^2(\Omega, \mathbb{R}^3))$, $\gamma \in W^{1,2}(0, T_e; H^1(\Omega, \mathbb{R}^3))$ and $z^{(0)} \in L^2(\Omega, \mathbb{R}^N)$. Suppose that the functions $\hat{\varepsilon}_p \in L^2(\Omega, \mathcal{S}^3)$, $\hat{b} \in H^{-1}(\Omega, \mathbb{R}^3)$ and $\hat{\gamma} \in H^1(\Omega, \mathbb{R}^3)$ are given and consider the problem

$$-\text{div}_x \tilde{T}(x) = \hat{b}(x), \tag{24}$$

$$\tilde{T}(x) = \mathcal{D}\big(\varepsilon(\nabla_x \tilde{u}(x)) - \hat{\varepsilon}_p(x)\big), \tag{25}$$

$$\tilde{u}(x) = \hat{\gamma}(x), \quad x \in \partial\Omega, \tag{26}$$

which has a unique solution in virtue of ellipticity theory. We drop the x-dependence of $\mathcal{D}$, L and χ ($g(z) = \partial_z \chi(z)$) for simplicity.

Definition 2. Let the linear operator $P : L^2(\Omega, \mathcal{S}^3) \to L^2(\Omega, \mathcal{S}^3)$ be defined by

$$P\hat{\varepsilon}_p = \varepsilon(\nabla_x \tilde{u}),$$

where $(\tilde{u}, \tilde{T})$ is the solution of (24)–(26) to $\hat{b} = 0$, $\hat{\gamma} = 0$ and $\hat{\varepsilon}_p \in L^2(\Omega, \mathcal{S}^3)$. With the identity operator I on $L^2(\Omega, \mathcal{S}^3)$ set $Q = I - P$.

Lemma 1. *(i) The operators P and Q are projectors on $L^2(\Omega, \mathcal{S}^3)$, which are orthogonal with respect to the scalar product $[\xi, \zeta]_\Omega := (\mathcal{D}\xi, \zeta)_\Omega$.*
(ii) The operator $B^T \mathcal{D} Q B : L^2(\Omega, \mathbb{R}^N) \to L^2(\Omega, \mathbb{R}^N)$ is selfajoint and non-negative with respect to the scalar product $(\xi, \zeta)_\Omega$.

Assume that the function $\chi : \mathbb{R}^N \to \bar{\mathbb{R}}$ is proper, convex, lower semi-continous and define $G : L^2(\Omega, \mathbb{R}^N) \to \bar{\mathbb{R}}$ by

$$H(z) = \begin{cases} \int_\Omega \chi(z(x)) dx & \text{if } \chi(z) \in L^1(\Omega) \\ +\infty & \text{otherwise} \end{cases}.$$

It is well known [27, p. 85] that H is proper, convex, lower semi-continous, and $z^* \in \partial H(z)$ if and only if

$$z \in L^2(\Omega, \mathbb{R}^N), \ z^* \in L^2(\Omega, \mathbb{R}^N) \text{ and } z^*(x) \in \partial \chi(z(x)), \text{ a.e. } x \in \Omega.$$

Note that ∂H is maximal monotone as a subdifferential of a proper, convex and lower semi-continous function.

Assume that $z(t)$ is known. Then the component $(u(t), T(t))$ of the solution is obtained as unique solution of the boundary value problem (1), (2), (5). Due to the linearity we have

$$(u(t), T(t)) = (\tilde{u}(t), \tilde{T}(t)) + (v(t), \sigma(t)), \tag{27}$$

where $(v(t), \sigma(t))$ is the solution of (24)–(26) to the data $\hat{b} = b(t)$, $\hat{\gamma} = \gamma(t)$, $\hat{\varepsilon}_p = 0$, and $(\tilde{u}(t), \tilde{T}(t))$ is the solution of (24)–(26) to the data $\hat{b} = \hat{\gamma} = 0$, $\hat{\varepsilon}_p = Bz(t)$. By definition of Q we have that $\tilde{T}(t) = -\mathcal{D} Q B z(t)$. Insertion of this equation into (3) yields

$$\frac{\partial}{\partial t} z(t) \in \partial H(z) \big(-Mz(t) + B^T \sigma(t)\big), \tag{28}$$

with the mappings $M : L^2(\Omega, \mathbb{R}^N) \to L^2(\Omega, \mathbb{R}^N)$ defined by

$$M = B^T \mathcal{D} Q B + L. \tag{29}$$

The operator M is symmetric and positive definite. Since σ is determined from the boundary value problem (24)–(26) to the data b, γ, it can be considered to be known. Therefore (28) is a non-autonomous evolution equation for z. We transform this equation to an autonomous equation with a maximal monotone evolution operator, since strong existence and perturbation theorems are mainly available for such equations. To this end, define a function $d : [0, T_e] \to L^2(\Omega, \mathbb{R}^N)$ by

$$d = -Mz + B^T \sigma. \tag{30}$$

We insert this function into (28) and use the initial condition (4) to obtain the initial boundary value problem

$$\frac{d}{dt}d(t) + A\,d(t) \ni B^T \sigma_t(t), \tag{31}$$

$$d(0) = -Mz^{(0)} + B^T \sigma(0), \tag{32}$$

for d, where the operator A is given by

$$A = M \partial H.$$

The relation between z and d given in (30) is one-to-one, and the evolution equation (31) is equivalent to the equation (28).

Since M^{-1} is selfadjoint and positive definite, the scalar product

$$\langle \xi, \zeta \rangle_\Omega := (M^{-1}\xi, \zeta)_\Omega$$

is well defined in $L^2(\Omega, \mathbb{R}^N)$. Let us denote by $\mathcal{L}^2(\Omega, \mathbb{R}^N)$ the Hilbert space $L^2(\Omega, \mathbb{R}^N)$ endowed with the scalar product $\langle \xi, \zeta \rangle_\Omega$. Then the operator A is a subdifferential of a proper, convex and lower semi-continuous function $\mathcal{H} : \mathcal{L}^2(\Omega, \mathbb{R}^N) \to \bar{\mathbb{R}}$ given by

$$\mathcal{H}(z) := H(z).$$

Indeed, one sees that from the equivalence

$$z^* \in M\partial H(z) \Leftrightarrow (M^{-1}z^*, y-z)_\Omega \leq H(y) - H(z)$$

$$\Leftrightarrow \langle z^*, y-z \rangle_\Omega \leq \mathcal{H}(y) - \mathcal{H}(z) \Leftrightarrow z^* \in \partial \mathcal{H}(z)$$

for all $y \in \mathcal{L}^2(\Omega, \mathbb{R}^N)$. Thus, due to Theorem 4.3 [27, p. 186] the problem (31)–(32) considered in $\mathcal{L}^2(\Omega, \mathbb{R}^N)$

$$\frac{d}{dt}d(t) + \partial \mathcal{H}\,d(t) \ni B^T \sigma_t(t),$$

$$d(0) = -Mz^{(0)} + B^T \sigma(0) \in D(\mathcal{H})$$

has a solution $d \in W^{1,2}(0, T_e, L^2(\Omega, \mathbb{R}^N))$ for $\sigma_t \in L^2(0, T_e, L^2(\Omega, \mathbb{R}^N))$. Note that the norms in $\mathcal{L}^2(\Omega, \mathbb{R}^N)$ and $L^2(\Omega, \mathbb{R}^N)$ are equivalent. Using the one-to-one relation between z and d we obtain that $z \in W^{1,2}(0, T_e, L^2(\Omega, \mathbb{R}^N))$. In virtue of ellipticity theory,

$$(u, T) \in W^{1,2}(0, T_e; H^1(\Omega, \mathbb{R}^3)) \times W^{1,2}(0, T_e; L^2(\Omega, \mathscr{S}^3)).$$

Thus we have the following existence result in the case of generalized standard materials.

Theorem 5 (Existence, $g = \partial \chi$). *Let the mapping $\chi : \mathbb{R}^N \to \bar{\mathbb{R}}$ be proper, convex, lower semi-continuous and satisfy the condition (33). Suppose that $b \in W^{1,2}(0, T_e; L^2(\Omega, \mathbb{R}^3))$, $\gamma \in W^{1,2}(0, T_e; H^1(\Omega, \mathbb{R}^3))$ and $z^{(0)} \in L^2(\Omega, \mathbb{R}^N)$. Assume further that the function $\chi : \mathbb{R}^N \to \bar{\mathbb{R}}$ satisfies the condition*

$$\int_\Omega \chi(B^T T^{(0)}(x) - L z^{(0)}(x)) dx < \infty, \tag{33}$$

where $(u^{(0)}, T^{(0)}) \in H^1(\Omega) \times L^2(\Omega)$ is the weak solution of the Dirichlet problem (1), (2), (5) of linear elasticity theory to the given data $b(0)$, $z(0) = z^{(0)}$, $\gamma(0)$. Then to every $T_e > 0$ there is a unique solution

$$(u, T) \in W^{1,2}(0, T_e; H^1(\Omega, \mathbb{R}^3)) \times W^{1,2}(0, T_e; L^2(\Omega, \mathscr{S}^3)), \tag{34}$$

$$z \in W^{1,2}(0, T_e; L^2(\Omega, \mathbb{R}^N)) \tag{35}$$

of the initial-boundary value problem (1)–(5) with $g = \partial \chi$.

Remark 2. We note that (33) is equivalent to the condition

$$B^T T^{(0)} - L z^{(0)} \in D(H).$$

The proof of Theorem 4 undergoes no essential changes in the case of generalized standard materials ($g = \partial \chi$), although the solution is of slightly weaker regularity compared to the general case. We formulate the result.

Theorem 6 (Boundary regularity, $g = \partial \chi$). *Let all conditions of Theorem 5 be satisfied. Assume additionally that $\gamma \in W^{1,2}(0, T_e; H^2(\Omega, \mathbb{R}^3))$ and $\partial \Omega \in C^2$. Then the solution of the problem (1)–(5) with $g = \partial \chi$ satisfies*

$$(u, T) \in L^\infty(0, T_e; B_{2,\infty}^{5/4}(\Omega, \mathbb{R}^3)) \times L^\infty(0, T_e; B_{2,\infty}^{1/4}(\Omega, \mathscr{S}^3)), \tag{36}$$

$$z \in L^\infty(0, T_e; B_{2,\infty}^{1/4}(\Omega, \mathbb{R}^N)). \tag{37}$$

For $H = B_{2,2}^0$ and $\delta > 0$ we have

$$(u, T) \in L^\infty(0, T_e; H^{4/3-\delta}(\Omega, \mathbb{R}^3)) \times L^\infty(0, T_e; H^{1/3-\delta}(\Omega, \mathscr{S}^3)), \tag{38}$$

$$z \in L^\infty(0, T_e; H^{1/3-\delta}(\Omega, \mathbb{R}^N)). \tag{39}$$

Remark 3. The model of Melan–Prager either with Prandtl–Reuss or Norton–Hoff flow rules meets all assumptions of Theorem 6.

References

1. H.-D. Alber (1998) *Materials with Memory – Initial-Boundary Value Problems for Constitutive Equations with Internal Variables.* Lecture Notes in Mathematics, Vol. 1682, Springer, Berlin.
2. H.-D. Alber, K. Chełmiński (2004) Quasistatic problems in viscoplasticity theory. I. Models with linear hardening. In: I. Gohberg, A.F. dos Santos, F.O. Speck, F.S. Teixeira, W. Wendland (Eds.), *Operator Theoretical Methods and Applications to Mathematical Physics*, Vol. 147. Birkhäuser, Basel.
3. H.-D. Alber, K. Chełmiński (2007) *Math. Models Meth. Appl. Sci.* **17**(2): 189–213.
4. H.-D. Alber, S. Nesenenko (2008) Submitted.
5. A. Bensoussan, J. Frehse (1993) Asymptotic behaviour of Norton-Hoff's law in plasticity theory and H^1 regularity. In: J.L. Lions (Ed.), *Boundary Value Problems for Partial Differential Equations and Applications*, Res. Notes Appl. Math., Vol. 29, Paris.
6. A. Bensoussan, J. Frehse (1996) *Comment. Math. Univ. Carolinae* **37**(2): 285–304.
7. A. Bensoussan, J. Frehse (2002) *Regularity Results for Nonlinear Elliptic Systems and Applications*, Applied Mathematical Sciences, Vol. 151. Springer, Berlin.
8. V.I. Burenkov (1998) *Sobolev Spaces on Domains*, Teubner-Texte zur Mathematik, Vol. 137. Teubner, Stuttgart/Leipzig.
9. C. Carstensen, S. Müller (2002) *SIAM J. Math. Anal.* **34**(2): 495–509.
10. A. Demyanov (2007) Regularity of the stresses in Prandtl–Reuss perfect plasticity. Preprint, SISSA Trieste http://hdl.handle.net/1963/1963
11. M. Fuchs, G. Seregin (2000) *Variational Methods for Problems from Plasticity Theory and for Generalized Newtonian Fluids*, Lecture Notes in Mathematics, Vol. 1749. Springer, New York.
12. B. Halphen, Nguyen Quoc Son (1975) Sur les matériaux standards généralisés. *J. Méc.* **14**: 39–63.
13. D. Knees (2004) Regularity results for quasilinear elliptic systems of power-law growth in nonsmooth domains. Boundary, transmission and crack problems. PhD thesis, Stuttgart
14. D. Knees (2006) *Math. Methods. Appl. Sci.* **29**: 1363–1391.
15. J. Lemaitre, J.-L. Chaboche (1990) *Mechanics of Solid Materials.* Cambridge University Press, Cambridge.
16. J. Lubliner (1990) *Plasticity Theory.* Macmillan, New York.
17. P. Neff, D. Knees (2008) *SIAM J. Math. Anal.*, to appear.
18. S. Nesenenko (2006) Homogenization and regularity in viscoplasticity. PhD Thesis, Darmstadt, Germany.
19. S. Nesenenko (2007) *SIAM J. Math. Anal.* **39**(1): 236–262.
20. L. Prandtl (1925) Spannungsverteilung in plastischen Körpern. In *Proc. Int. Congr. Appl. Mech. Delft 1924*: 43–54 (Gesammelte Abhandlungen. Springer, Berlin (1961), 133–148).
21. S.I. Repin (1996) *Math. Models Meth. Appl. Sci.* **6**(5): 587–604.
22. A. Reuss (1930) *Z. Angew. Math. Mech.* **10**: 266–274.
23. G.A. Seregin (1987) *Differents. Uravn.* **23**: 1981–1991. English translation in *Differential Equations* **23** (1987), 1349–1358.
24. G.A. Seregin (1988) A local Caccioppoli-type estimate for extremals of variational problems in Hencky plasticity. In: *Some Applications of Functional Analysis to Problems of Mathematical Physics*. Novosibirsk, pp. 127–138 [in Russian]
25. G.A. Seregin (1990) *Algebra Analiz* **2**: 121–140.
26. G.A. Seregin (1999) *J. Math. Sci.* **93**(5): 779–783.
27. E. Showalter (1997) *Monotone Operators in Banach Spaces and Nonlinear Partial Differential Equations.* Math. Surveys Monogr., Vol. 49, AMS, Providence.

Hamiltonian Theory of Viscoelasticity

A. Hanyga[1] and M. Seredyńska[2]

[1]*Department of Geosciences, University of Bergen, Allégaten 41,
5007 Bergen, Norway
E-mail: andrzej.hanyga@geo.uib.no*
[2]*Institute of Fundamental Technological Research, Polish Academy of Sciences,
ul. Świętokrzyska 21, 00-049 Warszawa, Poland
E-mail: msered@ippt.gov.pl*

Abstract. A conserved energy and a Hamiltonian is constructed for linear hereditary viscoelastic bodies whose relaxation moduli are positive definite. The Hamiltonian represents an elastic field interacting with a system of uncoupled oscillators.

Key words: viscoelasticity, Hamiltonian, relaxation, positive definite function.

1 Introduction and Definitions

Recent developments of nanotechnology have attracted some attention to physically motivated formulations of continuum mechanics. Hamiltonian formulation plays an important role in dielectric media as a prerequisite to quantization, which opens new areas, such as interactions of continua with molecules. Hamiltonian formulation also provides new tools for studying stability and chaos in continuous media.

Breuer and Onat noticed in [1] that the stored energy density U of a hereditary viscoelastic body is not uniquely determined by its defining equation

$$\dot{U} \leq \langle \sigma, \dot{\mathbf{e}} \rangle. \tag{1}$$

Since then several particular stored energy densities have been studied. We propose here to study the unique stored energy density U_c which satisfies the equation $\dot{U}_c = \langle \sigma, \dot{\mathbf{e}} \rangle$. In a viscoelastic body without external energy supply and with zero total energy flux U_c represents the contribution of strain energy to a conserved energy. From there on we obtain an infinite-dimensional Hamiltonian equation for such a system.

The state of a hereditary material is represented by the entire history of a set of independent observable variables such as strains in viscoelasticity. Since the state contains the information about its entire history one can expect that the conserved energy can somehow be constructed by adding the accumulated energy dissipation to the current energy. It turns out that the conserved energy can be explicitly expressed in terms of a set of auxiliary fields. The auxiliary variables replace the

B.D. Reddy (ed.), IUTAM Symposium on Theoretical, Modelling and Computational Aspects of Inelastic Media, 373–383.
© *Springer Science+Business Media B.V.* 2008

strain history in the constitutive equations for the other observables such as stress. The only precondition needed for the definition of the auxiliary variables and for the construction of the conserved energy is a very general property of material response functions.

The relaxation modulus $\mathsf{G}(t)$, $t \geq 0$, represents the relaxation of stress at fixed strains **e**

$$\sigma(t) = \mathsf{G}(t)\,\mathbf{e}; \quad \sigma_{ij}(t) = G_{ijkl}(t)\,e_{kl}. \tag{2}$$

It is convenient to regard $\mathsf{G}(t)$ as an operator on the space S of rank 2 symmetric tensors. The linear space of symmetric operators $S \to S$ is denoted by Σ. In view of the major symmetry the relaxation modulus $\mathsf{G}(t)$ is a function $\mathsf{G} : [0, \infty[\to \Sigma$. The stress response to an arbitrary strain history is given by the Volterra convolution

$$\sigma = \mathsf{G} * \dot{\mathbf{e}}. \tag{3}$$

The Volterra convolution of $\mathsf{F} : [0, \infty[\to \Sigma$ and $\mathbf{g} : \mathbb{R} \to S$ is defined by the formula

$$\mathsf{F} * \mathbf{g}(t) = \int_0^\infty \mathsf{F}(s)\,\mathbf{g}(t-s)\,\mathrm{d}s, \quad \forall t \in \mathbb{R}. \tag{4}$$

Definition 1. *A function* $\mathsf{F} : [0, \infty[\to \Sigma$ *is said to be* causal positive definite *if*

$$\int_{-\infty}^\infty \langle \mathbf{f}(t), \mathsf{F} * \mathbf{f}(t) \rangle\, \mathrm{d}t \geq 0 \tag{5}$$

for every integrable function $\mathbf{f} : \mathbb{R} \to S$ *with compact support.*

Relaxation moduli of real materials are causal positive definite. According to the Fluctuation-Dissipation Theorem [3, 11] the relaxation modulus is equal to the autocorrelation function of stress fluctuations,

$$\mathsf{G}(|s|) = \frac{\rho}{2\mathrm{k}T} \langle\langle \sigma(t) \otimes \sigma(t+s) + \sigma(t+s) \otimes \sigma(t) \rangle\rangle \tag{6}$$

which implies that G is a causal positive definite function.

The above physical justification is quite general. It also applies to the inverse dielectric modulus (relaxation of electric intensity at fixed electric displacement) [9]. Relaxation functions of real viscoelastic materials satisfy a more stringent condition: they have non-negative relaxation spectra and therefore they are completely monotone (Section 8 and [10, 12]). A completely monotone function which has a finite limit for $t \to 0$ is positive definite [6].

2 Spectral Decomposition

Let $S_\mathbb{C}$ denote the complex extension of S. We shall denote by $\Sigma_\mathbb{C}$ the space of symmetric operators on $S_\mathbb{C}$ and define the scalar product

$$\langle \mathbf{f}, \mathbf{g}\rangle_\mathbb{C} := \overline{f_{kl}}\, g_{kl} \qquad (7)$$

A spectral decomposition of relaxation modulus as well as stress can now be derived from the following extension of Bochner's theorem [8]:

Theorem 1. *If the Σ-valued function $\mathsf{F} : [0, \infty[\to \Sigma$ is causal positive definite and a finite limit $\mathsf{F}_0 := \lim_{t\to 0+} \mathsf{F}(t)$ exists, then there is a $\Sigma_\mathbb{C}$-valued Radon measure M on $\mathbb{R}$ which satisfies the inequalities:*

$$\langle \mathbf{y}, \mathsf{M}(E)\,\mathbf{y}\rangle_\mathbb{C} \geq 0 \quad \forall \mathbf{y} \in S_\mathbb{C} \qquad (8)$$

for every Borel subset E of $\mathbb{R}$, and the equation

$$\mathsf{F}(s) = \int_{-\infty}^{\infty} e^{i\xi s}\, \mathsf{M}(d\xi), \qquad \forall s \geq 0. \qquad (9)$$

The Radon measure M considered as a distribution is given by the inverse relation

$$\mathsf{M}(\xi) = \frac{1}{2\pi}\left(\hat{\mathsf{F}}(\xi) + \hat{\mathsf{F}}(\xi)^\dagger\right), \qquad (10)$$

where $\hat{\mathsf{F}}$ denotes the Fourier transform

$$\hat{\mathsf{F}}(\xi) := \int_0^{\infty} e^{-i\xi t}\, \mathsf{F}(t)\, dt. \qquad (11)$$

In view of our assumptions the theorem applies to the relaxation modulus G. The corresponding Radon measure will be denoted by M. Since G is real-valued, the Radon measure M has an additional symmetry with respect to reflections in $\mathbb{R}$: $\mathsf{M}(-E) = \mathsf{M}(E)$.

Note that both the transposition and the Hermitean conjugation are defined in $\Sigma_\mathbb{C}$. Symmetry with respect to the transposition reflects the major symmetry of viscoelastic moduli. On account of (10) the Radon measure M inherits the major symmetry from the relaxation modulus. On account of equation (10), $\mathsf{M}(E)$ is also a Hermitean operator on $S_\mathbb{C}$ for every Borel $E \subset \mathbb{R}$. It follows easily that $\mathsf{M}(E) \in \Sigma$ for every Borel $E \subset \mathbb{R}$.

Since $\mathsf{M}(E)$ is both real and positive semi-definite, an argument involving the determinant of a quadratic form implies the inequality

$$|\mathbf{v}^\top \mathsf{M}(E)\,\mathbf{w}| \leq |\mathbf{v}|\,|\mathbf{w}|\, m(E), \qquad (12)$$

where $m(E)$ is defined as the trace of $\mathsf{M}(E)$. By the Radon-Nikodym theorem the last inequality implies a very useful factorization of M:

$$\mathsf{M}(d\xi) = \mathsf{N}(\xi)\, m(d\xi) \qquad (13)$$

where m is a Borel measure on the real line, satisfying $m(-E) = m(E)$. It represents the density of the frequency spectrum. On the other hand $\mathsf{N}(\xi) \in \Sigma$ represents the anisotropic properties of the medium at the frequency ξ in the spectrum $\operatorname{supp} m$.

The function $\mathsf{N}: \mathbb{R} \to \Sigma$ is bounded ($|\mathsf{N}(\xi)| \leq 1$) and positive semi-definite for m-almost all $\xi \in \mathbb{R}$.

3 Auxiliary Fields

We shall now to replace strain rate history by a one-parameter family of auxiliary variables $\mathbf{w}$, defined for $\xi \geq 0$:

$$\mathbf{w}(t, x; \xi) := \int_{-\infty}^{t} \cos(\xi(t - t'))\, \dot{\mathbf{e}}(t', x)\, dt', \tag{14}$$

$$\mathbf{z}(t, x; \xi) := \int_{-\infty}^{t} \sin(\xi(t - t'))\, \dot{\mathbf{e}}(t', x)\, dt' \tag{15}$$

$$\mathbf{y}(t, x; \xi) := \int_{-\infty}^{t} e^{\mathrm{i}\xi(t-t')}\, \dot{\mathbf{e}}(t', x)\, dt', \tag{16}$$

For each ξ the auxiliary field $\mathbf{w}$ satisfies the differential equations of an oscillator

$$\frac{\partial^2 \mathbf{w}(t, x; \xi)}{\partial t^2} + \xi^2\, \mathbf{w}(t, x; \xi) = \ddot{\mathbf{e}}(t, x) \tag{17}$$

driven by the second strain rate, with the initial data

$$\mathbf{w}(0, x; \xi) = \int_{-\infty}^{0} \cos(\xi t')\, \dot{\mathbf{e}}(t', x)\, dt'. \tag{18}$$

The oscillators are not coupled to each other.

The stress can be expressed in terms of the auxiliary field $\mathbf{w}$:

$$\sigma(t, x) = \int_0^\infty \left[\int_{[0,\infty[} \cos(s\,\xi)\, \mathsf{N}(\xi, x)\, m(d\xi, x) \right] \dot{\mathbf{e}}(t - s)\, ds$$

$$= \int_{[0,\infty[} \mathsf{N}(\xi, x)\, \mathbf{w}(t, x; \xi)\, m(d\xi, x). \tag{19}$$

The stress can be decomposed into the elastic part (the contribution of $\xi = 0$) and the anelastic part

$$\sigma(t, x) = \mathsf{G}_0(x)\, \mathbf{e}(t, x) + \int_{]0,\infty[} \mathsf{N}(\xi, x)\, \mathbf{w}(t, x; \xi)\, m(d\xi, x) \tag{20}$$

where

$$G_0(x) := \int_{[0,\infty[} \mathsf{N}(\xi, x)\, m(\mathrm{d}\xi, x) = \lim_{t \to 0} G(t) \tag{21}$$

and $]0, \infty[:= [0, \infty[\setminus \{0\}$.

4 Construction of the Conserved Energy

We shall assume that the operator $\mathsf{N}(\xi)$ is invertible whenever defined, that is m-almost everywhere.

The stored energy density is now defined

$$U_c(t, x) := \frac{1}{2} \int_{[0,\infty[} \langle \mathbf{y}(t, x; \xi), \mathsf{N}(\xi, x)\, \mathbf{y}(t, x; \xi) \rangle_{\mathbb{C}}\, m(\mathrm{d}\xi)$$

$$= \frac{1}{2} \int_{[0,\infty[} [\langle \mathbf{w}(t, x; \xi), \mathsf{N}(\xi, x)\, \mathbf{w}(t, x; \xi) \rangle + \langle \mathbf{z}(t, x; \xi), \mathsf{N}(\xi, x)\, \mathbf{z}(t, x; \xi) \rangle]\, m(\mathrm{d}\xi)$$

$$= \frac{1}{2} \langle \mathbf{e}(t, x), G_\infty(x)\, \mathbf{e}(t, x) \rangle + \frac{1}{2} \int_{]0,\infty[} \langle \mathbf{w}(t, x; \xi), \mathsf{N}(\xi, x)\, \mathbf{w}(t, x; \xi) \rangle\, m(\mathrm{d}\xi)$$

$$+ \frac{1}{2} \int_{]0,\infty[} \langle \mathbf{z}(t, x; \xi), \mathsf{N}(\xi, x)\, \mathbf{z}(t, x; \xi) \rangle\, m(\mathrm{d}\xi) \tag{22}$$

where

$$G_\infty(x) := m(\{0\}, x)\, \mathsf{N}(0, x) \tag{23}$$

Note that U_c is a positive semi-definite form of the real variables $\mathbf{e}$ and $\{(\mathbf{w}(\xi), \mathbf{z}(\xi)) \mid \xi \in \mathbb{R}\}$.

Energy conservation follows from the following important theorem:

Theorem 2.
$$\frac{\mathrm{d}U_c}{\mathrm{d}t} = \langle \sigma, \dot{\mathbf{e}} \rangle \tag{24}$$

Proof. Substitute
$$\dot{\mathbf{y}} = \dot{\mathbf{e}} + i\xi \mathbf{y} \tag{25}$$

in
$$\frac{\mathrm{d}U_c}{\mathrm{d}t} = \mathrm{Re} \int_{[0,\infty[} \langle \dot{\mathbf{y}}, \mathsf{N}\mathbf{y} \rangle_{\mathbb{C}}\, m(\mathrm{d}\xi) \tag{26}$$

The contribution of the second term of (25) vanishes, hence

$$\frac{\mathrm{d}U_c}{\mathrm{d}t} = \int_{[0,\infty[} \langle \dot{\mathbf{e}}, \mathsf{N}(\xi)\, \mathbf{y}(\xi) \rangle_{\mathbb{C}}\, m(\mathrm{d}\xi) = \left\langle \int_{[0,\infty[} \mathsf{N}(\xi)\, \mathbf{w}(\xi)\, m(\mathrm{d}\xi), \dot{\mathbf{e}} \right\rangle \tag{27}$$

In view of equation (19) the last identity implies (24). □

A stored energy density satisfying (24) is uniquely defined by this equation. It is known as the maximum stored energy [2].

Substituting equation (24) in the well-known identity

$$\frac{d}{dt}\frac{\rho \mathbf{v}^2}{2} + \langle \sigma, \dot{\mathbf{e}} \rangle + \mathrm{div}\,\mathbf{j} = 0 \qquad (28)$$

where $\mathbf{j} = -\sigma \mathbf{v}$ is the energy flux density, yields the energy balance

$$\frac{d}{dt} E(t,x) + \mathrm{div}\,\mathbf{j} = 0 \qquad (29)$$

where

$$E := \frac{\rho \mathbf{v}^2}{2} + U_\mathrm{c} \qquad (30)$$

Hence follows the energy balance of the total energy $E_\mathrm{tot}(t) = \int_\Omega E(t,x)\,\lambda(\mathrm{d}x)$ of a viscoelastic body

$$\frac{dE_\mathrm{tot}}{dt} = -\int_{\partial\Omega} \mathbf{j}^\top \mathbf{n}\, da \qquad (31)$$

where $\mathbf{n}$ is the exterior unit normal on $\partial\Omega$, $\lambda(\mathrm{d}x)$ denotes the Lebesgue measure on Ω and a is the area on the external surface $\partial\Omega$ of Ω.

5 Hamiltonian Theory of Viscoelastic Bodies with Energy Conservation

The total energy is conserved if the total external energy flux vanishes. If $\partial\Omega = \Sigma_1 \cup \Sigma_2$, $\Sigma_1 \cap \Sigma_2 = \emptyset$ and $\mathbf{v} = 0$ on Σ_1, $\sigma\,\mathbf{n} = 0$ on Σ_2, then the total energy E_tot is conserved. In this case we proceed with the construction of the Hamiltonian by defining the generalized coordinates:

$$\mathbf{u}(x), \qquad (32)$$
$$\zeta(x;\xi) := \xi^{-1}\left(\mathbf{w}(x;\xi) - \mathbf{e}(x)\right) \qquad (33)$$

and the generalized momenta:

$$\mathbf{p}(x) = \rho(x)\,\mathbf{v}(x), \qquad (34)$$
$$\mathbf{q}(x;\xi) = \mathsf{N}(\xi,x)\,\dot{\zeta}(x;\xi). \qquad (35)$$

The Hamiltonian $H(\mathbf{u}, \zeta, \mathbf{p}, \mathbf{q})$ is the energy E_tot expressed in terms of the generalized coordinates and momenta:

$$H = \int_\Omega h(x)\,\lambda(\mathrm{d}x) \qquad (36)$$

where

$$h(x) := \frac{1}{2\rho} \mathbf{p}(x)^2 + \frac{1}{2} \langle \mathbf{e}(x), \mathsf{G}_0(x)\, \mathbf{e}(x) \rangle$$

$$+ \frac{1}{2} \int_{[0,\infty[} \xi^2 \, \langle \zeta(x,\xi), \mathsf{N}(\xi,x)\, \zeta(x,\xi) \rangle \, m(\mathrm{d}\xi, x)$$

$$+ \frac{1}{2} \int_{[0,\infty[} \langle \mathbf{q}(\xi), \mathsf{N}(\xi,x)^{-1} \mathbf{q}(\xi) \rangle \, m(\mathrm{d}\xi, x)$$

$$+ \left\langle \int_{[0,\infty[} \xi \, \mathsf{N}(\xi,x)\, \zeta(x,\xi)\, m(\mathrm{d}\xi, x), \mathbf{e}(x) \right\rangle. \tag{37}$$

Let $\mathsf{D}_\mathbf{u}$, $\mathsf{D}_\mathbf{p}$ denote the Gâteaux derivatives with respect to $\mathbf{u}, \mathbf{p} \in \mathcal{L}^2(\Omega; \mathbb{R}^d)$. D_ζ, $\mathsf{D}_\mathbf{q}$ denote the Gâteaux derivatives with respect to $\zeta, \mathbf{q} \in \mathcal{L}^2(\Omega \times [0,\infty[\,;\, S)$. The Hamiltonian equations assume the following form

$$\mathbf{v} = \mathsf{D}_\mathbf{p} H \equiv \frac{1}{\rho} \mathbf{p}(t,x,\xi), \tag{38}$$

$$\dot{\zeta} = \mathsf{D}_\mathbf{q} H \equiv \mathsf{N}(\xi,x)^{-1} \mathbf{q}(t,x;\xi), \tag{39}$$

$$\rho \dot{\mathbf{v}} \equiv \dot{\mathbf{p}} = -\mathsf{D}_\mathbf{u} H \equiv \mathrm{div}\, \sigma, \tag{40}$$

$$\mathsf{N}(\xi,x)\, \ddot{\zeta} \equiv \dot{\mathbf{q}} = -\mathsf{D}_\zeta H = -\xi\, \mathsf{N}(\xi,x)\, \mathbf{w}(t,x;\xi) \tag{41}$$

with $\mathbf{w} \equiv \xi \zeta + \mathbf{e}$. If $\mathsf{N}(\xi,x)$ is invertible ($m \times \lambda$)-almost everywhere, the last equation is equivalent to the equation $\ddot{\zeta} + \xi^2 \zeta = -\xi\, \mathbf{e}$.

6 Poisson Brackets

Let P denote the class of functionals of the form

$$F = \int_\Omega f_0(\mathbf{u}, \mathbf{p})\, \lambda(\mathrm{d}x) + \int_\Omega \int_{[0,\infty[} F_0(\zeta, \mathbf{q})\, m(\mathrm{d}\xi)\, \lambda(\mathrm{d}x).$$

The Poisson bracket of two functionals in P is defined by

$$\{F, H\} := \int_\Omega \left[\sum_{k=1}^d \left[\frac{\partial f_0}{\partial u_k} \frac{\partial g_0}{\partial p_k} - \frac{\partial f_0}{\partial p_k} \frac{\partial g_0}{\partial u_k} \right] \right] \lambda(\mathrm{d}x)$$

$$+ \int_\Omega \left[\int_{[0,\infty[} \sum_{k=1}^d \left[\mathsf{D}_{\zeta_k} F_0\, \mathsf{D}_{q_k} G_0 - \mathsf{D}_{q_k} F_0\, \mathsf{D}_{\zeta_k} G_0 \right] m(\mathrm{d}\xi) \right] \lambda(\mathrm{d}x). \tag{42}$$

It is proved in [8] that the Poisson bracket (42) satisfies the Jacobi identity.

The dynamics of an energetically isolated viscoelastic body can be expressed by the equation

$$\frac{dF}{dt} = \{F, H\} \tag{43}$$

where $F \in \mathsf{P}$ and H is the Hamiltonian defined by equation (36).

7 Interaction of a Field with Matter

The Hamiltonian (36) can be decomposed into two terms representing two subsystems (phonon or elastic field) and matter, as well as an interaction Hamiltonian:

$$H = H_0 + H_1 + H_{01}. \tag{44}$$

The Hamiltonian of the elastic subsystem has the familiar form

$$H_0 = \frac{1}{2} \int_\Omega \left[\frac{\mathbf{p}^2}{\rho} + \langle \mathbf{e}, \mathsf{G}_0\, \mathbf{e} \rangle \right] \lambda(dx). \tag{45}$$

The matter Hamiltonian represents a continuum of uncoupled oscillators specified by their frequencies $\xi \geq 0$:

$$H_1 = \frac{1}{2} \int_\Omega \left[\int_{[0,\infty[} \xi^2 \langle \zeta(x,\xi), \mathsf{N}(\xi, x)\, \zeta(x, \xi) \rangle\, m(d\xi, x) \right] \lambda(dx)$$

$$+ \frac{1}{2} \int_\Omega \left[\int_{[0,\infty[} \langle \mathbf{q}(\xi), \mathsf{N}(\xi, x)^{-1}\, \mathbf{q}(\xi) \rangle\, m(d\xi, x) \right] \lambda(dx). \tag{46}$$

The interaction Hamiltonian is

$$H_{01} = \int_\Omega \left[\left\langle \int_{[0,\infty[} \xi\, \mathsf{N}(\xi, x)\, \zeta(x,\xi)\, m(d\xi, x), \mathbf{e}(x) \right\rangle \right] \lambda(dx). \tag{47}$$

This interpretation provides an analogy with the dielectrics interacting with an external electric field.

8 A Monotonely Decreasing Energy

Inequality (1) does not ensure that the energy decays monotonely in an energetically isolated viscoelastic body [14]. In order to construct such an energy a more restrictive assumption about viscoelastic relaxation moduli is needed [7].

Definition 2. *A function* $\mathsf{F} : [0, \infty[\to \Sigma$ *is said to be* completely monotone *if it has derivatives of arbitrary order n on* $]0, \infty[$ *satisfying the inequalities*

$$(-1)^n \left\langle \mathbf{v}, \frac{d^n}{ds^n} \mathsf{F}(s) \, \mathbf{v} \right\rangle \geq 0 \quad \forall \mathbf{v} \in S \tag{48}$$

for all non-negative integers n.

We now assume that the relaxation modulus G is completely monotone.

A version of Bernstein's theorem (extended to tensor-valued functions) [10] yields the following spectral decomposition:

$$\mathsf{G}(t) = \int_{[0,\infty[} e^{-rt} \, \mathsf{M}_1(dr) = \int_{[0,\infty[} e^{-rt} \, \mathsf{N}_1(r) \, m_1(dr) \tag{49}$$

where m_1 is a Borel measure and the Σ-valued density N_1 is positive semi-definite m_1-almost everywhere.

Equation (49) is a generalization of a general property of real viscoelastic media. In rheological applications the Borel measure m_1 is commonly expressed in terms of a non-negative density $\chi(r)$ of the relaxation time spectrum

$$m_1(dr) = \chi(r) \, d\ln(r).$$

The auxiliary fields

$$\mathbf{y_1}(t, r) = \int_{-\infty}^{t} e^{-r(t-t')} \, \dot{\mathbf{e}}(t') \, dt' \tag{50}$$

satisfy the differential equations

$$\frac{d\mathbf{y_1}}{dt} + r \, \mathbf{y_1} = \dot{\mathbf{e}}, \quad r \geq 0. \tag{51}$$

We now define a different stored energy density

$$U_1 := \frac{1}{2} \int_{[0,\infty[} \langle \mathbf{y_1}(r), \mathsf{N}_1(r) \, \mathbf{y_1}(r) \rangle \, m_1(dr) \tag{52}$$

In an energetically isolated viscoelastic body the new energy, defined by the formula

$$E_{1,\text{tot}} := \int_{\Omega} \left[\frac{\rho \mathbf{v}^2}{2} + U_1 \right] \lambda(dx) \tag{53}$$

monotonely decreases with time

$$\frac{dE_{1,\text{tot}}}{dt} \leq 0, \quad t \geq 0. \tag{54}$$

If it is only assumed that the relaxation modulus is a positive definite function, then a generic energy density $E = \mathbf{v}^2/2 + U$ consistent with inequality (1) is neither conserved nor monotonely decaying [14].

9 A Non-Linear Example

Hamiltonian theory can be extended to some non-linear viscoelastic media. A relatively trivial non-linear single-integral model is obtained by replacing the driving term $\ddot{\mathbf{e}}(t, x)$ in equation (17) by $D^2 \mathbf{f}(\mathbf{e})(t, x)$, where $\mathbf{f} : S \to S$ is differentiable. The stress is then given by

$$\sigma = \mathsf{F}(\mathbf{e}) \, \mathsf{G} * D\mathbf{f}(\mathbf{e}) \tag{55}$$

with $F_{ijkl} := \partial f_{kl}/\partial e_{ij}$ and the stored energy is given by (22). The Hamiltonian, the generalized coordinates and momenta remain otherwise unchanged.

Equation (55) is similar to the equation of Quasi-Linear Viscoelasticity [5] except for the factor on the left. This factor ensures the existence of a Hamiltonian.

The same result has a different interpretation in geometrically non-linear viscoelasticity

$$\sigma(t) = \int_0^\infty \mathsf{G}(s) \, D\mathbf{C}_t(t-s) \, ds \tag{56}$$

where

$$\mathbf{C}_t(t-s) := \mathbf{F}(t) \, \mathbf{F}(t-s)^{-1} \left[\mathbf{F}(t) \, \mathbf{F}(t-s)^{-1} \right]^\top \tag{57}$$

and $\mathbf{F}$ is the deformation gradient with respect to an arbitrary reference configuration. Equations (16) and (24) are replaced by

$$\mathbf{y}(t, \xi) = \int_0^\infty e^{i\xi s} \, D\mathbf{C}_t(t-s) \, ds, \quad \mathbf{w} = \operatorname{Re} \mathbf{y}$$

and

$$\frac{dU}{dt} = \operatorname{Re} \int_{[0,\infty[} \mathbf{y}(t, \xi)^\dagger \, \mathsf{N}(d\xi) \, \mathbf{y}_t(t, \xi) \, m(d\xi) = 2 \left\langle \int_{[0,\infty[} \mathsf{N}(\xi) \, \mathbf{w}(\xi) \, m(d\xi), \mathbf{D} \right\rangle$$

respectively. The power transferred to the internal degrees of freedom on the right-hand side involves the geometrically non-linear strain rate (the first Rivlin–Ericksen tensor) $\mathbf{D} = D\mathbf{C}_t(t) = \left(\mathbf{L} + \mathbf{L}^\top \right)/2$, with $\mathbf{L} = \dot{\mathbf{F}} \mathbf{F}^{-1}$, while equations (19) and (22) remain unchanged.

10 Concluding Remarks

Dissipativity is not incompatible with existence of a conserved energy and Hamiltonian function. A linear hereditary viscoelastic body consists of two subsystems, an elastic subsystem and a one-parameter family of linear oscillators, which are driven by the elastic subsystem but otherwise uncoupled. This model resembles the model of a continuum coupled to a heat bath [4]. The oscillator dynamics in the viscoelastic case is deterministic. Thermal effects can however be introduced by randomizing the oscillator dynamics in accordance with a statistical model of driving forces.

References

1. S. Breuer and E.T. Onat. On the determination of free energy in viscoelastic solids. *ZAMP*, 15:185–191, 1964.
2. M. Fabrizio and A. Morro. *Mathematical Problems in Linear Viscoelasticity*. SIAM, Philadelphia, 1992.
3. B.U. Felderhof. On the derivation of the Fluctuation-Dissipation Theorem. *Journal of Physics A: Mathematical and General*, 11:921–927, 1978.
4. F.W. Ford, J.T. Lewis, and R.F. O'Connell. Independent oscillator model of a heat bath: Exact diagonalization of the Hamiltonian. *Journal of Statistical Physics*, 53:439–455, 1988.
5. Y.C. Fung. *Biomechanics. Mechanical Properties of Living Tissues*. Springer-Verlag, New York, 1981.
6. G. Gripenberg, S.O. Londen, and O.J. Staffans. *Volterra Integral and Functional Equations*. Cambridge University Press, Cambridge, 1990.
7. A. Hanyga. Viscous dissipation and completely monotone stress relaxation functions. *Rheologica Acta*, 44:614–621, 2005. doi:10.1007/s00397-005-0443-6.
8. A. Hanyga and M. Seredyńska. Hamiltonian and Lagrangian theory of viscoelasticity. *Continuum Mechanics and Thermodynamics*, 19:475–492, 2008. doi: 10.1007/s00161-007-0065-6.
9. A. Hanyga and M. Seredyńska. On a mathematical framework for dielectric relaxation functions. *Journal of Statistical Physics*, 131:269–303, 2008. doi: 10.1007/s10955-008-9501-7.
10. A. Hanyga and M. Seredyńska. Relations between relaxation modulus and creep compliance in anisotropic linear viscoelasticity. *Journal of Elasticity*, 88:41–61, 2007.
11. R. Kubo, N. Toda, and N. Hashitsune. *Statistical Physics II: Nonequilibrium Statistical Physics*. Springer-Verlag, Berlin, 1991. 2nd edition.
12. A. Molinari. Viscoélasticité linéaire and functions complètement monotones. *Journal de Mécanique*, 12:541–553, 1975.
13. P.J. Morrison. Hamiltonian description of an ideal fluid. *Reviews of Modern Physics*, 70:467–521, 1998.
14. M. Seredyńska and A. Hanyga. Nonlinear 2dof pendulum with fractional damping. *Acta Mechanica*, 176:169–183, 2005. doi:10.1007/s00707-005-0220-8.

Author Index

Acharya, A., 99
Alber, H.-D., 363
Armero, F., 251
Børvik, T., 329
Benallal, A., 329
Benzerga, A.A., 67
Berstad, T., 329
Brunssen, S., 155
Caddemi, S., 205
Caliò, I., 205
Carstensen, C., 41
Chevaugeon, N., 89
Chung Kim Yuen, S., 309
Cleary, P.W., 287
Das, R., 287
de Saxcé, G., 165
De Souza Neto, E.A., 3
Dettmer, W.G., 3
Dufour, F., 89
Ebobisse, F., 117
Garikipati, K., 217
Hachemi, A., 349
Hackl, K., 27
Hager, C., 155
Hain, M., 15
Hanyga, A., 373
Hjiaj, M., 165
Hopperstad, O., 329
Huth, R., 41
Karagiozova, D., 319
Kaunda, M.A.E., 263
Kochmann, D.M., 27
Langdon, G.S., 319
Lew, A., 227
Loehnert, S., 79
Logg, A., 195

Marotti de Sciarra, F., 107
Matthies, H.G., 185
McBride, A., 117, 237
Menzel, A., 275
Mielke, A., 53
Moës, N., 89
Mohr, R., 275
Mouhtamid, S., 349
Mueller-Hoeppe, D.S., 79
Needleman, A., 297
Neff, P., 129
Nesenenko, S., 363
Nguyen, A.D., 349
Nogueira de Codes, R., 329
Nurick, G.N., 309, 319
Ølgaard, K.B., 195
Ostien, J., 217
Perić, D., 3
Ponter, A.R.S., 341
Rangarajan, R., 227
Reddy, B.D., 117, 237
Reese, S., 175
Rosić, B.V., 185
Schmid, F., 155
Seredyńska, M., 373
Somer, D.D., 3
Steinmann, P., 275
Ten Eyck, A., 227
Tvergaard, V., 297
Uhlar, S., 275
Vladimirov, I.N., 175
Weichert, D., 349
Wells, G.N., 195
Wieners, C., 143
Wohlmuth, B., 155
Wriggers, P., 15

Subject Index

Γ-convergence, 53
assumed strain finite element methods, 251
automated modelling, 195
bi-potential, 165
brutal damage, 89
cement paste, 15
chaos expansion, 185
computational homogenization, 3
computational plasticity, 143
computer simulation, 297
concentrated damages, 205
consistent integration, 275
constrained dynamics, 275
continuum mechanics, 27
convex analysis, 165
coupled problems, 155
cracks, 79
crashworthiness, 309
cyclic loading, 341
damage, 15, 53, 287
damage identification, 205
debonding, 319
deep drawing, 175
difference quotient technique, 363
direct methods, 341, 349
discontinuous Galerkin, 227, 237
dislocation density, 129
dislocations, 99
distributions, 205
ductile fracture, 67
dynamic plasticity, 251
dynamic strain ageing, 329
dynamics, 263
elastic solid, 287
elastoplastic, 287
elastoplasticity, 155
energetic formulation, 117
energetic solutions, 53

energy absorbers, 309
energy-dissipative momentum-conserving time-stepping algorithms, 251
eXtended Finite Element Method (X-FEM), 79, 89
failure, 349
FEM, 41
Fenchel transform, 165
fibre-metal laminate, 319
finite deformations, 99, 237
finite elastoplasticity, 41
finite element analysis, 67
finite elements, 3, 15
Finite Elements in time, 275
forming process, 155
fracture, 79, 287
front curvature, 89
frost, 15
generalised functions, 205
gradient plasticity, 99, 129, 237
Hamiltonian, 373
heterogeneous materials, 3
homogenization, 67
immersed boundary methods, 227
imperfections, 309
incompressible elasticity, 195
induced anisotropy, 67
inelastic deformations, 275
inelastic irreversible behaviour, 185
inelastic solids, 3
inelasticity, 27
integration of algorithms, 263
interior penalty, 217
internal variables, 107, 165, 263
interpolation, 363
Izod test, 297
large deformations, 175
level set, 89

Lipunov function, 263
lifting operator, 217
localised blast loading, 319
locking, 227
maximal monotone operator, 363
Melan–Prager model, 363
microforces, 217
microstructures, 27
modelling, 319
modified Cam-clay model, 165
multi-scale, 15, 79
multi-scale analysis, 3
non-convex minimisation, 41
nonlinear elasticity, 227
nonlocal plasticity, 107
numerical relaxation, 41
optimisation, 349
overlapping domain decomposition, 155
panel, 319
plastic spin, 129
plasticity, 195, 263, 341, 349
PLC deformation bands, 329
polymer, 297
positive definite function, 373
quasiconvexity, 41
ratchetting, 341
rate-independent systems, 53

regularity, 363
relaxation, 27, 373
shakedown analysis, 349
shakedown, 341
singularities, 205
Smoothed Particle Hydrodynamics, 287
softening, 107 softening behaviour, 117
spinodal decomposition, 195
SQP methods, 143
stochastic generational return map, 185
stochastic model, 185
stochastic variational inequality, 185
strain gradient plasticity, 117
structural tensors, 175
thermodynamics, 107
thermography, 329
time-incremental minimization problems, 53
triggers, 309
tube crushing, 309
two-scale, 155
uncertainty, 185
variational inequality, 117, 165
variational methods, 195
viscoelasticity, 373
viscoplasticity, 297, 363
void coalescence, 67
void growth, 67